Potential Anti-SARS-CoV-2 Molecular Strategies

Potential Anti-SARS-CoV-2 Molecular Strategies

Editors

Giovanni N. Roviello
Caterina Vicidomini

MDPI • Basel • Beijing • Wuhan • Barcelona • Belgrade • Manchester • Tokyo • Cluj • Tianjin

Editors

Giovanni N. Roviello
Institute of Biostructure and Bioimaging (IBB)
Italian National Council for Research (CNR)
Naples
Italy

Caterina Vicidomini
Institute of Biostructure and Bioimaging (IBB)
Italian National Council for Research (CNR)
Naples
Italy

Editorial Office
MDPI
St. Alban-Anlage 66
4052 Basel, Switzerland

This is a reprint of articles from the Special Issue published online in the open access journal *Molecules* (ISSN 1420-3049) (available at: www.mdpi.com/journal/molecules/special_issues/molecules_AntiSARSCoV2_Molecular_Strategies).

For citation purposes, cite each article independently as indicated on the article page online and as indicated below:

LastName, A.A.; LastName, B.B.; LastName, C.C. Article Title. *Journal Name* **Year**, *Volume Number*, Page Range.

ISBN 978-3-0365-6963-5 (Hbk)
ISBN 978-3-0365-6962-8 (PDF)

Contents

About the Editors

Giovanni N. Roviello

Dr. Giovanni Roviello graduated with honors in chemistry from Federico II University (2002, Naples, Italy) and received his PhD in Biotechnology from the same university in 2006. He works as a senior researcher at the Institute of Biostructure and Bioimaging (IBB) of the Italian National Council for Research (CNR) in Naples, Italy. He was a visiting researcher in Germany (Institute of Molecular Biotechnology (IMB)- Jena 2003, Georg-August University of Goettingen 2005, FAU University–Erlangen 2018), the UK (University of Greenwich–Medway Campus in Chatham Maritime 2022), and Ireland (UCD, Dublin 2017) working in the field of biorganic chemistry. He has strong scientific and teaching connections with Yerevan State University (Armenia) with which he works on two funded research projects, AMU University of Poznan (Poland) for which he served as a lecturer for the nucleic acids course for Ph.D. students in 2021 and with Geomedi University of Tbilisi (Georgia) having been appointed the honorary professorship in medicinal chemistry for this university in 2012. He is currently a lecturer for the Georgian State University of Physical Education and Sport of Tbilisi (Georgia) helding a course on the 'Biotechnological Basis of the Physical Rehabilitation' and is the principal investigator for CNR for different international research activities including the EU-funded project NobiasFluors (ID 872331; 2020–2024), the Royal Society (UK)-funded project 'Design, synthesis and in vitro evaluation of TMPRSS2 inhibitor analogues as potential novel anti-COVID-19 drugs' (2022–2024), the project code 21T-1D057 funded by the Science Committee of the Republic of Armenia (2021–2024), and the MESRA (Armenia)-CNR-funded bilateral project 'Synthetic alpha-amino acids and peptide-based systems for anticancer strategies' (2023–2025) to cite only a few. He serves as an academic editor for different international scientific journals and is author of more than 100 scientific articles.

Caterina Vicidomini

Dr. Caterina Vicidomini Ph.D. is a researcher in the field of the pharmaceutical sciences applied to the research and development of new lead compounds for both pharmacological therapy and for diagnostic purposes. Dr. Vicidomini graduated in Pharmacy at the University of Salerno in 2002 and subsequently performed research activity in the frame of her Ph.D. internship at the Department of Pharmaceutical Sciences of the same university. During this period, she gained experience in organic synthesis, chromatography purification and characterization of new compounds of pharmaceutical interest. In July 2008 she completed a Postgraduate Master in Clinical Pharmacy (summa cum laude) at University of Salerno. Since 2011 she is working as permanent researcher at Institute of Biostructure and Bioimaging of Italian National Council for Research (IBB-CNR) in the screening of organic substances of natural and synthetic origin against different biological targets (e.g., neoplastic pathologies, taupathies) also by means of computational chemistry. Dr. Vicidomini is also working in the field of neuroscience and is involved in studies concerning the molecular diagnostics related to neurodegenerative diseases such as Parkinson's disease and others. Among the other skills of Dr. Vicidomini, the ability of coordinating the activities of research teams aimed at developing biologically active compounds is worth of mention together with the capability of working in highly multidisciplinary environments involving organic chemists, neurologists, radiologists, to cite only a few, collaborating with scientists having the most diverse backgrounds in order to set up and arrange strategies for the implementation of multidisciplinary projects.

Preface to "Potential Anti-SARS-CoV-2 Molecular Strategies"

The COVID-19 pandemic, associated with the SARS-CoV-2, has globally attracted scientific attention on the theme of human coronaviruses and, thus, the prophylaxis and therapy of infections caused by the most pathogenic members of this family have gained enormous relevance in the last three years. When SARS-CoV emerged approximately twenty years ago in China and other countries of Asia as the first deadly member of the group of beta coronaviruses, these RNA viruses were associated with simple colds, seasonal flu, and/or other mild forms of illness. With severe viral diseases such as SARS, MERS -a highly fatal disease in Middle East countries- and COVID-19, humanity has faced the most worrying effects of the beta coronaviruses. Consequently, the fundamental role of drugs such as synthetic and natural protease inhibitors, as well as prophylactic agents for, respectively, the therapy and prevention of the diseases caused by these respiratory viruses emerged in their full importance. This book contains some recent research studies conducted by scholars at the interface between chemistry, biomedicine, and pharmaceutical sciences that contributed to the development of innovative molecular approaches useful not only in the context of the COVID-19 disease, but also for the prevention and therapy of the diseases caused in general by the human beta coronaviruses.

Giovanni N. Roviello and Caterina Vicidomini
Editors

molecules

MDPI

Editorial

Potential Anti-SARS-CoV-2 Molecular Strategies

Caterina Vicidomini and Giovanni N. Roviello *

Institute of Biostructures and Bioimaging, Italian National Council for Research (IBB-CNR), Area di Ricerca Site and Headquarters, Via Pietro Castellino 111, 80131 Naples, Italy
* Correspondence: giroviel@unina.it

Abstract: Finding effective antiviral molecular strategies was a main concern in the scientific community when the severe acute respiratory syndrome coronavirus 2 (SARS-CoV-2) emerged at the end of 2019 as an easily transmissible and potentially deadly β-coronavirus able to cause the coronavirus disease 19 (COVID-19), which famously led to one of the most worrying pandemics in recent times. Other members of this zoonotic pathogenic family were already known before 2019, but apart from the SARS-CoV, which was responsible of severe acute respiratory syndrome (SARS) pandemic in 2002/2003, and Middle East respiratory syndrome coronavirus (MERS-CoV), whose main impact on humans is geographically restricted to Middle Eastern countries, the other human β-coronaviruses known at that time were those typically associated with common cold symptoms which had not led to the development of any specific prophylactic or therapeutic measures. Although SARS-CoV-2 and its mutations are still causing illness in our communities, COVID-19 is less deadly than before and we are returning to normality. Overall, the main lesson learnt after the past few years of pandemic is that keeping our bodies healthy and immunity defenses strong using sport, nature-inspired measures, and using functional foods are powerful weapons for preventing the more severe forms of illness caused by SARS-CoV-2 and, from a more molecular perspective, that finding drugs with mechanisms of action involving biological targets conserved within the different mutations of SARS-CoV-2—and possibly within the entire family of β-coronaviruses—gives more therapeutic opportunities in the scenario of future pandemics based on these pathogens. In this regard, the main protease (M^{pro}), having no human homologues, offers a lower risk of off-target reactivity and represents a suitable therapeutic target in the search for efficacious, broad-spectrum anti-β-coronavirus drugs. Herein, we discuss on the above points and also report some molecular approaches presented in the past few years to counteract the effects of β-coronaviruses, with a special focus on SARS-CoV-2 but also MERS-CoV.

Citation: Vicidomini, C.; Roviello, G.N. Potential Anti-SARS-CoV-2 Molecular Strategies. *Molecules* **2023**, *28*, 2118. https://doi.org/10.3390/molecules28052118

Received: 20 February 2023
Revised: 22 February 2023
Accepted: 22 February 2023
Published: 24 February 2023

1. Molecules against β-Coronaviruses

It is well known that the coronavirus disease 19 (COVID-19), caused by the severe acute respiratory syndrome coronavirus 2 (SARS-CoV-2), has caused enormous sanitary and socio-economic difficulties to the entire world due to the high transmission rates of the virus, the lack of effective treatments available when it first emerged, and the rapid genetic mutations of the virus observed in recent years. Among the biomolecules crucial for the viral processes, the spike protein is a key structural protein that mediates host infection by SARS-CoV-2, with the structure of the N-terminal signal peptide of this SARS-CoV-2 protein having an effect on the synthesis and secretion of the spike, which suggests that developing drugs which target the signaling peptide of the spike could be a winning anti-COVID-19 strategy [1]. Moreover, studies conducted on the S1 subunit of this protein have revealed the inactivation abilities of ozonated water and slightly acidic electrolyzed water and ethanol, and these were proposed as effective as SARS-CoV-2 disinfectants [2]. Another relevant SARS-CoV-2 protein target is nucleoprotein, also known as protein N or nucleocapsid protein, an abundant RNA-binding protein critical for viral genome packaging and essential for the replication machinery of SARS-CoV-2 [3]. It is a potential link between viral replication and the multiple signaling pathways which lead to long-lasting COVID-19 symptoms, and new ligands of the protein N were investigated by computational and experimental studies based on dynamic light scattering and surface plasmon resonance, which led to the identification of substance P(1-7) and enkephalins as effective binders of the major sites at the C-terminus or of β-sheets at the N-terminus of the viral protein. These studies led to the conclusion that antiviral drugs that target N can contribute to reducing brain fog and stroke risk and improving COVID-19 patient well-being [3]. One of the most widely explored strategies for lowering virus transmission and disease severity has been COVID-19 vaccination. Vaccines are a powerful weapon which can be used to prevent the worrying scenarios caused by a number of pathogens [4], and have been widely employed in the form of mass vaccination during the COVID-19 pandemic. Sadly, COVID-19 vaccines are not exempt from side effects, including, in some cases, thrombotic thrombocytopenia syndrome, which led to some specific COVID-19 vaccines being discouraged regarding their use in younger individuals [5]. Remarkably, vaccination responses vary significantly depending on the host, even though correlations with protection against SARS-CoV-2 were reported [6]. However, despite vaccinations, the recurrence of SARS-CoV-2 infection was frequently observed also in vaccinated individuals, while disease can still severely affect elderly and hospitalized patients who are fragile and vulnerable, which recalls the need for new efforts to be devoted to the discovery of potential global therapeutics, utilizable in the context of the infections of new mutations of SARS-CoV-2, but possibly also of other human β-coronaviruses. COVID-19 causes widespread respiratory and non-respiratory symptoms and, regardless of the underlying comorbid conditions, early treatment of COVID-19 has the potential to positively affect the clinical course of the disease. Moreover, even after the virus has been eradicated, the previous SARS-CoV-2 interaction with host cell receptors and consequent activation of pro-inflammatory pathways may still be responsible for endothelial and epithelial damage mechanisms. In this context, anti-inflammatory treatments, such as steroidal and non-steroidal drugs, as well as cytokine inhibitors, have been suggested in the treatment of COVID-19 patients, since inflammation treatment has been identified as a crucial step in the recovery process [7]. The complications observed in COVID-19 patients are numerous, especially in cases of the more severe forms of the diseases. For example, a strong correlation between the impaired renal function and in-hospital deaths of critically ill COVID-19 patients, particularly those with comorbidities and requiring renal replacement therapy, has been supported by data from numerous intensive care units around the world [8]. The common renal disorder known as acute kidney injury, characterized by a sudden and persistent decline in renal function, is sometimes experienced by COVID-19 patients during their hospitalization, with a significant impact on their survival. Catechins were also found to be effective therapeutics in COVID-19-associated acute kidney injury thanks to their wide range of pharmacological effects, which included anti-coronavirus, antioxidant, anti-inflammatory, and renoprotective effects against kidney damage [8]. More geographically confined than SARS-CoV-2, but also much more deadly, Middle East respiratory syndrome coronavirus (MERS-CoV) is another highly pathogenic zoonotic β-coronavirus, first discovered in patients living in the Arabian Peninsula in September 2012, that has an unusually high fatality rate in humans and causes severe, frequently fatal respiratory illness [9]. Natural products were investigated as MERS drugs using approaches employing quantum mechanics calculations, pharmacophore-based virtual screening, and molecular dynamics simulations, revealing the potential of compounds such as taiwanhomoflavone B, 2,3-dihydrohinokiflavone, and sophoricoside as effective natural anti-MERS-CoV candidates [9].

2. Drug Repositioning

Drug repositioning using previously FDA-approved drugs is a strategy typically employed when a new pathogen emerges in the attempt to treat patients who need urgent cures while waiting for more specific therapies to be developed, which clearly takes a longer time. This was especially true in the context of COVID-19, regarding which countless studies, often based on computational approaches using molecular docking and molecular dynamics simulations, were conducted on different molecular targets of SARS-CoV-2, including the papain-like protease (PL^{pro}), RNA-dependent RNA polymerase (RdRp), and SARS-CoV-2 main protease (M^{pro}), to name only a few. This led to the identification of salinomycin (Figure 1) from *Streptomyces albus* as a potential inhibitor of SARS-CoV-2 PL^{pro}, while the vegetal toxin ouabain was proposed as a dual inhibitor of the PL^{pro} and M^{pro} enzymes [10]. By screening 171 candidates obtained from the DrugBank database (http://www.drugbank.ca/ accessed on 17 February 2023), other in silico studies identified possible organic triazole compounds such as bemcentinib, and bisoctrizole, as M^{pro} inhibitors whose pharmacokinetic characteristics were also evaluated, and their complex stability and conformation were examined using molecular dynamics simulation [11]. Non-steroidal anti-inflammatory drugs, which are frequently used to treat upper airway infections symptomatically, are of crucial importance when administered in the early stages of SARS-CoV-2 infection and, in this context, ketoprofen lysine salt is a non-steroidal anti-inflammatory drug which was suggested to offer notable benefits in early COVID-19 therapy, based on the pharmacodynamic and pharmacokinetic characteristics of this drug [7].

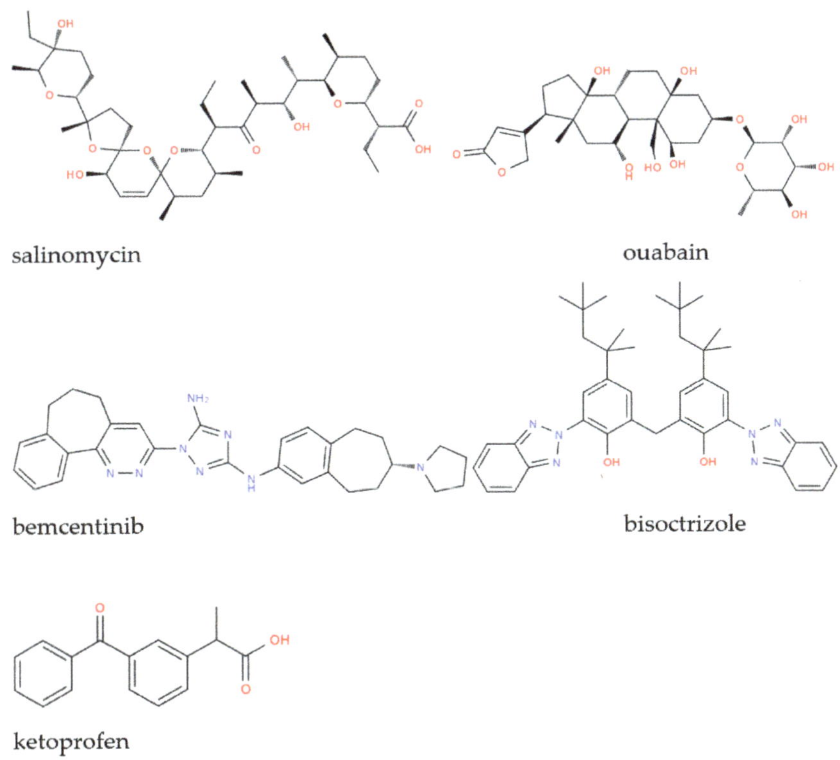

Figure 1. Structural representation of some of the drugs investigated as anti-COVID-19 therapeutics mentioned in this work.

3. Nature against β-Coronaviruses

Since nature is an inexhaustible source of remedies and therapeutic scaffolds, new potential drugs to be used in the fight against COVID-19 were often searched for from natural sources, especially plants [12–14], using highly diverse approaches such as transcriptomics, cheminformatics, and systems pharmacology with a holistic view [15]. Natural inhibitors of SARS-CoV-2 nonstructural protein (nsp10) were screened from a database containing 310 naturally isolated metabolites through various in silico selection methods including molecular similarity assessment, molecular fingerprint, docking studies, toxicity, ADMET (absorption, distribution, metabolism, excretion, and toxicity), and density-functional theory [16]. Polyphenol-rich tea leaf extracts containing concentrated theaflavins and several virucidal catechins were tested for their ability to inactivate SARS-CoV-2 in cellular studies and the virus structural proteins and viral RNA were also examined using Western blotting and real-time RT-PCR to assess the effects of the tea leaf extracts on viral proteins and the viral genome. Remarkably, cellular infection by SARS-CoV-2 was prevented by the treatment using natural preparations, which, from a molecular point of view, caused structural changes of the S2 subunit of the spike protein and damages to the viral genome, giving clues to the potential efficacy of tea leaf extracts as virucidal agents in vitro, which could improve the existing control measures adopted against β-coronaviruses [17]. In the context of the application of traditional and herbal medicine to the therapy and prevention of infections caused by β-coronaviruses, we have suggested the anti-COVID-19 effects of cloves (*Syzygium aromaticum* L.). This is a culinary spice with a long history of use in folk medicine for a variety of disorders, being used in traditional medicine to treat respiratory ailments since ancient times. Among its molecular ingredients, different clove phytochemicals have antiviral, anti-inflammatory, antibacterial, immunostimulatory, and antithrombotic properties, all useful aspects in the context of developing an effective anti-COVID-19 therapy [18]. In conclusion, more research on COVID-19 therapies making use of synthetic and natural molecules and their derivatives is unquestionably required. In fact, the scientific efforts regarding both therapy and the prophylaxis of β-coronavirus diseases should not be discontinued even if currently COVID-19 appears to be less deadly, because SARS-CoV-2 could re-emerge with more pathogenic variants and, moreover, it is not unlikely that, with the increased rate of human–wildlife contact due to the loss of intact ecosystems and forested areas, new zoonoses caused by β-coronaviruses will continue to emerge, and humanity could have to face new pandemics in the near future.

Author Contributions: C.V. and G.N.R. have contributed equally to this work. All authors have read and agreed to the published version of the manuscript.

Acknowledgments: Caterina Vicidomini and Giovanni N. Roviello, who served as Guest Editors for the Special Issue 'Potential Anti-SARS-CoV-2 Molecular Strategies' of *Molecules*, are grateful to all the authors for their contributions and to all the expert reviewers involved in the Special Issue.

Conflicts of Interest: The authors declare no conflict of interest.

References

1. Zhang, Z.; Wan, X.; Li, X.; Wan, C. Effects of a Shift of the Signal Peptide Cleavage Site in Signal Peptide Variant on the Synthesis and Secretion of SARS-CoV-2 Spike Protein. *Molecules* **2022**, *27*, 6688. [CrossRef] [PubMed]
2. Takeda, Y.; Jamsransuren, D.; Makita, Y.; Kaneko, A.; Matsuda, S.; Ogawa, H.; Oh, H. Inactivation Activities of Ozonated Water, Slightly Acidic Electrolyzed Water and Ethanol against SARS-CoV-2. *Molecules* **2021**, *26*, 5465. [CrossRef] [PubMed]
3. Henri, J.; Minder, L.; Mohanasundaram, K.; Dilly, S.; Goupil-Lamy, A.; Di Primo, C.; Slama Schwok, A. Neuropeptides, New Ligands of SARS-CoV-2 Nucleoprotein, a Potential Link between Replication, Inflammation and Neurotransmission. *Molecules* **2022**, *27*, 8094. [CrossRef] [PubMed]
4. Costanzo, V.; Roviello, G.N. The Potential Role of Vaccines in Preventing Antimicrobial Resistance (AMR): An Update and Future Perspectives. *Vaccines* **2023**, *11*, 333. [CrossRef]
5. Islam, A.; Bashir, M.S.; Joyce, K.; Rashid, H.; Laher, I.; Elshazly, S. An Update on COVID-19 Vaccine Induced Thrombotic Thrombocytopenia Syndrome and Some Management Recommendations. *Molecules* **2021**, *26*, 5004. [CrossRef] [PubMed]
6. Villar, M.; Urra, J.M.; Artigas-Jerónimo, S.; Mazuecos, L.; Contreras, M.; Vaz-Rodrigues, R.; Rodríguez-del-Río, F.J.; Gortázar, C.; de la Fuente, J. Correlates with Vaccine Protective Capacity and COVID-19 Disease Symptoms Identified by Serum Proteomics in Vaccinated Individuals. *Molecules* **2022**, *27*, 5933. [CrossRef] [PubMed]
7. Mariniello, D.F.; Allocca, V.; D'Agnano, V.; Villaro, R.; Lanata, L.; Bagnasco, M.; Aronne, L.; Bianco, A.; Perrotta, F. Strategies Tackling Viral Replication and Inflammatory Pathways as Early Pharmacological Treatment for SARS-CoV-2 Infection: Any Potential Role for Ketoprofen Lysine Salt? *Molecules* **2022**, *27*, 8919. [CrossRef] [PubMed]
8. Diniz, L.R.L.; Elshabrawy, H.A.; Souza, M.T.d.S.; Duarte, A.B.S.; Datta, S.; de Sousa, D.P. Catechins: Therapeutic Perspectives in COVID-19-Associated Acute Kidney Injury. *Molecules* **2021**, *26*, 5951. [CrossRef] [PubMed]

9. Bouback, T.A.; Pokhrel, S.; Albeshri, A.; Aljohani, A.M.; Samad, A.; Alam, R.; Hossen, M.S.; Al-Ghamdi, K.; Talukder, M.E.K.; Ahammad, F.; et al. Pharmacophore-Based Virtual Screening, Quantum Mechanics Calculations, and Molecular Dynamics Simulation Approaches Identified Potential Natural Antiviral Drug Candidates against MERS-CoV S1-NTD. *Molecules* **2021**, *26*, 4961. [CrossRef] [PubMed]
10. Qayed, W.S.; Ferreira, R.S.; Silva, J.R.A. In Silico Study towards Repositioning of FDA-Approved Drug Candidates for Anti-coronaviral Therapy: Molecular Docking, Molecular Dynamics and Binding Free Energy Calculations. *Molecules* **2022**, *27*, 5988. [CrossRef] [PubMed]
11. Sur, V.P.; Sen, M.K.; Komrskova, K. In Silico Identification and Validation of Organic Triazole Based Ligands as Potential Inhibitory Drug Compounds of SARS-CoV-2 Main Protease. *Molecules* **2021**, *26*, 6199. [CrossRef] [PubMed]
12. Falanga, A.P.; Terracciano, M.; Oliviero, G.; Roviello, G.N.; Borbone, N. Exploring the Relationship between G-Quadruplex Nucleic Acids and Plants: From Plant G-Quadruplex Function to Phytochemical G4 Ligands with Pharmaceutic Potential. *Pharmaceutics* **2022**, *14*, 2377. [CrossRef] [PubMed]
13. Greco, F.; Musumeci, D.; Borbone, N.; Falanga, A.P.; D'Errico, S.; Terracciano, M.; Piccialli, I.; Roviello, G.N.; Oliviero, G. Exploring the Parallel G-Quadruplex Nucleic Acid World: A Spectroscopic and Computational Investigation on the Binding of the c-myc Oncogene NHE III1 Region by the Phytochemical Polydatin. *Molecules* **2022**, *27*, 2997. [CrossRef] [PubMed]
14. Ricci, A.; Roviello, G.N. Exploring the Protective Effect of Food Drugs against Viral Diseases: Interaction of Functional Food Ingredients and SARS-CoV-2, Influenza Virus, and HSV. *Life* **2023**, *13*, 402. [CrossRef]
15. Adarshan, S.; Akassh, S.; Avinash, K.; Bharathkumar, M.; Muthuramalingam, P.; Shin, H.; Baskar, V.; Chen, J.-T.; Bhuvaneshwari, V.; Ramesh, M. Transcriptomics, Cheminformatics, and Systems Pharmacology Strategies Unveil the Potential Bioactives to Combat COVID-19. *Molecules* **2022**, *27*, 5955. [CrossRef] [PubMed]
16. Eissa, I.H.; Khalifa, M.M.; Elkaeed, E.B.; Hafez, E.E.; Alsfouk, A.A.; Metwaly, A.M. In Silico Exploration of Potential Natural Inhibitors against SARS-Cov-2 nsp10. *Molecules* **2021**, *26*, 6151. [CrossRef] [PubMed]
17. Takeda, Y.; Tamura, K.; Jamsransuren, D.; Matsuda, S.; Ogawa, H. Severe Acute Respiratory Syndrome Coronavirus-2 Inactivation Activity of the Polyphenol-Rich Tea Leaf Extract with Concentrated Theaflavins and Other Virucidal Catechins. *Molecules* **2021**, *26*, 4803. [CrossRef] [PubMed]
18. Vicidomini, C.; Roviello, V.; Roviello, G.N. Molecular Basis of the Therapeutical Potential of Clove (*Syzygium aromaticum* L.) and Clues to Its Anti-COVID-19 Utility. *Molecules* **2021**, *26*, 1880. [CrossRef] [PubMed]

molecules MDPI

Review

Molecular Basis of the Therapeutical Potential of Clove (*Syzygium aromaticum* L.) and Clues to Its Anti-COVID-19 Utility

Caterina Vicidomini [1,†], Valentina Roviello [2,†] and Giovanni N. Roviello [1,*]

1 Istituto di Biostrutture e Bioimmagini IBB-CNR, Via Tommaso De Amicis 95, 80145 Naples, Italy; caterina.vicidomini@ibb.cnr.it
2 Department of Chemical Materials and Industrial Production Engineering (DICMaPI), University of Naples Federico II, Piazzale V. Tecchio 80, 80125 Naples, Italy; valentina.roviello@unina.it
* Correspondence: giroviel@unina.it; Tel.: +39-08-1253-4585
† These authors equally contributed to this work.

Abstract: The current COronaVIrus Disease 19 (COVID-19) pandemic caused by SARS-CoV-2 infection is enormously affecting the worldwide health and economy. In the wait for an effective global immunization, the development of a specific therapeutic protocol to treat COVID-19 patients is clearly necessary as a short-term solution of the problem. Drug repurposing and herbal medicine represent two of the most explored strategies for an anti-COVID-19 drug discovery. Clove (*Syzygium aromaticum* L.) is a well-known culinary spice that has been used for centuries in folk medicine in many disorders. Interestingly, traditional medicines have used clove since ancient times to treat respiratory ailments, whilst clove ingredients show antiviral and anti-inflammatory properties. Other interesting features are the clove antithrombotic, immunostimulatory, and antibacterial effects. Thus, in this review, we discuss the potential role of clove in the frame of anti-COVID-19 therapy, focusing on the antiviral, anti-inflammatory, and antithrombotic effects of clove and its molecular constituents described in the scientific literature.

Keywords: coronavirus infections; pandemics; natural compounds; clove; *Syzygium aromaticum*; eugenol; eugeniin; SARS-CoV-2; COVID-19; phytochemicals; herbal medicine

Citation: Vicidomini, C.; Roviello, V.; Roviello, G.N. Molecular Basis of the Therapeutical Potential of Clove (*Syzygium aromaticum* L.) and Clues to Its Anti-COVID-19 Utility. *Molecules* **2021**, *26*, 1880. https://doi.org/10.3390/molecules26071880

Academic Editor: Diego Muñoz-Torrero

Received: 8 March 2021
Accepted: 22 March 2021
Published: 26 March 2021

Publisher's Note: MDPI stays neutral with regard to jurisdictional claims in published maps and institutional affiliations.

1. Introduction

Coronaviridae is a family of enveloped RNA viruses known as Coronaviruses (CoVs) that provoke infections in animals and humans [1–6]. Presently, seven human coronaviruses (HCoVs), commonly considered of zoonotic origin, are described in the scientific literature [7] that cause infections mainly associated with respiratory symptoms [8–10]. More in detail, HCoV-229E, HCoV-NL63, HCoV-OC43, and HCoV-HKU1 are "common cold" coronaviruses causing seasonal, usually mild, respiratory diseases [11,12]. Although, in most cases, these HCoVs do not lead to severe clinical symptoms, HCoV-NL63 and HCoV-HKU1 infections can provoke bronchiolitis and croup [13,14], whilst CoV 229E and OC43 can provoke pneumonia [15,16]. Nonetheless, three highly pathogenic HCoVs have emerged in the last two decades, i.e., Middle East Respiratory Syndrome (MERS)-CoV, Severe Acute Respiratory Syndrome (SARS)-CoV-1, and SARS-CoV-2 [17–19], which can lead to life-threatening pathologic events associated with the recent MERS, SARS, and the current COronaVIrus Disease 19 (COVID-19), which is causing enormous problems globally in both sanitary and socioeconomic terms [20]. SARS-CoV-1 and MERS-CoV are more lethal than SARS-CoV-2, but this latter is more transmissible, explaining the current pandemic status of COVID-19 [21]. In the first step of coronavirus infection, a specific molecular recognition between the virus particle, through the virus spike (S) protein, and the host cell takes place, involving different HCoV-specific receptors [22–26] that were identified for several CoVs and are considered one of the primary targets for anti-CoV biomedical strategies together with the SARS-CoV-2 main protease (M^{pro}) [27,28]. The receptors for the "common cold" HCoVs are human aminopeptidase N (APN), associated with the infection from HCoV-229E, and 9-O-acetylated sialic acid (9-O-Ac-Sia), used by HCoV-OC43 and HCoV-HKU1. On the other hand, the receptor for HCoV-NL63, i.e., angiotensin-converting enzyme 2 (ACE2), is also common to the more pathogenic SARS-CoV-1 and SARS-CoV-2, whilst dipeptidyl peptidase 4 (DPP4) was associated with MERS-CoV [29,30]. Once intracellular, all HCoVs replicate their RNA with the consequent expression of the viral proteins needed for the production of new viral particles inside the infected cell [31]. As anticipated, four out of the seven HCoVs are associated with usually mild upper respiratory infections, whilst MERS-CoV and SARS-CoV-1 and -2 can cause lethal events [32]. This latter, first emerging in China at the end of 2019 [32], can lead to severe pneumonia and, being easily transmissible, has spread worldwide rapidly, leading the World Health Organization (WHO) to declare COVID-19 a pandemic [33]. Currently, there are more than two million deaths (2,566,793, as found in Worldometers.info [34] accessed on 3 March 2021) worldwide due to COVID-19, with enormous consequences for the public health and the economy worldwide [35–37]. While the whole world is fighting against COVID19 and awaits an effective mass immunization, the scientific community is devoting immense efforts toward developing specific therapies for the treatment of SARS-CoV-2 infection. Moreover, since inflammatory cytokine storms together with immune system impairment are commonly observed in patients with severe COVID-19, several research studies have highlighted the advantages of dual therapies with antiviral and anti-inflammatory benefits [38,39]. Due to the urgent need for such a pharmacological treatment, drug repurposing [40–42] and herbal medicine are two of the most considered anti-COVID-19 approaches [43–47]. In fact, several plants such as mulberry, tea, and Dragon's Blood tree are known as remedies to treat respiratory ailments and for their anti-inflammatory and antithrombotic properties, which are useful aspects in the fight against COVID-19 [48–52].

2. Clove (*Syzygium aromaticum* L.) in Herbal Medicine and Its Active Constituents

Syzygium aromaticum L., also known as Eugenia caryophyllata L. [53], is an evergreen tree with sanguine flowers belonging to the family Myrtaceae that grows in tropical climates and has been widely used in Ayurveda and Chinese traditional medicines for over 2000 years. Arabic traders brought it to the Western world in the fourth century A.D., and in medieval Europe, it became very popular as a medicinal spice [54].

Indigenous to the Moluccas, this tree is cultivated in several countries of Asia and Africa, including India, Indonesia, Madagascar, Malaysia, Sri Lanka, and Zanzibar [55]. The dried flower bud of this plant is indicated by the English name "clove", derived from the Latin word "clavus" (nail), as the shape resembles that of a small-sized nail. Cloves are currently used in three different forms, as whole dried buds (commonly referred to as "cloves"), ground spice, and essential oil. Though all forms share similar biomedically-relevant properties, they differ in the degree of potency, with the oil showing the highest potency and, thus, often being diluted with almond oil. Whole cloves, containing a good amount of oil in their interiors, are still endowed with a medium potency, whilst ground cloves are the least potent form, as, in this form, the spice generally loses most of the essential oil [54].

Cloves have long been used in both traditional medicine and for culinary purposes and serve to produce an essential oil known since ancient times in food flavorings, traditional medicine, and perfume production [53]. Even though cloves are mostly used as a nutritional spice for food in the Western world, in the past, they have constituted a remedy for a variety of health concerns, with the clove anesthetic (due to eugenol), stimulating, antimicrobial, antifungal, antiviral, and antiseptic properties having been known for centuries [54].

On the other hand, the clove essential oil finds applications in dental care, including the treatment of gum infections [56], burns [57], and respiratory and digestive disorders [56,58]. The previous literature studies also evidenced other remarkable properties, such as antiangiogenic [53,59], anticancer [53,56,58], antioxidant [60], anti-inflammatory [61], and antimutagenic activities [62].

The American Food and Drug Administration (FDA) agency has confirmed the safety of clove buds, clove oil, and some clove ingredients as a food supplement [63], while the WHO has established the acceptable daily uptake of cloves in humans at 2.5 mg/kg body weight [64].

The spice contains a good amount of minerals like magnesium, manganese, potassium, iron, and selenium [54]. Among the others, potassium as an important electrolyte of the cell and body fluids has a key role in the heart rate and blood pressure control [65], while manganese is used by the body as a cofactor for the antioxidant enzyme superoxide dismutase [66,67]. Additionally, cloves are a good source of beta carotene vitamin B1, vitamin B6, vitamin C, vitamin K, riboflavin, and vitamin A, used by the body for maintaining healthy mucus membranes and skin [68]. Noteworthy, vitamin C sustains a resistance against infectious agents [69] and is used by cells to scavenge harmful oxygen-free radicals [70].

Several research studies have been carried out to identify the main clove phytochemicals [71–77]. Dried clove buds contain ~20% essential oil, which is rich in eugenol, accounting for 70–90%. The other main phytochemicals isolated from clove essential oil include eugenyl acetate, β-caryophyllene, and several sesquiterpenes [53,78], including α-cubebene, α-copaene, and γ- and δ-cadinene [79]. Crategolic acid, vanillin, gallotannic acid, methyl salicylate, eugeniin, rhamnetin, kaempferol, eugenitin, oleanolic acid, methyl amyl ketone, methyl salicylate, α- and β-humulene, benzaldehyde, chavicol, and β-ylangene are present in lesser amounts [74]. In particular, eugenol and minor constituents like methyl salicylate and methyl amyl ketone are responsible for the characteristic pleasant aroma of cloves. The extraction of phytochemicals, achievable with high efficiency by presoaking and the liquid ammonia treatment of plant materials [80], in the case of cloves was realized with different operating conditions, including using supercritical CO_2 [81].

2.1. *Clove as Herbal Remedy for Respiratory Ailments*

Traditional medicine uses cloves as respiratory aids, and in particular, the spice is one of the ingredients of teas used in tropical Asia to facilitate coughing [54]. Moreover, an aromatherapy procedure consisting of breathing in the aroma released from hot clove tea is another common way to use cloves for respiratory disorders like coughs, colds, asthma, bronchitis, and sinusitis [54]. Moreover, it is customary in Asia to chew cloves for treating soreness of throat and inflammation of the pharynx [54]. Chewing cloves after their thermal treatment is reported to bring relief from severe coughing [54]. Clove oil acts as an expectorant for treating respiratory disorders, including colds, bronchitis, cough, asthma, and upper-respiratory conditions [74]. In mixtures with honey, it helps in the case of chronic coughs and is mentioned to be specifically useful in the case of shortness of breath [82].

2.2. *Anti-Inflammatory, Immunostimulatory, and Antithrombotic Properties of Cloves*

Clove essential oil, often used in aromatherapy to treat inflammatory diseases, including arthritis and rheumatism [54], was found to have anti-inflammatory effects in animal models at doses of 0.05 and 0.20 mL/kg [83]. Interestingly, at this dosage, the anti-inflammatory effect of clove oil matches that of anti-inflammatory drugs like etodolac and indomethacin administered at 0.025 and 0.1, and 0.05 and 0.2 mL/kg doses, respectively [83]. The ethanol extracts of clove buds were also tested for anti-inflammatory effects at three doses (50, 100, and 200 mg/kg) in mice and Wistar rats using acetic acid-induced abdominal contractions in the former and formalin-induced hind paw edema in the latter animal models. The extract with an LD_{50} (50% Lethal Dose) of 565.7 mg/kg produced significant effects at all three doses, supporting the use of the clove extract in inflammatory conditions [84].

From a molecular point of view, clove buds contain flavonoids like β-caryophyllene, kaempferol, and rhamnetin, which contribute to clove anti-inflammatory properties [85–89]. In experimental animal models, eugenol (at 200 and 400 mg/kg doses) was shown to reduce the volume of pleural exudates without changing the total count of blood leukocytes, which indicates the anti-inflammatory activity of this molecule [90]. Eugenol is believed to regulate the cellular inflammatory cascades, including the NF-κB (nuclear factor kappa-light-chain-enhancer of activated B cells) and ERK (extracellular-signal-regulated kinase)/MAPK (mitogen-activated protein kinase) pathways, and the release of proinflammatory interleukins [82]. In other studies, LPS (lipopolysaccharide)-induced lung inflammation was relieved by the treatment with both whole clove aqueous extract and eugenol through a reduction of TNF-α (tumor necrosis factor alpha) and inhibition of NF-κB signaling, also with improvement in the alveolar damage [91,92]. Remarkably, clove aqueous extract showed protective effects on an animal model of pyelonephritis [93], a kidney inflammation reported in COVID-19 patients [94].

Traditional medicine attributes to clove the property of boosting the human immune system, improving disease resistance [54]. In experimental studies on animal models, clove oil improved the total white blood cell count and enhanced the delayed-type hypersensitivity response. Noteworthy, a dose-dependent restoration of both humoral and cellular immune responses was observed in cyclophosphamide-immunosuppressed mice treated with clove essential oil. The immunostimulatory activity was associated with improvement in the cell- and humor-mediated immune response mechanisms determined by clove essential oil [95].

Clove is mentioned to improve the blood supply to both the brain and the heart and is used as a tonic for the cardiovascular system [82]. Moreover, clove oil was shown to inhibit the platelet aggregation induced by the platelet-activating factor, arachidonic acid, and collagen, with a higher activity observed in the first two systems than the latter [74]. In vivo experiments carried out on rabbits showed that clove oil at 50–100 mg/kg doses afforded total protection against the platelet-activating factor and good (70%) protection against arachidonic acid-induced shock due to pulmonary platelet thrombosis [74]. Clove oil also inhibited thromboxane-A2 and 12-hydroxyeicosatetraenoic acid production by human platelets treated with C-14 arachidonic acid [96]. Antithrombotic and antiplatelet aggregation effects were also studied on clove extracts by ex vivo methods measuring the fibrinolytic activity and the inhibitory effect on thrombin-induced platelet aggregation [97]. The extracts showed remarkable fibrinolytic activity and inhibitory effects on platelet aggregation, suggesting clove anti-atherosclerotic potential [97].

Owing to the molecular basis for the clove antithrombotic effects, the main clove oil constituent, eugenol, has shown activity as a platelet inhibitor, thus preventing blood clots [87]. More in detail, the same compound was shown in vitro to inhibit arachidonic acid-induced platelet aggregation, as well prostaglandin biosynthesis and the formation of thromboxane B2 [98]. Together with acetyl eugenol, it was more effective than acetylsalicylic acid in inhibiting the platelet aggregation induced by arachidonic acid, adrenaline, and collagen, showing, in the first case, an anti-aggregation activity comparable to indomethacin [99]. Aside from the above-mentioned antithrombotic properties of eugenol, these were also revealed for rhamnetin, gallic acid, kaempferol, myricetin, and β-caryophyllene (Figure 1), as well for two polysaccharides isolated from the clove buds by chromatographic methods [100].

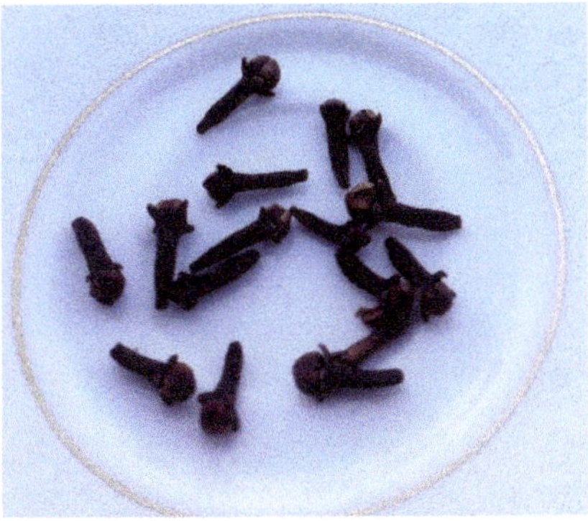

(a)

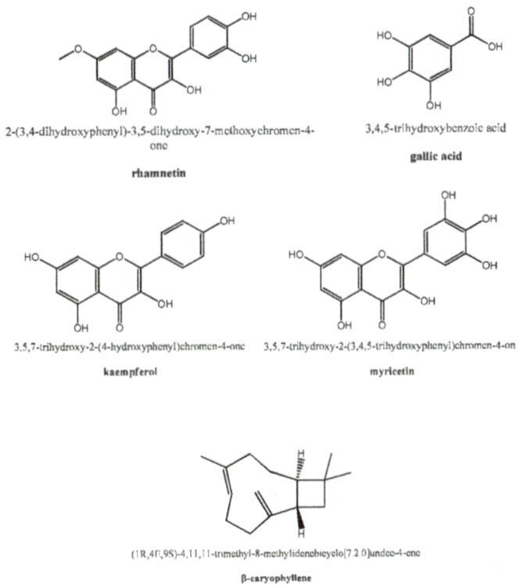

(b)

Figure 1. Clove, from culinary use to herbal medicine: (**a**) edible clove buds (photo taken by Giovanni N. Roviello). (**b**) Structure representation of some phytochemicals extracted from *Syzygium aromaticum* endowed with anti-inflammatory properties.

Both polysaccharides presented a backbone of type I rhamnogalacturonan and the side chain made of arabinan. However, one mainly composed of the sugars Ara, Gal, Glc, and Rha was endowed with a relatively high molecular weight (MW ~103,000), and the other mainly composed of Rha, Gal, GalA, and Ara showed a lower molecular weight (MW ~34,000). The high molecular weight polysaccharide showed antithrombotic activity with a plasma clotting time of 145 s in the activated partial thromboplastin time (APTT) assays, while the other displayed a lower activity with a plasma clotting time of 90 s in the APTT assay [100].

3. Clove Antiviral Properties

The whole clove antiviral activity was tested by Tragoolpua and Jatisatienr [101], who assayed an ethanol extract obtained from the plant flower buds for its anti-herpes simplex virus (HSV) properties. By a plaque reduction assay, the authors demonstrated that HSV was inhibited by the clove extract. Interestingly, the clove extract showed a direct inactivating action on the particles of the standard HSV strains. Moreover, the total HSV virus yield at 30 h declined after the treatment with the extract [101]. Another study performed on the methanol extracts of cloves showed a high in vitro activity of the extract in inhibiting the HCV protease, with a $\geq$90% protease inhibition at a dose of 100 µg/mL [102].

Antiviral Properties of Clove Phytochemicals

Eugenol (4-allyl-2-methoxyphenol; Figure 2), being the major constituent of cloves, was investigated for its antiviral activity by several research groups. The above-mentioned Tragoolpua and Jatisatienr [101] used pure eugenol as the reference compound in their anti-HSV studies and found that it exerted a higher antiviral activity than the ethanol extracts of whole clove buds. Similar findings were obtained by Benencia and Courreges [103], who reported the eugenol inhibition of HSV-1 and HSV-2 replication with inhibitory concentration 50% (IC_{50}) values of 25.6 μg/mL and 16.2 μg/mL, respectively. In the same study, eugenol was virucidal, whilst no compound-associated cytotoxicity was revealed at the concentrations tested [103]. Eugenol also showed antiviral activity against the influenza A virus (IAV), being able to inhibit IAV replication [104]. Finally, it was also found active as an inhibitor of the Ebola Virus in vitro [105].

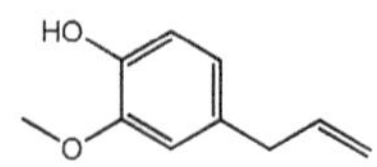

4-allyl-2-methoxyphenol

eugenol

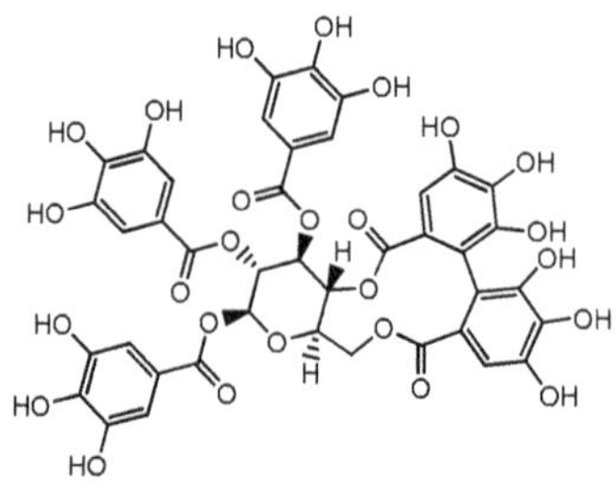

(11a*R*,13*S*,14*R*,15*S*,15a*R*)-2,3,4,5,6,7-hexahydroxy-9,17-dioxo-9,11,11a,13,14,15,15a,17-octahydrodibenzo[*g*,*i*]pyrano[3 *b*][1,5]dioxacycloundecine-13,14,15-triyl tris(3,4,5-trihydroxybenzoate)

eugeniin

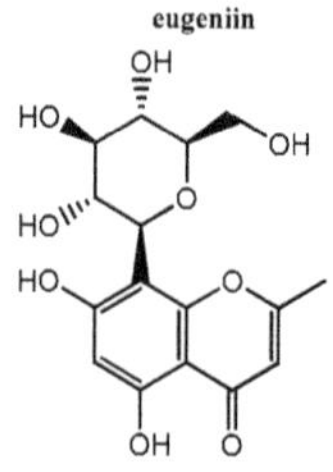

5,7-dihydroxy-2-methyl-8-[(2S,3R,4R,5S,6R)-3,4,5-trihydroxy-6-(hydroxymethyl)oxan-2-yl]chromen-4-one

isobiflorin

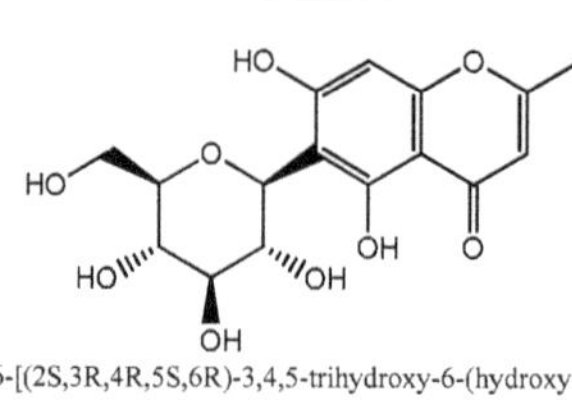

5,7-dihydroxy-2-methyl-6-[(2S,3R,4R,5S,6R)-3,4,5-trihydroxy-6-(hydroxymethyl)oxan-2-yl]chromen-4-one

biflorin

Figure 2. Structure representations of some clove phytochemicals with antiviral activity.

Other clove phytochemicals were investigated for their antiviral properties, and among them, eugeniin (Figure 2), isolated from the herbal extracts of cloves and, also, from *Geum japonicum*, showed anti-HSV activity at a 5-μg/mL concentration [106]. The HSV inhibitory activity of eugeniin was due to the inhibition of the viral DNA synthesis, as it acted as a selective inhibitor of the DNA polymerases of HSV-1 and HSV-2 [106].

Eugeniin was also found to act as a potent inhibitor of the protease of Dengue virus (DENV), which causes infections in tropical and subtropical regions of the world for which there are still no specific antiviral treatments available [107]. The IC_{50} values of eugeniin against the proteases of DENV serotype-2 and -3 were 94.7 nM and 7.5 μM, respectively. Thus, in consideration of the importance of DENV protease for the viral replication cycle, eugeniin was proposed as a promising drug in the context of anti-DENV therapeutics development [107]. The other investigated DENV protease inhibitors were isobiflorin and biflorin (Figure 2), even though their inhibitory activity was weaker than eugeniin [107]. The atomic-level details of the binding of these three clove phytochemicals to the viral protease were obtained by computational docking and saturation transfer difference (STD) NMR spectroscopy, which showed that the molecular recognition at the active site of the DENV protease involved networks of hydrophobic contacts and hydrogen bonds [107].

4. Clove in the Fight against COVID-19

The traditional therapeutic use of clove in respiratory disorders and its activity against different types of viruses, alongside its anti-inflammatory, immunostimulatory, and antithrombotic properties, are all attractive features highlighting its potential in the fight against the COVID-19 disease.

Clove is one of the medicinal plants currently employed to prevent and control the SARS-CoV-2-associated disease, together with *Eucalyptus globulus*, *Cymbopogon citratus*, *Zingiber officinale*, and other plants endowed with the advantage of being inexpensive and abundantly available around the globe [108]. More in detail, a protocol for the prevention and treatment of COVID-19 using cloves, as medicinal plant, was described by Kanyinda, J.N. M., who reported a proven effect for the treatment provided that it was carried out in the early stages of the disease [108]. The protocol included the preparation of a decoction in which cloves are boiled in water with other plant materials for 15 min. The released volatile active principles are then inhaled by patients for five minutes. The same protocol also included a drinkable decoction obtained with cloves and other plant materials [108]. Noteworthy, surveys have been conducted in India and Morocco, countries with low pandemic impacts [109,110], to identify the various home remedies used by the local populations during COVID-19, which have included many spices and herbs. Interestingly, more than 93% of the interviewed Indian people believed that spices are helpful in curing COVID-19 or other viral infections and can help in boosting the immunity. Cloves are mentioned as one of the most frequently used spices and herbs during the current COVID-19 pandemic in the areas under investigation, together with other plants like cinnamon, ginger, black pepper, garlic, neem, and basil [111]. Cloves are also being used in Morocco by herbalists from Salé Prefecture for the prevention and treatment of COVID-19 [112]. From a molecular point of view, some computational studies recommended phytocompounds extracted from cloves as potent anti-COVID-19 drugs [113,114], and one of them, kaempferol, was shown in silico to bind the substrate binding pocket of the main protease of SARS-CoV-2 with high affinity interacting with the active site residues such as Cys145 and His41 through hydrophobic interactions and hydrogen bonding, suggesting that natural compounds such as clove flavonoids could act as novel inhibitors of SARS-CoV-2 [115]. Molecular docking studies have also shown high affinities of clove compounds bicornin (−9.2 kcal/mol) and biflorin (−8.5 kcal/mol) for M^{pro}, suggesting their potential inhibitory activity [115].

5. Conclusions

The therapeutic use of cloves in traditional medicine to treat respiratory ailments and its experimentally proven activity against different types of viruses, as well its anti-inflammatory, immunostimulatory, and antithrombotic properties, all concur to compose a picture of the potential importance of cloves and their phytochemical constituents in the fight against the COVID-19 disease. Aside from the above-mentioned features, clove essential oil has shown remarkable antibacterial effects against the infections of immunosuppressed hospitalized patients [78], suggesting its utility to also prevent secondary bacterial infections in COVID-19 patients [82]. In conclusion, cloves, a precious spice largely used in countries where the impact of the novel coronavirus is lower than the Western world, are endowed with medicinal properties considered relevant in the prevention and therapy of COVID-19. Future clinical data on the activity of cloves and their constituents on COVID-19 patients and more molecular insights on the specific clove phytochemical interactions with SARS-CoV-2 protein targets are clearly desirable in order to realize the effective therapeutic protocols and design new drugs based on clove phytochemicals with optimized characteristics.

Author Contributions: All authors contributed equally to the literature collection, writing, literature data analysis, and editing and reviewing of the article. All authors have read and agreed to the published version of the manuscript.

Funding: This research received no external funding.

Acknowledgments: This review was written by the authors in smart-working mode (activated by Consiglio Nazionale delle Ricerche-CNR and the University of Naples Federico II-UNINA) during the COVID-19 crisis. We dedicate this study to all the healthcare workers and to all the people who suffered and still suffer around the world as a result of the COVID-19 pandemic.

Conflicts of Interest: The authors declare no conflict of interest.

References

1. V'kovski, P.; Kratzel, A.; Steiner, S.; Stalder, H.; Thiel, V. Coronavirus biology and replication: Implications for SARS-CoV-2. *Nat. Rev. Microbiol.* **2020**, *19*, 155–170. [CrossRef]
2. Chu, D.K.W.; Leung, C.Y.H.; Gilbert, M.; Joyner, P.H.; Ng, E.M.; Tse, T.M.; Guan, Y.; Peiris, J.S.M.; Poon, L.L.M. Avian Coronavirus in wild aquatic birds. *J. Virol.* **2011**, *85*, 12815–12820. [CrossRef]
3. Poon, L.L.M.; Chu, D.K.W.; Chan, K.H.; Wong, O.K.; Ellis, T.M.; Leung, Y.H.C.; Lau, S.K.P.; Woo, P.C.Y.; Suen, K.Y.; Yuen, K.Y.; et al. Identification of a novel coronavirus in bats. *J. Virol.* **2005**, *79*, 2001–2009. [CrossRef] [PubMed]
4. Wang, L.-F.; Anderson, D.E. Viruses in bats and potential spillover to animals and humans. *Curr. Opin. Virol.* **2019**, *34*, 79–89. [CrossRef]
5. Memish, Z.A.; Mishra, N.; Olival, K.J.; Fagbo, S.F.; Kapoor, V.; Epstein, J.H.; AlHakeem, R.; Durosinloun, A.; Al Asmari, M.; Islam, A.; et al. Middle East respiratory syndrome coronavirus in bats, Saudi Arabia. *Emerg. Infect. Dis.* **2013**, *19*, 1819–1823. [CrossRef]
6. Hofmann, H.; Pyrc, K.; van der Hoek, L.; Geier, M.; Berkhout, B.; Pohlmann, S. Human coronavirus NL63 employs the severe acute respiratory syndrome coronavirus receptor for cellular entry. *Proc. Natl. Acad. Sci. USA* **2005**, *102*, 7988–7993. [CrossRef] [PubMed]
7. Ye, Z.-W.; Yuan, S.; Yuen, K.-S.; Fung, S.-Y.; Chan, C.-P.; Jin, D.-Y. Zoonotic origins of human coronaviruses. *Int. J. Biol. Sci.* **2020**, *16*, 1686–1697. [CrossRef] [PubMed]
8. Gossner, C.; Danielson, N.; Gervelmeyer, A.; Berthe, F.; Faye, B.; Kaasik Aaslav, K.; Adlhoch, C.; Zeller, H.; Penttinen, P.; Coulombier, D. Human–dromedary camel interactions and the risk of acquiring zoonotic middle east respiratory syndrome Coronavirus infection. *Zoonoses Public Health* **2014**, *63*, 1–9. [CrossRef]
9. Sheahan, T.; Rockx, B.; Donaldson, E.; Sims, A.; Pickles, R.; Corti, D.; Baric, R. Mechanisms of zoonotic severe acute respiratory syndrome coronavirus host range expansion in human airway epithelium. *J. Virol.* **2008**, *82*, 2274–2285. [CrossRef] [PubMed]
10. Sheahan, T.; Rockx, B.; Donaldson, E.; Corti, D.; Baric, R. Pathways of cross-species transmission of synthetically reconstructed zoonotic severe acute respiratory syndrome coronavirus. *J. Virol.* **2008**, *82*, 8721–8732. [CrossRef]
11. Gaunt, E.R.; Hardie, A.; Claas, E.C.J.; Simmonds, P.; Templeton, K.E. Epidemiology and clinical presentations of the four Human Coronaviruses 229E, HKU1, NL63, and OC43 detected over 3 years using a novel multiplex real-time PCR method. *J. Clin. Microbiol.* **2010**, *48*, 2940–2947. [CrossRef]
12. Woldemeskel, B.A.; Kwaa, A.K.; Garliss, C.C.; Laeyendecker, O.; Ray, S.C.; Blankson, J.N. Healthy donor T cell responses to common cold coronaviruses and SARS-CoV-2. *J. Clin. Investig.* **2020**, *130*, 6631–6638. [CrossRef]
13. Abdul-Rasool, S.; Fielding, B.C. Understanding human Coronavirus. *Open Virol. J.* **2010**, *4*, 76–84. [CrossRef]
14. Esper, F.; Weibel, C.; Ferguson, D.; Landry, M.L.; Kahn, J.S. Coronavirus HKU1 infection in the United States. *Emerg. Infect. Dis.* **2006**, *12*, 775–779. [CrossRef]
15. Pene, F.; Merlat, A.; Vabret, A.; Rozenberg, F.; Buzyn, A.; Dreyfus, F.; Cariou, A.; Freymuth, F.; Lebon, P. Coronavirus 229E-related pneumonia in immunocompromised patients. *Clin. Infect. Dis.* **2003**, *37*, 929–932. [CrossRef] [PubMed]
16. Jordan, P.C.; Stevens, S.K.; Deval, J. Nucleosides for the treatment of respiratory RNA virus infections. *Antivir. Chem. Chemother.* **2018**, *26*, 204020661876448. [CrossRef] [PubMed]
17. Cevik, M.; Tate, M.; Lloyd, O.; Maraolo, A.E.; Schafers, J.; Ho, A. SARS-CoV-2, SARS-CoV-1 and MERS-CoV viral load dynamics, duration of viral shedding and infectiousness: A living systematic review and meta-analysis. *SSRN Electron. J.* **2020**. [CrossRef]
18. Andersen, K.G.; Rambaut, A.; Lipkin, W.I.; Holmes, E.C.; Garry, R.F. The proximal origin of SARS-CoV-2. *Nat. Med.* **2020**, *26*, 450–452. [CrossRef]

19. Rabaan, A.A.; Al-Ahmed, S.H.; Haque, S.; Sah, R.; Tiwari, R.; Malik, Y.S.; Dhama, K.; Yatoo, M.I.; Bonilla-Aldana, D.K.; Rodriguez-Morales, A.J. SARS-CoV-2, SARS-CoV, and MERS-COV: A comparative overview. *Infez. Med.* **2020**, *28*, 174–184.
20. Peeri, N.C.; Shrestha, N.; Rahman, M.S.; Zaki, R.; Tan, Z.; Bibi, S.; Baghbanzadeh, M.; Aghamohammadi, N.; Zhang, W.; Haque, U. The SARS, MERS and novel coronavirus (COVID-19) epidemics, the newest and biggest global health threats: What lessons have we learned? *Int. J. Epidemiol.* **2020**, *49*, 717–726. [CrossRef]
21. Gordon, D.E.; Hiatt, J.; Bouhaddou, M.; Rezelj, V.V.; Ulferts, S.; Braberg, H.; Jureka, A.S.; Obernier, K.; Guo, J.Z.; Batra, J.; et al. Comparative host-coronavirus protein interaction networks reveal pan-viral disease mechanisms. *Science* **2020**, *370*, eabe9403. [CrossRef]
22. Heurich, A.; Hofmann-Winkler, H.; Gierer, S.; Liepold, T.; Jahn, O.; Pohlmann, S. TMPRSS2 and ADAM17 Cleave ACE2 differentially and only proteolysis by TMPRSS2 augments entry driven by the severe acute respiratory syndrome coronavirus spike protein. *J. Virol.* **2013**, *88*, 1293–1307. [CrossRef]
23. Belouzard, S.; Millet, J.K.; Licitra, B.N.; Whittaker, G.R. Mechanisms of coronavirus cell entry mediated by the viral spike protein. *Viruses* **2012**, *4*, 1011–1033. [CrossRef] [PubMed]
24. Hulswit, R.J.G.; de Haan, C.A.M.; Bosch, B.J. Coronavirus spike protein and tropism changes. *Adv. Virus Res.* **2016**, *96*, 29–57. [PubMed]
25. Kirchdoerfer, R.N.; Cottrell, C.A.; Wang, N.; Pallesen, J.; Yassine, H.M.; Turner, H.L.; Corbett, K.S.; Graham, B.S.; McLellan, J.S.; Ward, A.B. Pre-fusion structure of a human coronavirus spike protein. *Nature* **2016**, *531*, 118–121. [CrossRef]
26. Pillay, T.S. Gene of the month: The 2019-nCoV/SARS-CoV-2 novel coronavirus spike protein. *J. Clin. Pathol.* **2020**, *73*, 366–369. [CrossRef]
27. Xia, S.; Liu, M.; Wang, C.; Xu, W.; Lan, Q.; Feng, S.; Qi, F.; Bao, L.; Du, L.; Liu, S.; et al. Inhibition of SARS-CoV-2 (previously 2019-nCoV) infection by a highly potent pan-coronavirus fusion inhibitor targeting its spike protein that harbors a high capacity to mediate membrane fusion. *Cell Res.* **2020**, *30*, 343–355. [CrossRef]
28. Roviello, V.; Musumeci, D.; Mokhir, A.; Roviello, G.N. Evidence of protein binding by a nucleopeptide based on a thymine-decorated L-diaminopropanoic acid through CD and in silico studies. *Curr. Med. Chem.* **2021**. [CrossRef] [PubMed]
29. Kim, C.-H. SARS-CoV-2 evolutionary adaptation toward host entry and recognition of receptor O-acetyl sialylation in virus–host interaction. *Int. J. Mol. Sci.* **2020**, *21*, 4549. [CrossRef]
30. Artese, A.; Svicher, V.; Costa, G.; Salpini, R.; Di Maio, V.C.; Alkhatib, M.; Ambrosio, F.A.; Santoro, M.M.; Assaraf, Y.G.; Alcaro, S.; et al. Current status of antivirals and druggable targets of SARS CoV-2 and other human pathogenic coronaviruses. *Drug Resist. Updates* **2020**, *53*, 100721. [CrossRef] [PubMed]
31. Schoeman, D.; Fielding, B.C. Coronavirus envelope protein: Current knowledge. *Virol. J.* **2019**, *16*, 69. [CrossRef] [PubMed]
32. Chen, B.; Tian, E.-K.; He, B.; Tian, L.; Han, R.; Wang, S.; Xiang, Q.; Zhang, S.; El Arnaout, T.; Cheng, W. Overview of lethal human coronaviruses. *Signal Transduct. Target. Ther.* **2020**, *5*, 89. [CrossRef]
33. Cucinotta, D.; Vanelli, M. WHO declares COVID-19 a pandemic. *Acta Biomed.* **2020**, *91*, 157–160. [PubMed]
34. Mercatelli, D.; Holding, A.N.; Giorgi, F.M. Web tools to fight pandemics: The COVID-19 experience. *Brief. Bioinform.* **2020**, bbaa261. [CrossRef] [PubMed]
35. Arthi, V.; Parman, J. Disease, downturns, and wellbeing: Economic history and the long-run impacts of COVID-19. *Explorat. Econ. History* **2020**, 101381.
36. Roviello, V.; Roviello, G.N. Lower COVID-19 mortality in Italian forested areas suggests immunoprotection by Mediterranean plants. *Environ. Chem. Lett.* **2020**, *19*, 699–710. [CrossRef] [PubMed]
37. Ibn-Mohammed, T.; Mustapha, K.B.; Godsell, J.; Adamu, Z.; Babatunde, K.A.; Akintade, D.D.; Acquaye, A.; Fujii, H.; Ndiaye, M.M.; Yamoah, F.A.; et al. A critical analysis of the impacts of COVID-19 on the global economy and ecosystems and opportunities for circular economy strategies. *Resour. Conserv. Recycl.* **2021**, *164*, 105169. [CrossRef]
38. Naveja, J.J.; Madariaga-Mazón, A.; Flores-Murrieta, F.; Granados-Montiel, J.; Maradiaga-Ceceña, M.; Alaniz, V.D.; Maldonado-Rodriguez, M.; García-Morales, J.; Senosiain-Peláez, J.P.; Martínez-Mayorga, K. Union is strength: Antiviral and anti-inflammatory drugs for COVID-19. *Drug Disc. Today* **2020**, *26*, 229–239. [CrossRef]
39. Zhang, W.; Zhao, Y.; Zhang, F.; Wang, Q.; Li, T.; Liu, Z.; Wang, J.; Qin, Y.; Zhang, X.; Yan, X. The use of anti-inflammatory drugs in the treatment of people with severe coronavirus disease 2019 (COVID-19): The Perspectives of clinical immunologists from China. *Clin. Immunol.* **2020**, *214*, 108393. [CrossRef]
40. Costanzo, M.; De Giglio, M.A.R.; Roviello, G.N. SARS CoV-2: Recent reports on antiviral therapies Based on lopinavir/ritonavir, darunavir/umifenovir, hydroxychloroquine, remdesivir, favipiravir and other drugs for the treatment of the new coronavirus. *Curr. Med. Chem.* **2020**, *27*, 4536–4541. [CrossRef]
41. Singh, T.U.; Parida, S.; Lingaraju, M.C.; Kesavan, M.; Kumar, D.; Singh, R.K. Drug repurposing approach to fight COVID-19. *Pharmacol. Rep.* **2020**, *72*, 1479–1508. [CrossRef]
42. Borbone, N.; Piccialli, G.; Roviello, G.N.; Oliviero, G. Nucleoside Analogs and Nucleoside Precursors as Drugs in the Fight against SARS-CoV-2 and Other Coronaviruses. *Molecules* **2021**, *26*, 986. [CrossRef]
43. Ang, L.; Lee, H.W.; Choi, J.Y.; Zhang, J.; Lee, M.S. Herbal medicine and pattern identification for treating COVID-19: A rapid review of guidelines. *Integr. Med. Res.* **2020**, *9*, 100407. [CrossRef]

44. Li, Y.; Liu, X.; Guo, L.; Li, J.; Zhong, D.; Zhang, Y.; Clarke, M.; Jin, R. Traditional Chinese herbal medicine for treating novel coronavirus (COVID-19) pneumonia: Protocol for a systematic review and meta-analysis. *Syst. Rev.* **2020**, *9*, 1–6. [CrossRef] [PubMed]
45. Ang, L.; Lee, H.W.; Kim, A.; Lee, J.A.; Zhang, J.; Lee, M.S. Herbal medicine for treatment of children diagnosed with COVID-19: A review of guidelines. *Complement. Ther. Clin. Pract.* **2020**, *14*, 101174. [CrossRef]
46. Shahrajabian, M.H.; Sun, W.; Shen, H.; Cheng, Q. Chinese herbal medicine for SARS and SARS-CoV-2 treatment and prevention, encouraging using herbal medicine for COVID-19 outbreak. *Acta Agric. Scand. Sect. B Soil Plant Sci.* **2020**, *70*, 437–443. [CrossRef]
47. Lee, D.Y.; Li, Q.Y.; Liu, J.; Efferth, T. Traditional Chinese herbal medicine at the forefront battle against COVID-19: Clinical experience and scientific basis. *Phytomedicine* **2021**, *80*, 153337. [CrossRef]
48. Wei, H.; Liu, S.; Liao, Y.; Ma, C.; Wang, D.; Tong, J.; Feng, J.; Yi, T.; Zhu, L. A systematic review of the medicinal potential of mulberry in treating diabetes mellitus. *Am. J. Chin. Med.* **2019**, *46*, 1743–1770. [CrossRef] [PubMed]
49. Ng, K.-W.; Cao, Z.-J.; Chen, H.-B.; Zhao, Z.-Z.; Zhu, L.; Yi, T. Oolong tea: A critical review of processing methods, chemical composition, health effects, and risk. *Crit. Rev. Food Sci. Nutr.* **2017**, *58*, 2957–2980. [CrossRef]
50. Yi, T.; Chen, H.-B.; Zhao, Z.-Z.; Yu, Z.-L.; Jiang, Z.-H. Comparison of the chemical profiles and anti-platelet aggregation effects of two "Dragon's Blood" drugs used in traditional Chinese medicine. *J. Ethnopharmacol.* **2011**, *133*, 796–802. [CrossRef]
51. Yi, T.; Tang, Y.; Zhang, J.; Zhao, Z.; Yang, Z.; Chen, H. Characterization and determination of six flavonoids in the ethnomedicine "Dragon's Blood" by UPLC-PAD-MS. *Chem. Cent. J.* **2012**, *6*, 116. [CrossRef]
52. Xue, Y.; Zhu, L.; Yi, T. Fingerprint analysis of Resina Draconis by ultra-performance liquid chromatography. *Chem. Cent. J.* **2017**, *11*. [CrossRef] [PubMed]
53. Zheng, G.-Q.; Kenney, P.M.; Lam, L.K. Sesquiterpenes from clove (*Eugenia caryophyllata*) as potential anticarcinogenic agents. *J. Nat. Prod.* **1992**, *55*, 999–1003. [CrossRef] [PubMed]
54. Bhowmik, D.; Kumar, K.S.; Yadav, A.; Srivastava, S.; Paswan, S.; Dutta, A.S. Recent trends in Indian traditional herbs *Syzygium aromaticum* and its health benefits. *J. Pharm. Phytochem.* **2012**, *1*, 13–22.
55. Cortés-Rojas, D.F.; de Souza, C.R.F.; Oliveira, W.P. Clove (*Syzygium aromaticum*): A precious spice. *Asian Pac. J. Trop. Biomed.* **2014**, *4*, 90–96. [CrossRef]
56. Aisha, A.F.; Abu-Salah, K.M.; Alrokayan, S.A.; Siddiqui, M.J.; Ismail, Z.; Majid, A.M.S.A. *Syzygium aromaticum* extracts as good source of betulinic acid and potential anti-breast cancer. *Rev. Bras. Farm.* **2012**, *22*, 335–343. [CrossRef]
57. Prashar, A.; Locke, I.C.; Evans, C.S. Cytotoxicity of clove (*Syzygium aromaticum*) oil and its major components to human skin cells. *Cell Prolif.* **2006**, *39*, 241–248. [CrossRef]
58. Banerjee, S.; Panda, C.K.; Das, S. Clove (*Syzygium aromaticum* L.), a potential chemopreventive agent for lung cancer. *Carcinogenesis* **2006**, *27*, 1645–1654. [CrossRef]
59. Aisha, A.; Nassar, Z.; Siddiqui, M.; Abu-Salah, K.; Alrokayan, S.; Ismail, Z.; Abdul Majid, A. Evaluation of antiangiogenic, cytotoxic and antioxidant effects of *Syzygium aromaticum* L. extracts. *Asian J. Biol. Sci.* **2011**, *4*, 282–290. [CrossRef]
60. Ogata, M.; Hoshi, M.; Urano, S.; Endo, T. Antioxidant activity of eugenol and related monomeric and dimeric compounds. *Chem. Pharm. Bull.* **2000**, *48*, 1467–1469. [CrossRef] [PubMed]
61. Darshan, S.; Doreswamy, R. Patented antiinflammatory plant drug development from traditional medicine. *Phytother. Res.* **2004**, *18*, 343–357. [CrossRef]
62. Miyazawa, M.; Hisama, M. Suppression of chemical mutagen-induced SOS response by alkylphenols from clove (*Syzygium aromaticum*) in the Salmonella typhimurium TA1535/pSK1002 umu test. *J. Agric. Food Chem.* **2001**, *49*, 4019–4025. [CrossRef]
63. Vijayasteltar, L.; Nair, G.G.; Maliakel, B.; Kuttan, R.; Krishnakumar, I. Safety assessment of a standardized polyphenolic extract of clove buds: Subchronic toxicity and mutagenicity studies. *Toxicol. Rep.* **2016**, *3*, 439–449. [CrossRef]
64. Ogunwande, I.; Olawore, N.; Ekundayo, O.; Walker, T.; Schmidt, J.; Setzer, W. Studies on the essential oils composition, antibacterial and cytotoxicity of *Eugenia uniflora* L. *Int. J. Aromather.* **2005**, *15*, 147–152. [CrossRef]
65. Udensi, U.K.; Tchounwou, P.B. Potassium homeostasis, oxidative stress, and human disease. *Int. J. Clin. Exp. Physiol.* **2017**, *4*, 111.
66. Horsburgh, M.J.; Wharton, S.J.; Karavolos, M.; Foster, S.J. Manganese: Elemental defence for a life with oxygen. *Trends Microbiol.* **2002**, *10*, 496–501. [CrossRef]
67. Fang, S.; Thomas, R.M.; Conklin, J.L.; Oberley, L.W.; Christensen, J. Co-localization of manganese superoxide dismutase and NADH diaphorase. *J. Histochem. Cytochem.* **1995**, *43*, 849–855. [CrossRef] [PubMed]
68. Cao, Y.; Qin, Y.; Bruist, M.; Gao, S.; Wang, B.; Wang, H.; Guo, X. Formation and dissociation of the interstrand i-motif by the sequences d(XnC4Ym) monitored with electrospray ionization mass spectrometry. *J. Am. Soc. Mass Spectrom.* **2015**, *26*, 994–1003. [CrossRef] [PubMed]
69. Wilson, J.X. Mechanism of action of vitamin C in sepsis: Ascorbate modulates redox signaling in endothelium. *Biofactors* **2009**, *35*, 5–13. [CrossRef]
70. Bagchi, D.; Garg, A.; Krohn, R.; Bagchi, M.; Tran, M.; Stohs, S. Oxygen free radical scavenging abilities of vitamins C and E, and a grape seed proanthocyanidin extract in vitro. *Res. Commun. Mol. Pathol. Pharmacol.* **1997**, *95*, 179–189.
71. Pino, J.A.; Marbot, R.; Agüero, J.; Fuentes, V. Essential oil from buds and leaves of clove (*Syzygium aromaticum* (L.) Merr. et Perry) grown in Cuba. *J. Essent. Oil Res.* **2001**, *13*, 278–279. [CrossRef]
72. Raina, V.; Srivastava, S.; Aggarwal, K.; Syamasundar, K.; Kumar, S. Essential oil composition of *Syzygium aromaticum* leaf from Little Andaman, India. *Flavour Fragr. J.* **2001**, *16*, 334–336. [CrossRef]

73. Zachariah, T.; Krishnamoorthy, B.; Rema, J.; Mathew, P. Oil constituents in bud and pedicel of clove (*Syzygium aromaticum*). *Ind. Perf.* **2005**, *49*, 313.
74. Mittal, M.; Gupta, N.; Parashar, P.; Mehra, V.; Khatri, M. Phytochemical evaluation and pharmacological activity of *Syzygium aromaticum*: A comprehensive review. *Int. J. Pharm. Pharm. Sci.* **2014**, *6*, 67–72.
75. Jimoh, S.O.; Arowolo, L.A.; Alabi, K.A. Phytochemical screening and antimicrobial evaluation of *Syzygium aromaticum* extract and essential oil. *Int. J. Curr. Microbiol. Appl. Sci.* **2017**, *6*, 4557–4567. [CrossRef]
76. El Ghallab, Y.; Al Jahid, A.; Eddine, J.J.; Said, A.A.H.; Zarayby, L.; Derfoufi, S. *Syzygium aromaticum* L.: Phytochemical investigation and comparison of the scavenging activity of essential oil, extracts and eugenol. *Adv. Tradit. Med.* **2020**, *20*, 153–158. [CrossRef]
77. Begum, S.; Siddiqui, B.S.; Khatoon, R.; Aftab, F. Phytochemical studies on *Syzygium aromaticum* Linn. *J. Chem. Soc. Pak.* **2014**, *36*, 512.
78. Chaieb, K.; Hajlaoui, H.; Zmantar, T.; Kahla-Nakbi, A.B.; Rouabhia, M.; Mahdouani, K.; Bakhrouf, A. The chemical composition and biological activity of clove essential oil, *Eugenia caryophyllata* (*Syzigium aromaticum* L. Myrtaceae): A short review. *Phytother. Res.* **2007**, *21*, 501–506. [CrossRef] [PubMed]
79. Gopalakrishnan, N.; Narayanan, C.; Mathew, A. Sesquiterpene hydrocarbons from clove oil. *Lebensmittel-Wissenschaft+ Technol.* **1984**, *17*, 42–43.
80. Zhao, C.; Qiao, X.; Shao, Q.; Hassan, M.; Ma, Z. Evolution of the lignin chemical structure during the bioethanol production process and its inhibition to enzymatic hydrolysis. *Energy Fuels* **2020**, *34*, 5938–5947. [CrossRef]
81. Frohlich, P.C.; Santos, K.A.; Palú, F.; Cardozo-Filho, L.; da Silva, C.; da Silva, E.A. Evaluation of the effects of temperature and pressure on the extraction of eugenol from clove (*Syzygium aromaticum*) leaves using supercritical CO_2. *J. Supercrit. Fluids* **2019**, *143*, 313–320. [CrossRef]
82. Bahramsoltani, R.; Rahimi, R. An evaluation of traditional Persian medicine for the management of SARS-CoV-2. *Front. Pharmacol.* **2020**, *11*. [CrossRef] [PubMed]
83. Öztürk, A.; Özbek, H. The anti-inflammatory activity of Eugenia caryophyllata essential oil: An animal model of anti-inflammatory activity. *Eur. J. Gen Med.* **2005**, *2*, 159–163.
84. Tanko, Y.; Mohammed, A.; Okasha, M.; Umah, A.; Magaji, R. Anti-nociceptive and anti-inflammatory activities of ethanol extract of *Syzygium aromaticum* flower bud in wistar rats and mice. *Afr. J. Trad. Complement. Altern. Med.* **2008**, *5*, 209–212. [CrossRef] [PubMed]
85. Martin, S.; Padilla, E.; Ocete, M.; Galvez, J.; Jimenez, J.; Zarzuelo, A. Anti-inflammatory activity of the essential oil of *Bupleurum fruticescens*. *Planta Med.* **1993**, *59*, 533–536. [CrossRef]
86. Rho, H.S.; Ghimeray, A.K.; Yoo, D.S.; Ahn, S.M.; Kwon, S.S.; Lee, K.H.; Cho, D.H.; Cho, J.Y. Kaempferol and kaempferol rhamnosides with depigmenting and anti-inflammatory properties. *Molecules* **2011**, *16*, 3338–3344. [CrossRef]
87. García-Mediavilla, V.; Crespo, I.; Collado, P.S.; Esteller, A.; Sánchez-Campos, S.; Tuñón, M.J.; González-Gallego, J. The anti-inflammatory flavones quercetin and kaempferol cause inhibition of inducible nitric oxide synthase, cyclooxygenase-2 and reactive C-protein, and down-regulation of the nuclear factor kappaB pathway in Chang Liver cells. *Eur. J. Pharmacol.* **2007**, *557*, 221–229. [CrossRef]
88. Jnawali, H.N.; Lee, E.; Jeong, K.-W.; Shin, A.; Heo, Y.-S.; Kim, Y. Anti-inflammatory activity of rhamnetin and a model of its binding to c-Jun NH2-terminal kinase 1 and p38 MAPK. *J. Nat. Prod.* **2014**, *77*, 258–263. [CrossRef]
89. Novo Belchor, M.; Hessel Gaeta, H.; Fabri Bittencourt Rodrigues, C.; Ramos da Cruz Costa, C.; de Oliveira Toyama, D.; Domingues Passero, L.F.; Dalastra Laurenti, M.; Hikari Toyama, M. Evaluation of rhamnetin as an inhibitor of the pharmacological effect of secretory phospholipase A2. *Molecules* **2017**, *22*, 1441. [CrossRef]
90. Daniel, A.N.; Sartoretto, S.M.; Schmidt, G.; Caparroz-Assef, S.M.; Bersani-Amado, C.A.; Cuman, R.K.N. Anti-inflammatory and antinociceptive activities A of eugenol essential oil in experimental animal models. *Rev. Bras. Farm.* **2009**, *19*, 212–217. [CrossRef]
91. Magalhães, C.B.; Riva, D.R.; DePaula, L.J.; Brando-Lima, A.; Koatz, V.L.G.; Leal-Cardoso, J.H.; Zin, W.A.; Faffe, D.S. In vivo anti-inflammatory action of eugenol on lipopolysaccharide-induced lung injury. *J. Appl. Physiol.* **2010**, *108*, 845–851. [CrossRef]
92. Chniguir, A.; Zioud, F.; Marzaioli, V.; El-Benna, J.; Bachoual, R. *Syzygium aromaticum* aqueous extract inhibits human neutrophils myeloperoxidase and protects mice from LPS-induced lung inflammation. *Pharm. Biol.* **2019**, *57*, 55–63. [CrossRef]
93. Nassan, M.; Mohamed, E.; Abdelhafez, S.; Ismail, T. Effect of clove and cinnamon extracts on experimental model of acute hematogenous pyelonephritis in albino rats: Immunopathological and antimicrobial study. *Int. J. Immunopathol. Pharmacol.* **2015**, *28*, 60–68. [CrossRef]
94. Su, H.; Yang, M.; Wan, C.; Yi, L.-X.; Tang, F.; Zhu, H.-Y.; Yi, F.; Yang, H.-C.; Fogo, A.B.; Nie, X. Renal histopathological analysis of 26 postmortem findings of patients with COVID-19 in China. *Kidney Int.* **2020**, *98*, 219–227. [CrossRef]
95. Carrasco, F.R.; Schmidt, G.; Romero, A.L.; Sartoretto, J.L.; Caparroz-Assef, S.M.; Bersani-Amado, C.A.; Cuman, R.K.N. Immunomodulatory activity of *Zingiber officinale Roscoe*, *Salvia officinalis* L. and *Syzygium aromaticum* L. essential oils: Evidence for humor-and cell-mediated responses. *J. Pharm. Pharmacol.* **2009**, *61*, 961–967. [CrossRef] [PubMed]
96. Saeed, S.A.; Gilani, A.H. Antithrombotic activity of clove oil. *J. Pak. Med. Assoc.* **1994**, *44*, 112. [PubMed]
97. Yang, Y.-Y.; Lee, M.-J.; Lee, H.-S.; Park, W.-H. Screening of antioxidative, anti-platelet aggregation and anti-thrombotic effects of clove extracts. *J. Physiol. Pathol. Korean Med.* **2011**, *25*, 471–481.
98. Rasheed, A.; Laekeman, G.; Totte, J.; Vlietinck, A.; Herman, A. Eugenol and prostaglandin biosynthesis. *N. Engl. J. Med.* **1984**, *310*, 50–51. [PubMed]

99. Srivastava, K. Antiplatelet principles from a food spice clove (*Syzgium aromaticum* L). *Prostag. Leukot. Essent. Fat. Acids* **1993**, *48*, 363–372. [CrossRef]
100. Im LEE, J.; Lee, H.S.; Jun, W.J.; Yu, K.W.; Shin, D.H.; Hong, B.S.; Cho, H.Y.; Yang, H.C. Purification and Characterization of Antithrombotics from Syzygium aromaticum (L.) MERR. & PERRY. *Biol. and Pharm. Bull.* **2001**, *24*, 181–187. [CrossRef]
101. Tragoolpua, Y.; Jatisatienr, A. Anti-herpes simplex virus activities of *Eugenia caryophyllus* (Spreng.) Bullock & SG Harrison and essential oil, eugenol. *Phytother. Res.* **2007**, *21*, 1153–1158. [PubMed]
102. Hussein, G.; Miyashiro, H.; Nakamura, N.; Hattori, M.; Kakiuchi, N.; Shimotohno, K. Inhibitory effects of Sudanese medicinal plant extracts on hepatitis C virus (HCV) protease. *Phytother. Res.* **2000**, *14*, 510–516. [CrossRef]
103. Benencia, F.; Courreges, M. In vitro and in vivo activity of eugenol on human herpesvirus. *Phytother. Res.* **2000**, *14*, 495–500. [CrossRef]
104. Dai, J.-P.; Zhao, X.-F.; Zeng, J.; Wan, Q.-Y.; Yang, J.-C.; Li, W.-Z.; Chen, X.-X.; Wang, G.-F.; Li, K.-S. Drug screening for autophagy inhibitors based on the dissociation of Beclin1-Bcl2 complex using BiFC technique and mechanism of eugenol on anti-influenza A virus activity. *PLoS ONE* **2013**, *8*, e61026. [CrossRef]
105. Lane, T.; Anantpadma, M.; Freundlich, J.S.; Davey, R.A.; Madrid, P.B.; Ekins, S. The natural product eugenol is an inhibitor of the ebola virus in vitro. *Pharm. Res.* **2019**, *36*, 1–6. [CrossRef] [PubMed]
106. Kurokawa, M.; Hozumi, T.; Basnet, P.; Nakano, M.; Kadota, S.; Namba, T.; Kawana, T.; Shiraki, K. Purification and Characterization of Eugeniin as an Anti-herpesvirus Compound from *Geum japonicum* and *Syzygium aromaticum*. *J. Pharmacol. Exp. Ther.* **1998**, *284*, 728–735.
107. Saleem, H.N.; Batool, F.; Mansoor, H.J.; Shahzad-ul-Hussan, S.; Saeed, M. Inhibition of dengue virus protease by Eugeniin, Isobiflorin, and Biflorin Isolated from the Flower Buds of *Syzygium aromaticum* (Cloves). *ACS Omega* **2019**, *4*, 1525–1533. [CrossRef]
108. Kanyinda, J.N.M. Coronavirus (COVID-19): A protocol for prevention and treatment (Covalyse®). *Eur. J. Med. Health Sci.* **2020**, *2*, 1–4. [CrossRef]
109. Samaddar, A.; Gadepalli, R.; Nag, V.L.; Misra, S. The enigma of low COVID-19 fatality rate in India. *Front. Genet.* **2020**, *11*, 854. [CrossRef]
110. Rhodes, J.M.; Subramanian, S.; Laird, E.; Kenny, R.A. Low population mortality from COVID-19 in countries south of latitude 35 degrees North supports vitamin D as a factor determining severity. *Aliment. Pharmacol. Ther.* **2020**, *51*, 1434–1437. [CrossRef]
111. Singh, N.A.; Kumar, P.; Kumar, N. Spices and herbs: Potential antiviral preventives and immunity boosters during COVID-19. *Phytother. Res.* **2021**, *29*. [CrossRef]
112. Chaachouay, N.; Douira, A.; Zidane, L. COVID-19, prevention and treatment with herbal medicine in the herbal markets of Salé Prefecture, North-Western Morocco. *Eur. J. Integrat. Med.* **2021**, *42*, 101285. [CrossRef] [PubMed]
113. Pandey, P.; Singhal, D.; Khan, F.; Arif, M. An in silico screening on *Piper nigrum*, *Syzygium aromaticum* and *Zingiber officinale* roscoe derived compounds against SARS-CoV-2: A drug repurposing approach. *Biointerface Res. Appl. Chem.* **2020**, *11*, 11122–11134.
114. Joshi, T.; Joshi, T.; Sharma, P.; Mathpal, S.; Pundir, H.; Bhatt, V.; Chandra, S. In silico screening of natural compounds against COVID-19 by targeting Mpro and ACE2 using molecular docking. *Eur. Rev. Med. Pharmacol. Sci.* **2020**, *24*, 4529–4536. [PubMed]
115. Rehman, M.; AlAjmi, M.F.; Hussain, A. Natural compounds as inhibitors of SARS-CoV-2 main protease (3CLpro): A molecular docking and simulation approach to combat COVID-19. *Curr. Pharm. Des.* **2020**, *16*. [CrossRef]

molecules MDPI

Review

Strategies Tackling Viral Replication and Inflammatory Pathways as Early Pharmacological Treatment for SARS-CoV-2 Infection: Any Potential Role for Ketoprofen Lysine Salt?

Domenica Francesca Mariniello [1,2], Valentino Allocca [1,2,†], Vito D'Agnano [1,2,†], Riccardo Villaro [3], Luigi Lanata [4], Michela Bagnasco [4], Luigi Aronne [1,2], Andrea Bianco [1,2] and Fabio Perrotta [1,2,*]

1 Department of Translational Medical Sciences, University of Campania "L. Vanvitelli", 80131 Naples, Italy
2 U.O.C. Clinica Pneumologica "L. Vanvitelli", A.O. dei Colli, Ospedale Monaldi, 80131 Naples, Italy
3 Section of Infectious Diseases, Department of Clinical Medicine and Surgery, University of Naples Federico II, 80131 Naples, Italy
4 Medical Deptartment, Dompé Farmaceutici SpA, 20122 Milan, Italy
* Correspondence: fabio.perrotta@unicampania.it
† These authors contributed equally to this work.

Abstract: COVID-19 is an infective disease resulting in widespread respiratory and non-respiratory symptoms prompted by SARS-CoV-2 infection. Interaction between SARS-CoV-2 and host cell receptors prompts activation of pro-inflammatory pathways which are involved in epithelial and endothelial damage mechanisms even after viral clearance. Since inflammation has been recognized as a critical step in COVID-19, anti-inflammatory therapies, including both steroids and non-steroids as well as cytokine inhibitors, have been proposed. Early treatment of COVID-19 has the potential to affect the clinical course of the disease regardless of underlying comorbid conditions. Non-steroidal anti-inflammatory drugs (NSAIDs), which are widely used for symptomatic relief of upper airway infections, became the mainstay of early phase treatment of COVID-19. In this review, we discuss the current evidence for using NSAIDs in early phases of SARS-CoV-2 infection with focus on ketoprofen lysine salt based on its pharmacodynamic and pharmacokinetic features.

Keywords: NSAIDs; COVID-19; SARS-CoV-2; inflammation; ketoprofen

Citation: Mariniello, D.F.; Allocca, V.; D'Agnano, V.; Villaro, R.; Lanata, L.; Bagnasco, M.; Aronne, L.; Bianco, A.; Perrotta, F. Strategies Tackling Viral Replication and Inflammatory Pathways as Early Pharmacological Treatment for SARS-CoV-2 Infection: Any Potential Role for Ketoprofen Lysine Salt? *Molecules* **2022**, *27*, 8919. https://doi.org/10.3390/molecules27248919

Academic Editors: Giovanni N. Roviello and Caterina Vicidomini

Received: 16 November 2022
Accepted: 12 December 2022
Published: 15 December 2022

Publisher's Note: MDPI stays neutral with regard to jurisdictional claims in published maps and institutional affiliations.

1. Introduction

The severe acute respiratory syndrome coronavirus 2 (SARS-CoV-2), responsible for the novel coronavirus disease 2019 (COVID-19), has been rapidly transmitted around the world during the last three years, causing a global public health emergency [1]. Coronaviruses are pathogens that largely affect the respiratory system but the expression of host SARS-CoV-2 receptor, angiotensin-converting enzyme 2 (ACE2), is not lung-specific, and its presence in a variety of tissues, including the brain, the intestine, the blood vessels, and the kidney, could subject these organs to direct infection by SARS-CoV-2 making COVID-19 a systemic disease [2]. Most patients commonly present with fever, myalgia, shortness of breath, malaise, and dry cough, although patients may present with asymptomatic, mild, moderate, or severe disease. High systemic levels of cytokines, referred to as "cytokine storm", have frequently been found in severe COVID-19 disease. Therefore, targeting inflammation is one of the key strategies in the management of COVID-19 disease. In this scenario, recent data suggest that NSAIDs may represent a safe strategy in the treatment of SARS-CoV-2 infection [3].

SARS-CoV-2 Pathogenesis and Associated Damage Mechanisms

SARS-CoV-2 enters the human body via the upper airways. In concordance with other coronaviruses, it is transmitted mainly through respiratory droplets, although other ways of transmission are reported such as aerosol, contact with contaminated surfaces, and fecal–oral transmission [4]. Initially, the virus proliferates in the upper respiratory tract and then the virus moves into the lower respiratory airway. The virus is made up of six accessory proteins and four main structural proteins: nucleocapsid (N), membrane (M), envelope (E), and spike (S), the latter one responsible for viral entry [5]. SARS-CoV2 binds to the host receptor membrane angiotensin-converting enzyme 2 (ACE2) [6,7]. To enable viral entry, the spike (S) protein is cleaved into two subunits, S1 and S2; the S1 subunit is responsible for binding to ACE2 [8]. Type 2 transmembrane cellular proteases (TMPRSS2) mediate the proteolytic cleavage of S-protein [9]. The viral genome is translated by host machinery into a polyprotein, virions are assembled and then released by exocytosis, ready to infect other cells. After viral entry, ACE2 receptor is internalized and degraded [10]. Within a few hours of infection, the recognition of viral components called pathogen-associated molecular patterns (PAMPs) by pattern recognition receptors (PRRs) in the alveolar cells and site of invasion, such as Toll-like receptors and retinoic acid-inducible gene-1 (RIG-1)-like receptors, activate the innate immune response [11]. Nuclear factor kB (NF-κB) and interferon regulatory factor 3 (IRF3) initiate the expression of interferon type 1 and cytokines such as tumor necrosis factor alpha (TNF-a), interleukin-1 (IL-1), and IL-6. These cytokines represent the first line of host defense and subsequently induce adaptive immune responses [12]. The upregulation of the NF-κB signaling pathway leads to cytokine storm and hyperinflammation, with a greater risk of severe COVID-19 disease [13,14]. Regarding adaptive immune response, T lymphocytes attempt to eliminate virus by killing infected cells and secrete cytokines to amplify T lymphocytes immune response [15]. Severe COVID-19 patients often exhibit lymphopenia with a reduction in the number of T-cells, probably due to chemotaxis of these cells toward the site of inflammation or to direct cell damage provoked by SARS-CoV-2 [16].

B lymphocytes produce antibodies specific to SARS-CoV-2, in particular antibodies versus spike protein, which may neutralize the virus and stimulate systemic immunity. Reduction in the B cell count has been observed in patients with severe COVID-19.

Early autoptic studies documented in patients deceased from COVID-19 diffuse pulmonary microthrombi that appear to form directly within pulmonary vessels and do not result from embolic propagation of peripheral venous thrombi [17–19]. In severe COVID-19 patients, reduction in platelet count with augmentation of fibrinogen and dimer levels is commonly observed [1]. There are several molecular/cellular pathways that potentially can explain this hypercoagulable state in COVID-19: it may involve ACE2, a carboxypeptidase responsible for the conversion of angiotensin II (Ang II) to angiotensin 1-7. As mentioned above, after binding with the spike protein of the virus, ACE2 is internalized and degraded [20], leading to up-regulation of Ang II signaling and consequentially of pro-thrombotic pathways. On the other hand, the level of angiotensin 1-7 is diminished [21]. Angiotensin 1-7 mediates an anti-thrombotic activity by the production of nitric oxide and prostacyclin and the inhibition of platelet activation [22]. In addition, the hyperinflammation in COVID-19 disease mediated by cytokines such as IL-1, TNF-α, and IL-6 also leads to a hypofibrinolytic state through an increase of fibrinogen and plasminogen activator inhibitor-1 (PAI-1) [23]. Inflammatory cytokines promote the activation and aggregation of platelets, the activation of vascular endothelial cells (EC) and the expression of tissue factor (TF) on the surface of endothelium cells and leucocytes [24]. Furthermore, during inflammation, the production of natural anticoagulants such as antithrombin III, tissue factor inhibitor, and Protein C is diminished [25]. A vicious circle is established as the coagulation cascade can in turn promote inflammation. In fact, thrombin is a major activator of protease-activated receptor 1 (PAR 1) that contributes to the release of IL-1, IL-2, IL-6, IL-8, TNFα and the expression of adhesion molecules such as E- and P-selectin and ICAM-1 on the endothelial surface involved in immune cell recruitment [26]. This review aims to summarize the current therapeutic scenario for the early phases of COVID-19 focusing on the role of anti-inflammatory treatment that appears still controversial.

2. Pharmacological Agents with Selective Activity against SARS-CoV-2

2.1. Antivirals Targeting SARS-CoV-2

The group of antiviral drugs comprises several molecules directly targeting the pathogen to hinder its growth. According to their different mechanisms of action, antiviral drugs may be categorized into three main subgroups. (1) Inhibitors of S protein; (2) inhibitors of viral proteases; (3) inhibitors of viral RNA dependent RNA polymerase; (4) host-oriented.

2.1.1. Entry Inhibitors

As stated above, S protein is responsible for virus entry. The S protein is a trans-membrane protein with N-exo and C-endo terminals. The N terminal S1 subunit contains receptor binding domain (RBD), known to interact with the peptidase domain of ACE2 and to be the main target of neutralizing antibodies, while the C terminal S2 subunit induces membrane fusion. Monoclonal antibodies (mAbs) are laboratory-produced molecules which derive from natural B cells of subjects who have experienced or been injected with the antigen of interest. As a result, mAbs are able to mimic a normal immune response against a predetermined antigen [27]. During SARS-CoV-2 pandemic, various mABs have been progressively approved worldwide, whilst other are currently under investigation [28]. Target population for treatment with Abs is represented by high-risk patients with symptomatic mild to moderate infection, not requiring supplemental oxygen due to COVID-19. High risk features, among others, include older age, chronic kidney disease, obesity, cardiovascular and metabolic disease, and chronic lung diseases.

Bamlanivimab-Etesevimab

In the BLAZE-1 trial (*NCT04427501*), the cocktail of bamlanivimab and etesevimab, has been reported to significantly reduce hospitalizations and death rate in high-risk patients affected by COVID-19. At the end of 29 days observational period, in the bamlanivimab-etesevimab group, a total of 11 of 518 patients (2.1%) experienced COVID-19 related hospitalization or death from any cause, as compared with 36 of 517 patients (7.0%) in the placebo group (absolute risk difference, −4.8 percentage points; 95% confidence interval [CI], −7.4 to −2.3; relative risk difference, 70%; $p < 0.001$). Contrary to placebo group, where nine of the ten occurred deaths have been designed as COVID-19 related, no patients died in the bamlanivimab-etesevimab group [29].

Casirivimab and Imdevimab

REGN-COV2 (casirivimab and imdevimab; *NCT04452318*) represents a mixture of the human Abs, casirivimab, and imdevimab. Although identified by different methods, they both target the S protein RBD. Subcutaneous casirivimab and imdevimab, 1200 mg, succeeded in preventing progression to symptomatic disease compared to placebo; odds ratio, 0.54 [95% CI, 0.30–0.97]; $p = 0.04$; absolute risk difference, −13.3% [95% CI, −26.3% to −0.3%]). Furthermore, the combination of casirivimab and imdevimab has been reported to reduce the number of symptomatic weeks per 1000 participants (895.7 weeks vs. 1637.4 weeks with placebo; $p = 0.03$) with an approximately 5.6-day reduction in symptom duration per symptomatic participant. With respect to adverse events, the proportion of participants receiving casirivimab and imdevimab who experienced one or more treatment-emergent adverse events was 33.5% compared to 48.1% for placebo, including events related (25.8% vs. 39.7%) or not related (11.0% vs. 16.0%) to COVID-19 [30].

Sotrovimab

Sotrovimab, a pan-sarbecovirus monoclonal antibody was shown to significantly reduce the risk of disease progression, leading to hospitalization (for >24 h) for any cause or death, as demonstrated in a phase 3, multicenter, randomized, doubleblind, placebo-controlled trial. A total of 3 of 291 patients in the sotrovimab group (1%) experienced hospitalization for >24 h for any cause or death, in comparison to 21 of 292 patients in the placebo group (7%) (relative risk reduction 85%; 97.24% confidence interval [CI], 44 to 96; $p = 0.002$) [31].

Tixagevimab-Cilgavimab

Tixagevimab-cilgavimab is a neutralizing monoclonal antibody combination whose capability to improve outcomes for patients hospitalized with COVID-19 has lately been investigated. Although tixagevimab–cilgavimab did not improve the primary outcome of time to sustained recovery versus placebo, (89% for tixagevimab–cilgavimab and 86% for placebo group at day 90 [(recovery rate ratio [RRR] 1.08 [95% CI 0.97–1.20]; $p = 0.21$), it must be noted that mortality was lower in the tixagevimab–cilgavimab group (61 [9%]) versus placebo group (86 [12%]; hazard ratio [HR] 0.70 [95% CI 0.50–0.97]; $p = 0.032$) [32].

Other Anti-SARS-CoV-2 Monoclonal Antibodies

SARS-CoV-2 may mutate over time making certain treatments less useful and allowing the pandemic to spread. In this respect, variants of concern (VOCs) are continuously monitored due to their great impact on decreasing efficacy of treatment with mAbs. As omicron VOC has quickly spread becoming the dominant variant in US, bebtelovimab has been considered the only monoclonal antibody-based treatment approved by the Food and Drug Administration (FDA) at present [33]. Based on its capabilities to link S protein amino acids that are reported to be rarely mutated, bebtelovimab, a fully human immunoglobulin G1, may represent a long-term solution for COVID-19 treatment as suggested by Westendorf K et al. [34].

2.1.2. Inhibitors of Viral Proteases

SARS-CoV-2 is constituted by four conserved structural proteins—spike (S), envelope (E), membrane (M), and nucleocapsid (N)—and six accessory proteins. Among them, there are two recognized cysteine proteases—Mpro (3CLpro) and PLpro—which are essential for viral replication. Given the absence of human homolog as well as its important role in the viral gene expression, Mpro has been utilized as a potential molecular target.

Lopinavir/Ritonavir

Lopinavir is an anti-retroviral protease inhibitor employed in combination with ritonavir, in the treatment of HIV infection. Although lopinavir/ritonavir administration was not associated with a significant difference in the time to clinical improvement, in a post-hoc analysis, 28-day mortality was lower in treated population compared to control, albeit not significantly (19.2% vs. 25%) [35]. At present, lopinavir/ritonavir is not recommended for COVID-19 treatment and it can only be considered for patients included in clinical trials [33,36].

Nirmatrelvir/Ritonavir

Nirmatrelvir, a novel orally active inhibitor of 3CL protease inhibitor, in combination with ritonavir, has been investigated in a phase III trial in a cohort of 2246 symptomatic, unvaccinated, non-hospitalized adults at high risk for progression. Authors reported that the incidence of COVID-19-related hospitalization or death by day 28 was 0.77% (3 of 389 patients) in the nirmatrelvir group compared to 7.01% (27 of 385 patients) in the placebo group, with 7 deaths, and a 89.1% relative risk reduction. Results were confirmed in the final analysis involving the 1379 patients with a difference in terms of hospitalization of −5.81 percentage points (95% CI, −7.78 to −3.84; $p < 0.001$; relative risk reduction, 88.9%) [34]. Currently, ritonavir boosted-nirmatrelvir is recommended within 5 days of symptoms onset, for non-hospitalized adults and pediatric patients aged ≥12 years and weighing ≥40 kg affected by mild to moderate COVID-19 with high risk of disease progression [33].

2.1.3. Inhibitors of Viral RNA Dependent RNA Polymerase (RdRp)

Remdesivir

Originally developed for the treatment of Ebola and Marburg virus infections, remdesivir (GS-5734) was considered early on as a potential candidate for COVID-19 treatment due to its capacity to cause premature termination of SARS-CoV-2 viral RNA transcription. By acting as nucleotide analog, it is incorporated by the RdRp, and RNA synthesis is consequently inhibited. In the Adaptive COVID-19 Treatment Trial (ACTT-1), in a cohort of 1062 randomized patients, subjects who received remdesivir had a median recovery time of 10 days compared to 15 days of the control group (rate ratio for recovery, 1.29; 95% CI, 1.12 to 1.49; $p < 0.001$, by a log-rank test). With regard to mortality, remdesivir has shown superiority compared to placebo both at day 15 and day 29 (6.7% and 11.4% with remdesivir vs. 11.9 and 15.2% in control, respectively; hazard ratio, 0.73; 95% CI, 0.52 to 1.03). Currently, remdesivir has been recommended for hospitalized patients who require (BIIa) or do not require (BIII) conventional oxygen supplementation. Likewise, the use of remdesivir might be considered for hospitalized patients who require High Flow Nasal Cannula (HFNC) Oxygen or Non-Invasive Ventilation (NIV) in association with dexamethasone plus per os (PO) baricitinib (AI) or IV tocilizumab (BIIa) [37]. Remdesivir has also been recommended for nonhospitalized patients with mild to moderate COVID-19 who were at high risk of progression to severe disease.

Molnupiravir

Molnupiravir, the biological prodrug of NHC (β-D-N(4)-hydroxycytidine), represents another ribonucleoside analogue with activity against SARS-CoV-2 and other RNA viruses. Started within 5 days of the onset of signs or symptoms, molnupiravir has showed superiority in decreasing the risk of hospitalization for any cause or death through to day 29 compared to controls in a cohort of 1433 nonhospitalized adults with mild-to-moderate COVID-19 with at least one risk factor for severe COVID-19 (28 of 385 participants [7.3%]) than with placebo (53 of 377 [14.1%]) (difference, −6.8 percentage points; 95% confidence interval [CI], −11.3 to −2.4; $p = 0.001$) [38]. On December 2021, the Food and Drug Administration (FDA) approved monlupiravir for the treatment of adults with mild or moderate COVID-19, within five days of symptom onset, with no alternative antiviral therapies available.

2.1.4. Host-Oriented Therapies for SARS-CoV-2 Infection

Novel drugs targeting host immune factors named host-oriented therapies are currently under development. Virus-host cell interaction prompts the innate immune system response which may be ineffective in determining a complete viral clearance [39,40]. Therefore, the possibility to restore the altered immune responses with these host-oriented therapies offers a great opportunity against viral infections [41–43]. SNG001 is an inhaled drug containing INF-β, an antiviral protein produced during viral spread and it is under evaluation in clinical trial (NCT04385095). SARS-CoV-2 might weaken the immune system response also through the inhibition of IFN-β expression [44,45]. In this phase 2 trial, COVID-19 patients receiving inhaled SNG001 had greater improvement in clinical symptoms and recovered more rapidly than patients who received placebo [45]. Another host-oriented strategy in evaluation for COVID-19 is the possibility to use IL-7 to support the host's immune system. IL-7 promotes lymphocytic count increase counteracting the lymphocytopenia, a pathologic hallmark of severe COVID-19. In a recent study with a small number of patients with COVID-19, it was shown that IL-7 can be safely administered and it was associated with an increase in lymphocytes count, appearing to counteract a pathologic hallmark of COVID-19 [46]. These results were under evaluation in another trial (NCT04379076), actually terminated for poor accrual. The administration of IL-7 seems to improve clinical outcome as already demonstrated in septic patients [47,48]. However, for the lack of strong evidence to support the use of IL-7 in COVID-19 patients, more studies are needed to approve its clinical use. Other drugs which directly affect viral entry are under evaluation. Aprotinin, a serine protease inhibitor that could inhibit TMPRSS2, seems to have potential role in the control of SARS-CoV-2 replication, especially in early stages, and in the prevention of COVID-19 progression to a severe disease [47]. Host factors assisting in viral replication are other targets for host-oriented therapies [49]. Signaling via tyrosine kinases play a crucial role in viral replications and receptor tyrosine kinase inhibitors such as genistein could have a role to treat COVID-19 [50]. Genistein is an isoflavone with potent anti-inflammatory and immunomodulators effects due to its ability to modulate intracellular pathways such as PI3K, Akt, mTOR, NF-κB, PPARγ, AMPK, and Nrf2, preventing viral entry and therefore reducing lung injuries [42,50]. Host-oriented therapies also could be used to mitigate the cytokine storm induced by SARS-CoV-2 and other viral infection due hyper-activation of TLR. Several TLR-4 antagonists such as eritoran are under evaluation for its use in viral infection for their ability to mitigate host damage due to excessive inflammation [51].

3. Anti-Inflammatory Drugs in COVID-19

3.1. Corticosteroids Use in COVID-19 Patients

Corticosteroids (CCS) are steroid hormones implicated in several physiological processes such as the control of inflammatory response, protein catabolism, gluconeogenesis, antiallergy proprieties, and potent immunomodulator effects [52]. In clinical practice, there are two major classes of corticosteroids that are commonly used, glucocorticoids (e.g., dexamethasone, prednisolone and methylprednisolone) named for their gluconeogenic proprieties, and mineralcorticoids (e.g., fludrocortisone) named for their role in salt-water balance [53]. The potent anti-inflammatory effect of these drugs is due to the inhibition of NF-κB pathway, reducing IL-6 and TNF-α expression. Corticosteroids also act in cellular immunity inhibiting $CD8^+$ T cells, TH1 cells, and NK cells [54,55].

Despite these proprieties, the potential side effects such as risk of secondary infection, long-term complications, and delayed viral clearance initially discouraged the use of systemic corticosteroids in COVID-19 patients [54]. Subsequently, the RECOVERY trial, published in 2021, an open-label trial comparing a range of possible treatments in patients who were hospitalized with COVID-19, has demonstrated that the use of dexamethasone 6 mg reduced 28-days mortality in hospitalized patients who were receiving ventilatory support or oxygen alone but not in patients who were receiving no respiratory support [55]. The role of corticosteroids is controversial in mild-moderate COVID-19 and their use in the early stage of the disease. A systematic review showed that patients with mild-moderate COVID-19 treated with CCS have a longer hospitalization and more days of viral shedding [56]. In another randomized controlled trial (RCT) of adult patients with acute hypoxemic failure related to SARS-CoV-2 infection who received corticosteroids versus any comparator, the use of corticosteroids increased mortality in the subgroup of patients not requiring respiratory support [57]. In addition, the National Institutes of Health (NIH) guidelines for the treatment of COVID-19 established that there are no data to support the use of systemic corticosteroids in non-hospitalized patients with COVID-19 [34] and Infectious Diseases Society of America (IDSA) guidelines stated that there was no evidence for benefit with dexamethasone in patients who were not on supplemental oxygen [34]. Another study on the use of dexamethasone especially in early stage of disease appears to show it to be associated with severe COVID-19 due, probably, to an increase in viral load [58]. The increase of mortality related to the early use of corticosteroids is shown in the Liu and Zhang study that showed that patients who received corticosteroid therapy early have poorer outcomes due to a delayed viral clearance [59,60]. For the use in the early phase of SARS-CoV-2 infection, inhaled corticosteroids are being tested in several RCTs in the subset of patients with mild-moderate disease [54]. In the OIC (Steroid in COVID) trial, budesonide dry powder, administered at dose of 800 mcg twice daily (bid) within 7 days of onset of mild symptom, has demonstrated a reduction in urgent medical care and reduced time of recovery with low rate of adverse events [61]. These data were confirmed in the largest PRINCIPLE trial that investigated the use of inhaled budesonide (800 mcg bid) in outpatients aged 65 years old or 50 years old with comorbidities within 14 days of symptoms onset. This study has demonstrated that inhaled budesonide improves the time of recovery reducing the probability of hospital admissions or deaths in people who are at higher risk of complications although the combined endpoint was not achieved [62].

These benefits are not shown with all inhaled corticosteroids. The COVERAGE study, which tested inhaled ciclesonide in 217 outpatients with COVID-19, risk factors for severe disease, symptoms onset $\leq$ 7 days, and no criteria for hospitalization, revealed no efficacy to reducing the need for oxygen therapy, hospitalization, and/or death [63]. Another study investigated the role of a combination of inhaled ciclesonide and intranasal ciclesonide in young patients with COVID-19 and could not show a statistical improvement in respiratory symptoms compared to the control group [64]. Inhaled steroids are relatively safe drugs with a low risk of side-effects, but their utility in the early stage of COVID-19 requires further studies to better define their role [54].

3.2. *Non-Steroidal Anti-Inflammatory Drugs in the Early Stage of the Therapeutic Scenario of COVID-19*

Non-steroidal anti-inflammatory drugs (NSAIDs) are very widely used to alleviate fever, pain, and inflammation (common symptoms of COVID-19 patients) through effectively blocking production of prostaglandins (PGs) via inhibition of cyclooxygenase enzymes, namely COX-1 and COX-2, that catalyze the two-step conversion of arachidonic acid into thromboxane, prostaglandins, and prostacyclins. Prostaglandins are key inflammatory mediators. The use of NSAIDs during the COVID-19 pandemic, especially in the first wave, was controversial and NSAIDs were largely avoided in this phase of COVID-19 pandemic to favor the analgesic antipyretic paracetamol with no anti-inflammatory effect. Later, on the 18th of March 2020 European Medicines Agency (EMA) clarified and concluded that there is no clinical reason to withdraw the use of NSAIDs during the SARS-CoV-2 pandemic [65]. Moreover, some evidence suggested that early use of non-steroidal anti-inflammatory drugs in COVID-19 might interfere with the disease progression in patients with mild-to-moderate COVID-19 [66,67]. Similarly, in a Danish cohort study, the use of NSAIDs in subjects positive for SARS-CoV-2 was not associated with 30-day mortality, hospitalization, or complications [68]. Non-steroidal anti-inflammatory drugs (NSAIDs) form a large heterogeneous pharmacological family which inhibit COX-1 or COX-2; COX-1, constitutively expressed in the body, plays a pivotal role in platelet aggregation, gastrointestinal barrier integrity, and maintenance of renal function, while COX-2 is expressed during an inflammatory response where pro-inflammatory cytokines or growth factors stimulation are produced, thus being considered the most relevant mediator in promoting inflammation, fever, and pain. In particular, in SARS-CoV-2 infection it has been demonstrated that proteins S and N of the viral nucleocapsid induce the overexpression of COX-2, responsible for the progression of the inflammatory storm, as well as of the disease [69,70].

NSAIDs are defined nonselective when they inhibit both COX-1 and COX-2, and COX-2 selective. Their main indications are to alleviate fever, inflammation, and pain, including in patients with chronic inflammatory disorders such as rheumatoid arthritis and osteoarthritis. To reduce gastrointestinal side effects mediated essentially by inhibition of COX-1, highly selective COX-2 inhibitors (-coxib) were developed, although a number of recent studies have highlighted more serious potential cardiovascular side effects. The cardiovascular risk is not limited to the use of COX-2 inhibitors but is also correlated with the use of non-selective NSAIDs.

It is now recognized that an excessive inflammation is a key feature of severe COVID-19. Since PGs contribute to strengthen the inflammatory cytokines production, NSAIDs may mitigate this pathologic feature. As demonstrated by Chen and co-workers, although NSAIDs did not influence the expression of ACE2 and did not affect both SARS-CoV-2 entry and replication, meloxicam dampened the production of a subgroup of proinflammatory cytokines induced by infection, such as IL-6, CCL2, GM-CSF, CXCL10, IL-2, and TNF-α [71]. Moreover, NSAIDs may promote a more efficient virus clearance through inhibition of PGD_2 and PGE_2, which have been reported to impair both innate and adaptive immunity [72,73]. On the other hand, a reduced production of PGI_2 may be potentially detrimental considering its capacity of modulate virally induced illness [74]. In a recent metanalysis, the risk of severe COVID-19 has been demonstrated to be significantly decreased in subject receiving NSAIDs before admission (adjusted OR = 0.79, 95% CI 0.71–0.89, I2 = 0%) as well as the risk of death compared to controls (adjusted OR = 0.68, 95% CI 0.52–0.89, I2 = 85%) [75]. However, NSAIDs usage is not associated with a reduced risk of either SARS-CoV-2 infection or hospitalization due to COVID-19 [75]. With respect to adverse effects, prior usage of NSAIDs has been reported to significantly increase the risk of stroke in COVID-19 patients (OR = 2.32, 95% CI 1.04–5.2, I2 = 0%), whilst not thrombotic events (including deep vein thrombosis, pulmonary embolism, myocardial infarction) or renal failure [75].

3.3. *Ketoprofen Lysine Salt in the Therapeutic Scenario of SARS-CoV-2 Infection*

3.3.1. Ketoprofen Lysine Salt Mechanism of Action

An optimal NSAID should give relief and prevent complications and worsening of the disease. Ketoprofen is a NSAID belonging to the family of propionic derivates, with analgesic, anti-inflammatory, antipyretic, and antiplatelet properties [76,77].

Salification of ketoprofen with the lysine amino acid allows for higher solubility that facilitates a more rapid and complete absorption of the drug with a high peak plasma concentration reached after 15 min vs. 60 min [78–81]. Ketoprofen lysine salt (KLS) is a non-selective NSAID but also demonstrates high activity on COX-2 and inhibits the lipoxygenase pathway of the arachidonic acid cascade leading to a decrease in the synthesis of leukotrienes [82]. Ketoprofen, in common with other NSAIDs, has both peripheral and central sites of action [75] through the inhibition of both nitric oxide (NO) and COX synthase in the brain [83] and is rapidly and readily distributed into the central nervous system passing the blood brain barrier within 15 min, thanks to its high level of liposolubility (Figure 1) [79]. KLS penetrates extensively into the tonsillar tissue and in the adipose tissue; this is particularly interesting since recent data demonstrated the ability of SARS-CoV-2 to infect human adipocytes and that subjects with severe COVID-19 have higher local visceral adipose tissue inflammation [84]. Moreover, the use of ketoprofen lysine salt is associated with low prevalence of gastrointestinal adverse effects due to the improved solubility and bioavailability that determine a better gastrointestinal safety profile [80]. Compared with ibuprofen, ketoprofen lysine salt has minor effects on the gastric mucosa and therefore a better gastrointestinal safety profile; probably due to the protective effect exerted by lysine in the ketoprofen molecule, for its antioxidant effect and for its capacity to stimulate the production of endogenous gastro-protective proteins [85,86].

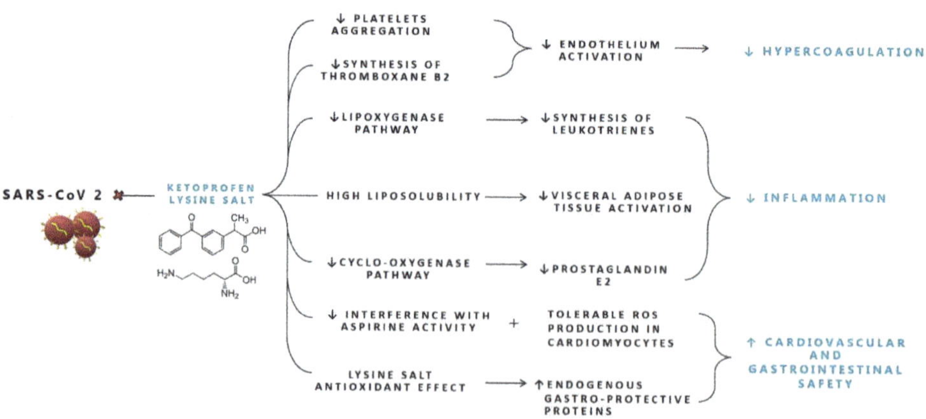

Figure 1. Proposed mechanisms of action of Ketoprofen lysine salt in SARS-CoV-2 infection. ROS: reactive oxygen species.

Several clinical studies and meta-analysis have demonstrated that ketoprofen lysine salt provided better and longer analgesic control as well as greater and faster pain relief than other NSAIDs, showing that it has the highest ratio between antinflammatory and analgesic effect, being the best molecule among the different NSAIDs compared, such as ibuprofen and diclofenac [78,87]. This strong efficacy has been confirmed in a recent meta-analysis underlying the better analgesic effect of ketoprofen compared with ibuprofen and diclofenac [88,89]. Another interesting effect of ketoprofen lysine salt is its role in reducing platelet aggregation and the synthesis of thromboxane B_2 with a potency superimposable to that of aspirin [87,90]. This is particularly important in the management of COVID-19 disease as during the cytokine storm vessels' endothelium is activated reducing prostacyclin and NO production, important antiaggregant mediators, causing diffuse coagulopathy as part of the systemic inflammatory response syndrome [91]. For this reason, in the early phases of the SARS-CoV-2 infection it is particularly important to choose NSAIDs with antiplatelet activity. This effect should be considered as potentially differentiating in the choice of an antinflammatory drug, since not all NSAIDs share this antiaggregant activity, for example ibuprofen has a negligible effect [90,92]. Moreover, another important aspect that should be considered when treating patients with COVID-19 is background cardioprotective therapy with low-dose aspirin. For these patients, the appropriate choice of NSAIDs should be carefully evaluated, since pharmacodynamic interactions between NSAIDs and low-dose aspirin are not classified as class effect, because not all NSAIDs interact with aspirin to the same extent: ketoprofen lysine salt does not interfere with antiplatelet activity, while clinical dosing regimen of ibuprofen or naproxen competitively inhibit aspirin's antiplatelet effect [93–95]. Altogether, while the above-reported pharmacokinetic and pharmacodynamic proprieties of KLS suggest potential for use in COVID-19 patients, at this time no significant evidence from studies in COVID-19 are currently available.

3.3.2. Ketoprofen Lysine Salt Cardiovascular Safety

In the last two years, real life studies and clinical experience have clarified that SARS-CoV-2 can impair cardiovascular system infecting heart and vascular tissues via ACE2 (angiotensin-converting enzyme 2), highlighting that cardiovascular diseases highly influence the susceptibility to and the outcomes of SARS-CoV-2 infection.

The systemic inflammation related to COVID-19 could accelerate the worsening of subclinical disorders or cause de novo cardiovascular diseases [96,97] such as myocardial injury or acute coronary syndrome largely linked to advanced systemic inflammation. Attention should be paid to the potential drug–disease interactions, preexisting cardiovascular diseases and drug cardiovascular safety profile in COVID-19 patients [98]. Some non-selective NSAIDs could increase the cardiovascular (CV) risk [99] and actually a study to investigate CV safety in NSAIDs funded by the European Commission (EMA) is still in progress [100,101]. Among the selected NSAIDs, the cardiovascular risk increased for seven traditional NSAIDs (diclofenac, ibuprofen, indomethacin, ketorolac, naproxen, nimesulide, and piroxicam) while ketoprofen showed the lowest relative risk for heart failure (OR 1.04 (0.96–1.12) and for acute myocardial infarction (OR_{pool} 1.00 (0.86 to 1.16) [100,101], showing a cardiovascular safety profile better than many other NSAIDs. Furthermore, a recent study applying a very elegant in vitro model, strongly supports the hypothesis that immortalized human cardiomyocytes exposed to ketoprofen are subjected to tolerable stress events while diclofenac exposition leads to cell death, dramatically increasing ROS (reactive oxygen species) production. Increased ROS levels cause proteasome dysfunction and the opening of mitochondrial permeability transition pores (mPTPs) allowing the release of cytochrome c and the activation of caspase-3/9, thus inducing intrinsic apoptotic pathway. This study shows an alteration of mPTP only in cells exposed to diclofenac; conversely cells exposed to ketoprofen show the same behavior of untreated cells [102].

4. Future Perspective

Inflammation represents a key factor of SARS-CoV-2 infection and a growing body of literature is focusing on this issue. By exploiting their anti-inflammatory proprieties, NSAIDs have been regarded as safe drugs to be utilized in COVID-19. However, new approaches, from diagnosis to treatment, are currently under the spotlight to tackle this issue. Geromichalou et al. have investigated the potential antiviral activity of copper (II) complexes with non-steroidal anti-inflammatory drugs on various SARS-CoV-2 proteins [72]. Although in vitro model, authors demonstrated that these molecular complexes may potentially impact on SARS-CoV-2 replication [72]. Nanomaterials and nanotechnologies may play a significant role not only for a quick and accurate diagnosis; they also may improve COVID-19 treatment as well as new strategies for vaccines [103,104].

5. Conclusions

Targeting pro-inflammatory pathways in early phases of COVID-19 is one of the key strategies in avoiding progression of the disease. Current literature should however be interpreted with caution as the research has been conducted during different COVID-19 waves and the presence of SARS-CoV-2 variants is not negligible. However, NSAIDs, because of their interaction with different pro-inflammatory mediators, may play pleiotropic effects in COVID-19. Available evidence about the role of NSAIDs in SARS-CoV-2 are based on retrospective studies. In this respect, large prospective randomized clinical trials are needed to better investigate the efficacy and safety of these drugs also exploring the different pharmacokinetic properties existing among the drugs. Finally, the evolution of therapeutic strategies for the early stages of SARS-CoV-2 infection should be oriented in developing combination of complementary drugs able to interact with this complex network between virus, host cell, and innate immune cells.

Author Contributions: Conceptualization, A.B. and F.P.; data curation, D.F.M., V.A., V.D., L.L., M.B. and L.A.; writing—original draft preparation, D.F.M., L.A., V.A., V.D., R.V., L.L. and M.B.; writing—review and editing, A.B. and F.P.; supervision, A.B., D.F.M. and F.P.; project administration, D.F.M., V.A. and V.D. All authors have read and agreed to the published version of the manuscript.

Funding: This research received no external funding.

Institutional Review Board Statement: Not applicable.

Informed Consent Statement: Not applicable.

Data Availability Statement: Not applicable.

Conflicts of Interest: The authors declare no conflict of interest.

References

1. Wang, Y.; Zhou, Y.; Yang, Z.; Xia, D.; Hu, Y.; Geng, S. Clinical Characteristics of Patients with Severe Pneumonia Caused by the SARS-CoV-2 in Wuhan, China. *Respiration* **2020**, *99*, 649–657. [CrossRef] [PubMed]
2. Wadman, J.C.-F.M.; Kaiser, C. How Does Coronavirus Kill? Clinicians Trace a Ferocious Rampage through the Body, from Brain to Toes. *Science* **2020**.
3. Zhou, Q.; Zhao, S.; Gan, L.; Wang, Z.; Peng, S.; Li, Q.; Liu, H.; Liu, X.; Wang, Z.; Shi, Q.; et al. Use of Non-Steroidal Anti-Inflammatory Drugs and Adverse Outcomes during the COVID-19 Pandemic: A Systematic Review and Meta-Analysis. *EClinicalMedicine* **2022**, *46*, 101373. [CrossRef] [PubMed]
4. Hasan, S.S.; Capstick, T.; Ahmed, R.; Kow, C.S.; Mazhar, F.; Merchant, H.A.; Zaidi, S.T.R. Mortality in COVID-19 Patients with Acute Respiratory Distress Syndrome and Corticosteroids Use: A Systematic Review and Meta-Analysis. *Expert Rev. Respir. Med.* **2020**, *14*, 1149–1163. [CrossRef]
5. Li, F. Structure, Function, and Evolution of Coronavirus Spike Proteins. *Annu. Rev. Virol.* **2016**, *3*, 237–261. [CrossRef]
6. Perrotta, F.; Matera, M.G.; Cazzola, M.; Bianco, A. Severe Respiratory SARS-CoV-2 Infection: Does ACE2 Receptor Matter? *Respir. Med.* **2020**, *168*, 105996. [CrossRef]
7. Harrison, A.G.; Lin, T.; Wang, P. Mechanisms of SARS-CoV-2 Transmission and Pathogenesis. *Trends Immunol.* **2020**, *41*, 1100–1115. [CrossRef]

8. Hatmal, M.M.; Alshaer, W.; Al-Hatamleh, M.A.I.; Hatmal, M.; Smadi, O.; Taha, M.O.; Oweida, A.J.; Boer, J.C.; Mohamud, R.; Plebanski, M. Comprehensive Structural and Molecular Comparison of Spike Proteins of SARS-CoV-2, SARS-CoV and MERS-CoV, and Their Interactions with ACE2. *Cells* **2020**, *9*, 2638. [CrossRef]
9. Glowacka, I.; Bertram, S.; Müller, M.A.; Allen, P.; Soilleux, E.; Pfefferle, S.; Steffen, I.; Tsegaye, T.S.; He, Y.; Gnirss, K.; et al. Evidence That TMPRSS2 Activates the Severe Acute Respiratory Syndrome Coronavirus Spike Protein for Membrane Fusion and Reduces Viral Control by the Humoral Immune Response. *J. Virol.* **2011**, *85*, 4122–4134. [CrossRef]
10. Imai, Y.; Kuba, K.; Penninger, J.M. The Discovery of Angiotensin-Converting Enzyme 2 and Its Role in Acute Lung Injury in Mice. *Exp. Physiol.* **2008**, *93*, 543–548. [CrossRef]
11. Okamoto, M.; Tsukamoto, H.; Kouwaki, T.; Seya, T.; Oshiumi, H. Recognition of Viral RNA by Pattern Recognition Receptors in the Induction of Innate Immunity and Excessive Inflammation During Respiratory Viral Infections. *Viral Immunol.* **2017**, *30*, 408–420. [CrossRef] [PubMed]
12. Koop, A.; Lepenies, I.; Braum, O.; Davarnia, P.; Scherer, G.; Fickenscher, H.; Kabelitz, D.; Adam-Klages, S. Novel Splice Variants of Human IKKε Negatively Regulate IKKε-Induced IRF3 and NF-KB Activation. *Eur. J. Immunol.* **2011**, *41*, 224–234. [CrossRef] [PubMed]
13. Bianco, A.; Conte, S.; Mariniello, D.F.; Allocca, V.; Matera, M.G.; D'Agnano, V.; Lanata, L.; Cazzola, M.; Perrotta, F. Mucolytic and Antioxidant Properties of Carbocysteine as a Strategy in COVID-19 Therapy. *Life* **2022**, *12*, 1824. [CrossRef]
14. Perrotta, F.; Corbi, G.; Mazzeo, G.; Boccia, M.; Aronne, L.; D'Agnano, V.; Komici, K.; Mazzarella, G.; Parrella, R.; Bianco, A. COVID-19 and the Elderly: Insights into Pathogenesis and Clinical Decision-Making. *Aging Clin. Exp. Res.* **2020**, *32*, 1599–1608. [CrossRef] [PubMed]
15. Maloir, Q.; Ghysen, K.; von Frenckell, C.; Louis, R.; Guiot, J. Acute respiratory distress revealing antisynthetase syndrome. *Rev. Med. Liege* **2018**, *73*, 370–375. [PubMed]
16. Zhang, X.; Tan, Y.; Ling, Y.; Lu, G.; Liu, F.; Yi, Z.; Jia, X.; Wu, M.; Shi, B.; Xu, S.; et al. Viral and Host Factors Related to the Clinical Outcome of COVID-19. *Nature* **2020**, *583*, 437–440. [CrossRef] [PubMed]
17. Carfora, V.; Spiniello, G.; Ricciolino, R.; Di Mauro, M.; Migliaccio, M.G.; Mottola, F.F.; Verde, N.; Coppola, N. Anticoagulant Treatment in COVID-19: A Narrative Review. *J. Thromb. Thrombolysis* **2021**, *51*, 642–648. [CrossRef]
18. Cattaneo, M.; Bertinato, E.M.; Birocchi, S.; Brizio, C.; Malavolta, D.; Manzoni, M.; Muscarella, G.; Orlandi, M. Pulmonary Embolism or Pulmonary Thrombosis in COVID-19? Is the Recommendation to Use High-Dose Heparin for Thromboprophylaxis Justified? *Thromb. Haemost.* **2020**, *120*, 1230–1232. [CrossRef]
19. Boccia, M.; Aronne, L.; Celia, B.; Mazzeo, G.; Ceparano, M.; D'Agnano, V.; Parrella, R.; Valente, T.; Bianco, A.; Perrotta, F. COVID-19 and Coagulative Axis: Review of Emerging Aspects in a Novel Disease. *Monaldi Arch. Chest Dis.* **2020**, *90*. [CrossRef]
20. Scialo, F.; Daniele, A.; Amato, F.; Pastore, L.; Matera, M.G.; Cazzola, M.; Castaldo, G.; Bianco, A. ACE2: The Major Cell Entry Receptor for SARS-CoV-2. *Lung* **2020**, *198*, 867–877. [CrossRef]
21. Verdecchia, P.; Cavallini, C.; Spanevello, A.; Angeli, F. The Pivotal Link between ACE2 Deficiency and SARS-CoV-2 Infection. *Eur. J. Intern. Med.* **2020**, *76*, 14–20. [CrossRef] [PubMed]
22. Mehta, P.K.; Griendling, K.K. Angiotensin II Cell Signaling: Physiological and Pathological Effects in the Cardiovascular System. *Am. J. Physiol. Cell Physiol.* **2007**, *292*, C82–C97. [CrossRef] [PubMed]
23. Santos, R.A.S.; Sampaio, W.O.; Alzamora, A.C.; Motta-Santos, D.; Alenina, N.; Bader, M.; Campagnole-Santos, M.J. The ACE2/Angiotensin-(1-7)/MAS Axis of the Renin-Angiotensin System: Focus on Angiotensin-(1-7). *Physiol. Rev.* **2018**, *98*, 505–553. [CrossRef] [PubMed]
24. Schmitt, F.C.F.; Manolov, V.; Morgenstern, J.; Fleming, T.; Heitmeier, S.; Uhle, F.; Al-Saeedi, M.; Hackert, T.; Bruckner, T.; Schöchl, H.; et al. Acute Fibrinolysis Shutdown Occurs Early in Septic Shock and Is Associated with Increased Morbidity and Mortality: Results of an Observational Pilot Study. *Ann. Intensive Care* **2019**, *9*, 19. [CrossRef] [PubMed]
25. Branchford, B.R.; Carpenter, S.L. The Role of Inflammation in Venous Thromboembolism. *Front. Pediatr.* **2018**, *6*, 142. [CrossRef]
26. José, R.J.; Williams, A.; Manuel, A.; Brown, J.S.; Chambers, R.C. Targeting Coagulation Activation in Severe COVID-19 Pneumonia: Lessons from Bacterial Pneumonia and Sepsis. *Eur. Respir. Rev.* **2020**, *29*, 200240. [CrossRef]
27. Chambers, R.C.; Scotton, C.J. Coagulation Cascade Proteinases in Lung Injury and Fibrosis. *Proc. Am. Thorac. Soc.* **2012**, *9*, 96–101. [CrossRef]
28. Kreuzberger, N.; Hirsch, C.; Chai, K.L.; Tomlinson, E.; Khosravi, Z.; Popp, M.; Neidhardt, M.; Piechotta, V.; Salomon, S.; Valk, S.J.; et al. SARS-CoV-2-Neutralising Monoclonal Antibodies for Treatment of COVID-19. *Cochrane Database Syst. Rev.* **2021**, *9*, CD013825. [CrossRef]
29. Hwang, Y.-C.; Lu, R.-M.; Su, S.-C.; Chiang, P.-Y.; Ko, S.-H.; Ke, F.-Y.; Liang, K.-H.; Hsieh, T.-Y.; Wu, H.-C. Monoclonal Antibodies for COVID-19 Therapy and SARS-CoV-2 Detection. *J. Biomed. Sci.* **2022**, *29*, 1. [CrossRef]
30. Dougan, M.; Nirula, A.; Azizad, M.; Mocherla, B.; Gottlieb, R.L.; Chen, P.; Hebert, C.; Perry, R.; Boscia, J.; Heller, B.; et al. Bamlanivimab plus Etesevimab in Mild or Moderate COVID-19. *N. Engl. J. Med.* **2021**, *385*, 1382–1392. [CrossRef]
31. O'Brien, M.P.; Forleo-Neto, E.; Sarkar, N.; Isa, F.; Hou, P.; Chan, K.-C.; Musser, B.J.; Bar, K.J.; Barnabas, R.V.; Barouch, D.H.; et al. Effect of Subcutaneous Casirivimab and Imdevimab Antibody Combination vs Placebo on Development of Symptomatic COVID-19 in Early Asymptomatic SARS-CoV-2 Infection: A Randomized Clinical Trial. *JAMA* **2022**, *327*, 432–441. [CrossRef] [PubMed]

32. Gupta, A.; Gonzalez-Rojas, Y.; Juarez, E.; Crespo Casal, M.; Moya, J.; Falci, D.R.; Sarkis, E.; Solis, J.; Zheng, H.; Scott, N.; et al. Early Treatment for COVID-19 with SARS-CoV-2 Neutralizing Antibody Sotrovimab. *N. Engl. J. Med.* **2021**, *385*, 1941–1950. [CrossRef] [PubMed]
33. Holland, T.L. Tixagevimab–Cilgavimab for Treatment of Patients Hospitalised with COVID-19: A Randomised, Double-Blind, Phase 3 Trial. *Lancet Respir. Med.* **2022**, *10*, 972–984. [CrossRef] [PubMed]
34. National Institutes of Health. *Treatment Guidelines Panel. Coronavirus Disease 2019 (COVID-19). COVID-19 Treatment Guidelines Panel Coronavirus Disease 2019 (COVID-19) Treatment Guidelines*; National Institutes of Health: Bethesda, MD, USA, 2019.
35. Westendorf, K.; Žentelis, S.; Wang, L.; Foster, D.; Vaillancourt, P.; Wiggin, M.; Lovett, E.; van der Lee, R.; Hendle, J.; Pustilnik, A.; et al. LY-CoV1404 (Bebtelovimab) Potently Neutralizes SARS-CoV-2 Variants. *Cell Rep.* **2022**, *39*, 110812. [CrossRef]
36. Cao, B.; Wang, Y.; Wen, D.; Liu, W.; Wang, J.; Fan, G.; Ruan, L.; Song, B.; Cai, Y.; Wei, M.; et al. A Trial of Lopinavir-Ritonavir in Adults Hospitalized with Severe COVID-19. *N. Engl. J. Med.* **2020**, *382*, 1787–1799. [CrossRef]
37. Hammond, J.; Leister-Tebbe, H.; Gardner, A.; Abreu, P.; Bao, W.; Wisemandle, W.; Baniecki, M.; Hendrick, V.M.; Damle, B.; Simón-Campos, A.; et al. Oral Nirmatrelvir for High-Risk, Nonhospitalized Adults with COVID-19. *N. Engl. J. Med.* **2022**, *386*, 1397–1408. [CrossRef]
38. Kokic, G.; Hillen, H.S.; Tegunov, D.; Dienemann, C.; Seitz, F.; Schmitzova, J.; Farnung, L.; Siewert, A.; Höbartner, C.; Cramer, P. Mechanism of SARS-CoV-2 Polymerase Stalling by Remdesivir. *Nat. Commun.* **2021**, *12*, 279. [CrossRef]
39. Scialo, F.; Vitale, M.; Daniele, A.; Nigro, E.; Perrotta, F.; Gelzo, M.; Iadevaia, C.; Cerqua, F.S.; Costigliola, A.; Allocca, V.; et al. SARS-CoV-2: One Year in the Pandemic. What Have We Learned, the New Vaccine Era and the Threat of SARS-Cov-2 Variants. *Biomedicines* **2021**, *9*, 611. [CrossRef]
40. D'Agnano, V.; Scialò, F.; Perna, F.; Atripaldi, L.; Sanduzzi, S.; Allocca, V.; Vitale, M.; Pastore, L.; Bianco, A.; Perrotta, F. Exploring the Role of Krebs von Den Lungen-6 in Severe to Critical COVID-19 Patients. *Life* **2022**, *12*, 1141. [CrossRef]
41. NIH. COVID-19 Treatment Guidelines Therapeutic Management of Hospitalized Adults With COVID-19. Available online: https://www.covid19treatmentguidelines.nih.gov/management/clinical-management/hospitalized-adults--therapeutic-management/ (accessed on 11 December 2022).
42. Jayk Bernal, A.; Gomes da Silva, M.M.; Musungaie, D.B.; Kovalchuk, E.; Gonzalez, A.; Delos Reyes, V.; Martín-Quirós, A.; Caraco, Y.; Williams-Diaz, A.; Brown, M.L.; et al. Molnupiravir for Oral Treatment of COVID-19 in Nonhospitalized Patients. *N. Engl. J. Med.* **2022**, *386*, 509–520. [CrossRef]
43. Kaufmann, S.H.E.; Dorhoi, A.; Hotchkiss, R.S.; Bartenschlager, R. Host-Directed Therapies for Bacterial and Viral Infections. *Nat. Rev. Drug Discov.* **2018**, *17*, 35–56. [CrossRef] [PubMed]
44. Paschos, K.; Allday, M.J. Epigenetic Reprogramming of Host Genes in Viral and Microbial Pathogenesis. *Trends Microbiol.* **2010**, *18*, 439–447. [CrossRef] [PubMed]
45. Tripathi, D.; Sodani, M.; Gupta, P.K.; Kulkarni, S. Host Directed Therapies: COVID-19 and Beyond. *Curr. Res. Pharmacol. Drug Discov.* **2021**, *2*, 100058. [CrossRef] [PubMed]
46. Monk, P.D.; Marsden, R.J.; Tear, V.J.; Brookes, J.; Batten, T.N.; Mankowski, M.; Gabbay, F.J.; Davies, D.E.; Holgate, S.T.; Ho, L.-P.; et al. Safety and Efficacy of Inhaled Nebulised Interferon Beta-1a (SNG001) for Treatment of SARS-CoV-2 Infection: A Randomised, Double-Blind, Placebo-Controlled, Phase 2 Trial. *Lancet Respir. Med.* **2021**, *9*, 196–206. [CrossRef]
47. Laterre, P.F.; François, B.; Collienne, C.; Hantson, P.; Jeannet, R.; Remy, K.E.; Hotchkiss, R.S. Association of Interleukin 7 Immunotherapy With Lymphocyte Counts Among Patients With Severe Coronavirus Disease 2019 (COVID-19). *JAMA Netw. Open* **2020**, *3*, e2016485. [CrossRef]
48. Lee, J.S.; Shin, E.-C. The Type I Interferon Response in COVID-19: Implications for Treatment. *Nat. Rev. Immunol.* **2020**, *20*, 585–586. [CrossRef]
49. Venet, F.; Foray, A.-P.; Villars-Méchin, A.; Malcus, C.; Poitevin-Later, F.; Lepape, A.; Monneret, G. IL-7 Restores Lymphocyte Functions in Septic Patients. *J. Immunol.* **2012**, *189*, 5073–5081. [CrossRef]
50. Bojkova, D.; Bechtel, M.; McLaughlin, K.-M.; McGreig, J.E.; Klann, K.; Bellinghausen, C.; Rohde, G.; Jonigk, D.; Braubach, P.; Ciesek, S.; et al. Aprotinin Inhibits SARS-CoV-2 Replication. *Cells* **2020**, *9*, 2377. [CrossRef]
51. Kumar, N.; Liang, Y.; Parslow, T.G.; Liang, Y. Receptor Tyrosine Kinase Inhibitors Block Multiple Steps of Influenza a Virus Replication. *J. Virol.* **2011**, *85*, 2818–2827. [CrossRef]
52. Jafari, A.; Esmaeilzadeh, Z.; Khezri, M.R.; Ghasemnejad-Berenji, H.; Pashapour, S.; Sadeghpour, S.; Ghasemnejad-Berenji, M. An Overview of Possible Pivotal Mechanisms of Genistein as a Potential Phytochemical against SARS-CoV-2 Infection: A Hypothesis. *J. Food Biochem.* **2022**, *46*, e14345. [CrossRef]
53. Olejnik, J.; Hume, A.J.; Mühlberger, E. Toll-like Receptor 4 in Acute Viral Infection: Too Much of a Good Thing. *PLOS Pathog.* **2018**, *14*, e1007390. [CrossRef] [PubMed]
54. Riedemann, T.; Patchev, A.V.; Cho, K.; Almeida, O.F.X. Corticosteroids: Way Upstream. *Mol. Brain* **2010**, *3*, 2. [CrossRef] [PubMed]
55. Perico, N.; Cortinovis, M.; Suter, F.; Remuzzi, G. Home as the New Frontier for the Treatment of COVID-19: The Case for Anti-Inflammatory Agents. *Lancet. Infect. Dis.* **2022**. [CrossRef] [PubMed]
56. Scialò, F.; Mariniello, D.F.; Nigro, E.; Komici, K.; Allocca, V.; Bianco, A.; Perrotta, F.; D'Agnano, V. Effects of Different Corticosteroid Doses in Elderly Unvaccinated Patients with Severe to Critical COVID-19. *Life* **2022**, *12*, 1924. [CrossRef]
57. Horby, P.; Lim, W.S.; Emberson, J.R.; Mafham, M.; Bell, J.L.; Linsell, L.; Staplin, N.; Brightling, C.; Ustianowski, A.; Elmahi, E.; et al. Dexamethasone in Hospitalized Patients with COVID-19. *N. Engl. J. Med.* **2021**, *384*, 693–704. [CrossRef] [PubMed]

58. Ordinola Navarro, A.; Lopez Luis, B.A. Corticosteroids Prescription for Mild-Moderate COVID-19 in Primary Care. *J. Infect. Dev. Ctries.* **2021**, *15*, 1813–1815. [CrossRef]
59. Pasin, L.; Navalesi, P.; Zangrillo, A.; Kuzovlev, A.; Likhvantsev, V.; Hajjar, L.A.; Fresilli, S.; Lacerda, M.V.G.; Landoni, G. Corticosteroids for Patients With Coronavirus Disease 2019 (COVID-19) With Different Disease Severity: A Meta-Analysis of Randomized Clinical Trials. *J. Cardiothorac. Vasc. Anesth.* **2021**, *35*, 578–584. [CrossRef]
60. Callejas Rubio, J.L.; Aomar Millan, I.; Moreno-Higueras, M.; Martín Ripoll, L.; Yuste Osorio, E.; Ríos-Fernández, R. Caution with the use of dexamethasone in patients with COVID-19 in its initial phases. *Rev. Clin. Esp.* **2021**, *221*, 592–595. [CrossRef]
61. Liu, J.; Zhang, S.; Dong, X.; Li, Z.; Xu, Q.; Feng, H.; Cai, J.; Huang, S.; Guo, J.; Zhang, L.; et al. Corticosteroid Treatment in Severe COVID-19 Patients with Acute Respiratory Distress Syndrome. *J. Clin. Investig.* **2020**, *130*, 6417–6428. [CrossRef]
62. Matthay, M.A.; Wick, K.D. Corticosteroids, COVID-19 Pneumonia, and Acute Respiratory Distress Syndrome. *J. Clin. Investig.* **2020**, *130*, 6218–6221. [CrossRef]
63. Ramakrishnan, S.; Nicolau, D.V.J.; Langford, B.; Mahdi, M.; Jeffers, H.; Mwasuku, C.; Krassowska, K.; Fox, R.; Binnian, I.; Glover, V.; et al. Inhaled Budesonide in the Treatment of Early COVID-19 (STOIC): A Phase 2, Open-Label, Randomised Controlled Trial. *Lancet. Respir. Med.* **2021**, *9*, 763–772. [CrossRef] [PubMed]
64. Yu, L.-M.; Bafadhel, M.; Dorward, J.; Hayward, G.; Saville, B.R.; Gbinigie, O.; Van Hecke, O.; Ogburn, E.; Evans, P.H.; Thomas, N.P.B.; et al. Inhaled Budesonide for COVID-19 in People at High Risk of Complications in the Community in the UK (PRINCIPLE): A Randomised, Controlled, Open-Label, Adaptive Platform Trial. *Lancet* **2021**, *398*, 843–855. [CrossRef]
65. Duvignaud, A.; Lhomme, E.; Onaisi, R.; Sitta, R.; Gelley, A.; Chastang, J.; Piroth, L.; Binquet, C.; Dupouy, J.; Makinson, A.; et al. Inhaled Ciclesonide for Outpatient Treatment of COVID-19 in Adults at Risk of Adverse Outcomes: A Randomised Controlled Trial (COVERAGE). *Clin. Microbiol. Infect. Off. Publ. Eur. Soc. Clin. Microbiol. Infect. Dis.* **2022**, *28*, 1010–1016. [CrossRef] [PubMed]
66. Ezer, N.; Belga, S.; Daneman, N.; Chan, A.; Smith, B.M.; Daniels, S.-A.; Moran, K.; Besson, C.; Smyth, L.Y.; Bartlett, S.J.; et al. Inhaled and Intranasal Ciclesonide for the Treatment of COVID-19 in Adult Outpatients: CONTAIN Phase II Randomised Controlled Trial. *BMJ* **2021**, *375*, e068060. [CrossRef]
67. EMA. EMA Gives Advice on the Use of Non-Steroidal Anti-Inflammatories for COVID-19. *Eur. Med. Agency* **2020**, 1–4.
68. Kelleni, M.T. Early Use of Non-Steroidal Anti-Inflammatory Drugs in COVID-19 Might Reverse Pathogenesis, Prevent Complications and Improve Clinical Outcomes. *Biomed. Pharmacother.* **2021**, *133*, 110982. [CrossRef] [PubMed]
69. Fazio, S.; Bellavite, P.; Zanolin, E.; McCullough, P.A.; Pandolfi, S.; Affuso, F. Retrospective Study of Outcomes and Hospitalization Rates of Patients in Italy with a Confirmed Diagnosis of Early COVID-19 and Treated at Home Within 3 Days or After 3 Days of Symptom Onset with Prescribed and Non-Prescribed Treatments Between November 2020 and August 2021. *Med. Sci. Monit. Int. Med. J. Exp. Clin. Res.* **2021**, *27*, e935379. [CrossRef]
70. Lund, L.C.; Hallas, J.; Nielsen, H.; Koch, A.; Mogensen, S.H.; Brun, N.C.; Christiansen, C.F.; Thomsen, R.W.; Pottegård, A. Post-Acute Effects of SARS-CoV-2 Infection in Individuals Not Requiring Hospital Admission: A Danish Population-Based Cohort Study. *Lancet Infect. Dis.* **2021**, *21*, 1373–1382. [CrossRef]
71. Yan, X.; Hao, Q.; Mu, Y.; Timani, K.A.; Ye, L.; Zhu, Y.; Wu, J. Nucleocapsid Protein of SARS-CoV Activates the Expression of Cyclooxygenase-2 by Binding Directly to Regulatory Elements for Nuclear Factor-Kappa B and CCAAT/Enhancer Binding Protein. *Int. J. Biochem. Cell Biol.* **2006**, *38*, 1417–1428. [CrossRef]
72. Liu, S.-J.; Leng, C.-H.; Lien, S.-P.; Chi, H.-Y.; Huang, C.-Y.; Lin, C.-L.; Lian, W.-C.; Chen, C.-J.; Hsieh, S.-L.; Chong, P. Immunological Characterizations of the Nucleocapsid Protein Based SARS Vaccine Candidates. *Vaccine* **2006**, *24*, 3100–3108. [CrossRef]
73. Chen, J.S.; Alfajaro, M.M.; Chow, R.D.; Wei, J.; Filler, R.B.; Eisenbarth, S.C.; Wilen, C.B. Non-Steroidal Anti-Inflammatory Drugs Dampen the Cytokine and Antibody Response to SARS-CoV-2 Infection. *J. Virol.* **2021**, *95*, e00014-21. [CrossRef]
74. Geromichalou, E.G.; Trafalis, D.T.; Dalezis, P.; Malis, G.; Psomas, G.; Geromichalos, G.D. In Silico Study of Potential Antiviral Activity of Copper(II) Complexes with Non-Steroidal Anti-Inflammatory Drugs on Various SARS-CoV-2 Target Proteins. *J. Inorg. Biochem.* **2022**, *231*, 111805. [CrossRef] [PubMed]
75. Kalinski, P. Regulation of Immune Responses by Prostaglandin E2. *J. Immunol.* **2012**, *188*, 21–28. [CrossRef]
76. Dorris, S.L.; Peebles, R.S.J. PGI2 as a Regulator of Inflammatory Diseases. *Mediat. Inflamm.* **2012**, *2012*, 926968. [CrossRef] [PubMed]
77. Zhao, H.; Huang, S.; Huang, S.; Liu, F.; Shao, W.; Mei, K.; Ma, J.; Jiang, Y.; Wan, J.; Zhu, W.; et al. Prevalence of NSAID Use among People with COVID-19 and the Association with COVID-19-Related Outcomes: Systematic Review and Meta-Analysis. *Br. J. Clin. Pharmacol.* **2022**, *88*, 5113–5127. [CrossRef] [PubMed]
78. Arienzo, D.M.; Pennisi, M.; Zanolo, G.; Borsa, M. Ketoprofen Lysine: Ketoprofen Serum Levels and Analgesic Activity. *Drugs Under Exp. Clin. Res.* **1984**, *10*, 2–3. [CrossRef]
79. McCormack, K.; Brune, K. Dissociation between the Antinociceptive and Anti-Inflammatory Effects of the Nonsteroidal Anti-Inflammatory Drugs. A Survey of Their Analgesic Efficacy. *Drugs* **1991**, *41*, 533–547. [CrossRef]
80. Sarzi-Puttini, P.; Atzeni, F.; Lanata, L.; Bagnasco, M.; Colombo, M.; Fischer, F.; D'Imporzano, M. Pain and Ketoprofen: What Is Its Role in Clinical Practice? *Reumatismo* **2010**, *62*, 172–188. [CrossRef]
81. Varrassi, G.; Alon, E.; Bagnasco, M.; Lanata, L.; Mayoral-Rojals, V.; Paladini, A.; Pergolizzi, J.V.; Perrot, S.; Scarpignato, C.; Tölle, T. Towards an Effective and Safe Treatment of Inflammatory Pain: A Delphi-Guided Expert Consensus. *Adv. Ther.* **2019**, *36*, 2618–2637. [CrossRef]

82. Rusca, A. Two Way Crossover, Randomised, Single Dose Comparative Bioavailability Study of Ketoprofen Lysine Salt after Oral Admin-Istration to Healthy Volunteers of Both Sexes. (IPAS–KETO–025–94), Data on File 1994. *File* **1994**, 1–8.
83. Marmo, E.; Ottavo, R.; Giordano, L.; Paone, G.; Falcone, O.; Spaziante, G.; Visone, C.; Campidonico, U. Experimental assessment of some pharmacodynamic features of ketoprofen lysine. Pain relief activity, antipyretic effects, anti-inflammatory activity, anti-platelet aggregation activity and interference with the biosynthesis of prostaglandins. *Arch. Sci. Med.* **1980**, *137*, 387–404.
84. Ossipov, M.H.; Jerussi, T.P.; Ren, K.; Sun, H.; Porreca, F. Differential Effects of Spinal (R)-Ketoprofen and (S)-Ketoprofen against Signs of Neuropathic Pain and Tonic Nociception: Evidence for a Novel Mechanism of Action of (R)-Ketoprofen against Tactile Allodynia. *Pain* **2000**, *87*, 193–199. [CrossRef] [PubMed]
85. Nigro, E.; Perrotta, F.; Polito, R.; D'Agnano, V.; Scialò, F.; Bianco, A.; Daniele, A. Metabolic Perturbations and Severe COVID-19 Disease: Implication of Molecular Pathways. *Int. J. Endocrinol.* **2020**, *2020*, 8896536. [CrossRef]
86. Díaz-Reval, M.I.; Ventura-Martínez, R.; Déciga-Campos, M.; Terrón, J.A.; Cabré, F.; López-Muñoz, F.J. Evidence for a Central Mechanism of Action of S-(+)-Ketoprofen. *Eur. J. Pharmacol.* **2004**, *483*, 241–248. [CrossRef] [PubMed]
87. Cimini, A.; Brandolini, L.; Gentile, R.; Cristiano, L.; Menghini, P.; Fidoamore, A.; Antonosante, A.; Benedetti, E.; Giordano, A.; Allegretti, M. Gastroprotective Effects of L-Lysine Salification of Ketoprofen in Ethanol-Injured Gastric Mucosa. *J. Cell. Physiol.* **2015**, *230*, 813–820. [CrossRef] [PubMed]
88. Scaglione, F. Utilizzo Dei FANS Nelle Infezioni Virali Respiratorie, Incluso COVID 19. Focus Su Ketoprofene Sale Di Lisina. *6 Giugno 2022* **2022**.
89. Brandolini, L.; d'Angelo, M.; Antonosante, A.; Villa, S.; Cristiano, L.; Castelli, V.; Benedetti, E.; Catanesi, M.; Aramini, A.; Luini, A.; et al. Differential Protein Modulation by Ketoprofen and Ibuprofen Underlines Different Cellular Response by Gastric Epithelium. *J. Cell. Physiol.* **2018**, *233*, 2304–2312. [CrossRef]
90. Sarzi-Puttini, P.; Atzeni, F.; Lanata, L.; Bagnasco, M. Efficacy of Ketoprofen vs. Ibuprofen and Diclofenac: A Systematic Review of the Literature and Meta-Analysis. *Clin. Exp. Rheumatol.* **2013**, *31*, 731–738.
91. Atzeni, F.; Masala, I.F.; Bagnasco, M.; Lanata, L.; Mantelli, F.; Sarzi-Puttini, P. Comparison of Efficacy of Ketoprofen and Ibuprofen in Treating Pain in Patients with Rheumatoid Arthritis: A Systematic Review and Meta-Analysis. *Pain Ther.* **2021**, *10*, 577–588. [CrossRef]
92. Van Solingen, R.M.; Rosenstein, E.D.; Mihailescu, G.; Drejka, M.L.; Kalia, A.; Cohen, A.J.; Kramer, N. Comparison of the Effects of Ketoprofen on Platelet Function in the Presence and Absence of Aspirin. *Am. J. Med.* **2001**, *111*, 285–289. [CrossRef]
93. Stichtenoth, D.O.; Tsikas, D.; Gutzki, F.M.; Frölich, J.C. Effects of Ketoprofen and Ibuprofen on Platelet Aggregation and Prostanoid Formation in Man. *Eur. J. Clin. Pharmacol.* **1996**, *51*, 231–234. [CrossRef] [PubMed]
94. Hohlfeld, T.; Saxena, A.; Schrör, K. High on Treatment Platelet Reactivity against Aspirin by Non-Steroidal Anti-Inflammatory Drugs–Pharmacological Mechanisms and Clinical Relevance. *Thromb. Haemost.* **2013**, *109*, 825–833. [CrossRef] [PubMed]
95. Canzano, P.; Brambilla, M.; Porro, B.; Cosentino, N.; Tortorici, E.; Vicini, S.; Poggio, P.; Cascella, A.; Pengo, M.F.; Veglia, F.; et al. Platelet and Endothelial Activation as Potential Mechanisms Behind the Thrombotic Complications of COVID-19 Patients. *JACC Basic Transl. Sci.* **2021**, *6*, 202–218. [CrossRef] [PubMed]
96. Saxena, A.; Balaramnavar, V.M.; Hohlfeld, T.; Saxena, A.K. Drug/Drug Interaction of Common NSAIDs with Antiplatelet Effect of Aspirin in Human Platelets. *Eur. J. Pharmacol.* **2013**, *721*, 215–224. [CrossRef]
97. Catella-Lawson, F.; Reilly, M.P.; Kapoor, S.C.; Cucchiara, A.J.; DeMarco, S.; Tournier, B.; Vyas, S.N.; FitzGerald, G.A. Cyclooxygenase Inhibitors and the Antiplatelet Effects of Aspirin. *N. Engl. J. Med.* **2001**, *345*, 1809–1817. [CrossRef]
98. Madjid, M.; Safavi-Naeini, P.; Solomon, S.D.; Vardeny, O. Potential Effects of Coronaviruses on the Cardiovascular System: A Review. *JAMA Cardiol.* **2020**, *5*, 831–840. [CrossRef]
99. Nishiga, M.; Wang, D.W.; Han, Y.; Lewis, D.B.; Wu, J.C. COVID-19 and Cardiovascular Disease: From Basic Mechanisms to Clinical Perspectives. *Nat. Rev. Cardiol.* **2020**, *17*, 543–558. [CrossRef]
100. Chung, M.K.; Zidar, D.A.; Bristow, M.R.; Cameron, S.J.; Chan, T.; Harding, C.V., 3rd; Kwon, D.H.; Singh, T.; Tilton, J.C.; Tsai, E.J.; et al. COVID-19 and Cardiovascular Disease: From Bench to Bedside. *Circ. Res.* **2021**, *128*, 1214–1236. [CrossRef]
101. McGettigan, P.; Henry, D. Cardiovascular Risk and Inhibition of Cyclooxygenase: A Systematic Review of the Observational Studies of Selective and Nonselective Inhibitors of Cyclooxygenase 2. *JAMA* **2006**, *296*, 1633–1644. [CrossRef]
102. Arfè, A.; Scotti, L.; Varas-Lorenzo, C.; Nicotra, F.; Zambon, A.; Kollhorst, B.; Schink, T.; Garbe, E.; Herings, R.; Straatman, H.; et al. Non-Steroidal Anti-Inflammatory Drugs and Risk of Heart Failure in Four European Countries: Nested Case-Control Study. *BMJ* **2016**, *354*, i4857. [CrossRef]
103. Eftekhari, A.; Alipour, M.; Chodari, L.; Maleki Dizaj, S.; Ardalan, M.; Samiei, M.; Sharifi, S.; Zununi Vahed, S.; Huseynova, I.; Khalilov, R.; et al. A Comprehensive Review of Detection Methods for SARS-CoV-2. *Microorganisms* **2021**, *9*, 232. [CrossRef] [PubMed]
104. Rashidzadeh, H.; Danafar, H.; Rahimi, H.; Mozafari, F.; Salehiabar, M.; Rahmati, M.A.; Rahamooz-Haghighi, S.; Mousazadeh, N.; Mohammadi, A.; Ertas, Y.N.; et al. Nanotechnology against the Novel Coronavirus (Severe Acute Respiratory Syndrome Coronavirus 2): Diagnosis, Treatment, Therapy and Future Perspectives. *Nanomedicine* **2021**, *16*, 497–516. [CrossRef] [PubMed]

molecules

MDPI

Article

Neuropeptides, New Ligands of SARS-CoV-2 Nucleoprotein, a Potential Link between Replication, Inflammation and Neurotransmission

Julien Henri [1], Laetitia Minder [2], Kevin Mohanasundaram [3], Sébastien Dilly [3], Anne Goupil-Lamy [4], Carmelo Di Primo [5] and Anny Slama Schwok [3,*]

1 Laboratoire de Biologie Computationnelle et Quantitative, Institut de Biologie Paris-Seine, UMR-CNRS 7238, Sorbonne Université, F-75005 Paris, France
2 Institut Européen de Chimie et Biologie (IECB), CNRS, INSERM UAR 3033, US001, Univ. Bordeaux, F-33000 Bordeaux, France
3 Saint Antoine Hospital, Centre de Recherche Saint Antoine, Sorbonne Université, Biology and Cancer Therapeutics, INSERM U938, F-75231 Paris, France
4 Biovia, Dassault Systèmes, 10 Rue Marcel Dassault, CS40501, CEDEX, F-78946 Vélizy-Villacoublay, France
5 CNRS, INSERM, ARNA, UMR 5320, U1212, IECB, Univ. Bordeaux, F-33000 Bordeaux, France
* Correspondence: anny.slama-schwok@inserm.fr or anny.schwok@gmail.com

Abstract: This work identifies new ligands of the nucleoprotein N of SARS-CoV-2 by in silico screening, which used a new model of N, built from an Alphafold model refined by molecular dynamic simulations. The ligands were neuropeptides, such as substance P (1-7) and enkephalin, bound at a large site of the C-terminal or associated with the N-terminal β−sheet. The BA4 and BA5 Omicron variants of N also exhibited a large site as in wt N, and an increased flexibility of the BA5 variant, enabling substance P binding. The binding sites of some ligands deduced from modeling in wt N were assessed by mutation studies in surface plasmon resonance experiments. Dynamic light scattering showed that the ligands impeded RNA binding to N, which likely inhibited replication. We suggest that the physiological role of these neuropeptides in neurotransmission, pain and vasodilation for cholecystokinin and substance P could be altered by binding to N. We speculate that N may link between viral replication and multiple pathways leading to long COVID-19 symptoms. Therefore, N may constitute a "danger hub" that needs to be inhibited, even at high cost for the host. Antivirals targeted to N may therefore reduce the risk of brain fog and stroke, and improve patients' health.

Keywords: nucleocapsid or nucleoprotein; SARS-CoV-2; structure-based drug discovery; alphafold model and molecular dynamics; neuroinflammation; brain fog; neuropeptides; neurotransmission; metabolism

Citation: Henri, J.; Minder, L.; Mohanasundaram, K.; Dilly, S.; Goupil-Lamy, A.; Di Primo, C.; Slama Schwok, A. Neuropeptides, New Ligands of SARS-CoV-2 Nucleoprotein, a Potential Link between Replication, Inflammation and Neurotransmission. *Molecules* **2022**, *27*, 8094. https://doi.org/10.3390/molecules27228094

Academic Editors: Giovanni N. Roviello and Caterina Vicidomini

Received: 20 September 2022
Accepted: 15 November 2022
Published: 21 November 2022

1. Introduction

The nucleoprotein N is one of the four structural proteins of SARS-CoV-2 virus [1–4]. N is present in large number of copies. As such, N is one of the major targets for antibody development and has been widely used for COVID-19 detection in the present pandemic outbreak. N binds the long viral RNA genome and is associated in the ribonucleoprotein (RNP) complex required for viral replication [5,6]. N is also involved in the formation of new virions through its interactions with the membrane protein M. Besides these functions, N acts as a mediator of inflammation. N represses the host antiviral response (as RNA interference and RIG-I mediated interferon) [7,8]. N targets the stress granule protein G3BP1, an essential antiviral protein known to induce innate immune response [9]. Importantly, N is associated with long-term SARS-CoV-2-specific immune and inflammatory responses, since the frequency of N-specific interferon-γ-producing $CD8^+$ T cells decline more rapidly in long-hauler COVID-19 patients [10]. N also activates endothelial cells dysfunction, leading to vasculopathy and coagulopathy observed in some COVID-19 patients [11]. N is thus an important target for the development of antivirals [12].

It is recognized that SARS-CoV-2 induces an early host inflammatory response that activates a cyclooxygenase-2 (COX-2) inflammatory cascade associated with NF-κB activation [13,14]. Indeed, in a mouse model, N protein promoted the expression of pro-inflammatory cytokines and triggered lung injury via NF-κB activation [11,15]. The use of anti-inflammatory drugs, especially non-steroidal anti-inflammatory drugs (NSAIDs), in the initial outpatient stage of COVID-19 appears to be a valuable therapeutic strategy [16]. We identified, using a structure-based approach a cyclooxygenase (COX) inhibitor, naproxen, as an antiviral against SARS-CoV-2 and Influenza A virus that combines anti-inflammatory properties [17–20]. We showed that naproxen bound to SARS-CoV-2 N-terminal domain (NTD) in vitro and reduced the viral load of infected cells. Naproxen protected lung cells against viral injury in a model of lung epithelium, in contrast with celecoxib or paracetamol [17]. NTD was also shown to bind AMP, NADPH and single-stranded RNA [3,12].

In this paper, we extend our previous work and identify by in silico screening new ligands of the full-length N protein (FL). We report that the nucleocapsid can sequester in a large cavity of its C-terminal various host neuropeptides involved in neurotransmission, vasodilation, inflammation and ligands perturbing cell metabolism. This capture, although likely decreasing viral replication, may contribute to perturbations in brain function and metabolism. We discuss these hypotheses in light of the recent literature. Our study sheds light on the need to block N by antivirals which may overcome some long COVID-19 symptoms [21,22].

2. Results

2.1. Generation of Models of Full-Length Nucleoprotein N

Recent structures of both N amino-terminal domain (NTD) and C-terminal domain (CTD) have been solved by X-ray crystallography, Nuclear Magnetic Resonance (NMR) and electron microscopy [2,3,23–25]. N also bears dynamic, disordered domains: an SR-rich domain between the NTD and CTD, as well as N- and C-terminal tails (Figure 1A). The isolated NTD formed a monomer which bound RNA and other single-stranded nucleic acid fragments [2].

The N NTD structure presented a central antiparallel sheet of 3–5 β-strands ("the hand palm") with a characteristic extrusion loop ("basic finger") that can close on RNA upon binding [2,3]. The CTD also bound RNA and forms oligomers, mostly dimers [2]. The N protein also underwent liquid–liquid phase separation when mixed with RNA [1]. The full length (FL) N was reported to be a tetramer by size exclusion chromatography coupled to light scattering [2], but may also form a mixture of monomer and dimer [26]. However, the structure of the full-length N protein of SARS-CoV-2 is not yet available, partly because this recombinant protein is resistant to RNase treatments [26].

We present here a model of the full-length N protein initially generated by Alphafold and further optimized by MD simulations as schematized in Figure 1B, and described in the experimental section [27]. Figure 1B also shows the comparison of our model with the NMR structure of the NTD (PDB 7ACT [2]) shown in brown and the X-ray structure of the CTD (PDB 6WZO [3]) shown in blue, with root mean square deviations (RMSD) of 4.0 and 4.7 Å, respectively (see also Supplementary Figure S1 for the superimposition of the structures and the RMSD of two dynamic trajectories). Figure 1C shows that the protein presented a large cavity at its C-terminal domain, highlighted by a star, which could bind large ligands.

While the NTD presented a number of suitable sites for ligands binding, these sites were small compared to the large cavity found at the C-terminal (Figure 2A), also detected in the isolated CTD X-ray structure [4,28]. Moreover, as this model of N was obtained based on the initial Wuhan strain of SARS-CoV-2 (wt FL), we also built the models of two Omicron variants of N to make sure that the detected cavity in wt FL would also bind ligands in more recent variants of N. Of the two Omicron variants of N, BA4 carried five mutations—P13L, P151S, R203K, G204R, S413R—and a deletion—31-33del—and BA5 had the same modifications but no mutation in P151 [29]. Figure 2A shows modifications of the cavities in the two variants, which only differed by the mutation of one residue. The latter cavity in BA5 was shifted apart from the β-sheet. The RMSD of the main chain atoms with respect to the wt ones were 1.57 Å and 1.28 Å in BA4 and BA5, respectively.

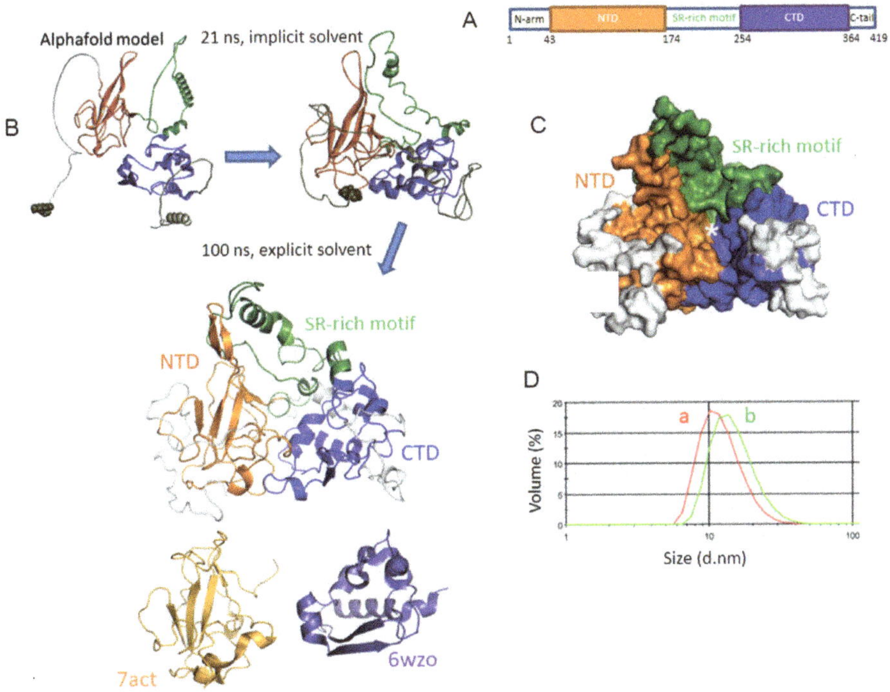

Figure 1. Sequence and model of the SARS-CoV-2 nucleoprotein wt N. (**A**) Schematic sequence of N, containing two structured domains at the N-terminal and C-terminal, NTD (brown) and CTD (blue), respectively. (**B**) Model of N, based on Alphafold, refined with MD simulations in implicit solvent, then in explicit solvent: the NTD and the RMN structure of the NTD (PDB 7ACT) [3] just below are shown in brown, and the CTD and the X-ray structure of the CTD (PDB 6WZO) just below are shown in blue, the SR-rich motif is presented in green, the other linkers in grey; the RMSD of the model compared to 7ACT is 4.7 Å and with one unit of the dimer of 6VYO [30] is 6.2 Å, the RMSD of the model with 6WZO [2] and 7DE1 [24] are 4.0 and 5.1 Å, respectively; the superimposition of the model and the experimental structures is shown in Supplementary Figure S1; (**C**) FL is shown as a surface with each domain colored with the same color code as depicted in A; note the large, mainly hydrophobic cavity, highlighted by the white star, at the C-terminal; see also Figure 2 for a visualization of the sites detected by Discovery Studio; (**D**) size distribution of the N protein (2 μM) in 20 mM Tris buffer pH = 7.9 containing 100 mM NaCl, (a) N alone and (b) N in the presence of RNA (TAR-polyA 6 μM) determined by dynamic light scattering (DLS).

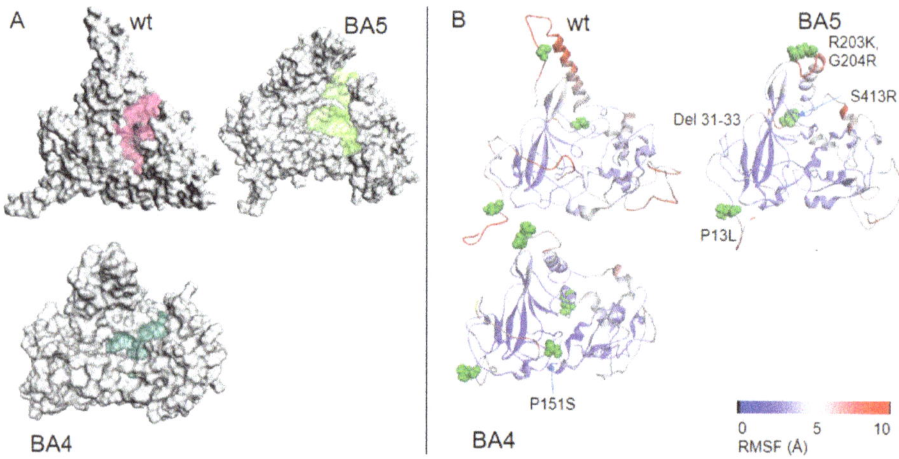

Figure 2. Models of wt N and variants. (**A**) Comparison of the modeled structures of N-FL shown as solvent-accessible surfaces: wt (Wuhan sequence); Omicron BA4 variant; Omicron BA5 variant. The C-terminal binding cavities are highlighted in colors. (**B**) 3D representation of the RMSF, as a measure of the protein flexibility of the main chain atoms (see also Supplementary Figure S1 for a 2D representation of the RMSF). The position of the deletion of the 31–33 residues is highlighted, the other mutations are shown in CPK. The RMSF follows a color code from blue (less flexible) to red (most flexible), note, in wt, BA4 and BA5 variants, the modifications of the fluctuations of the α-helix close to the R203K, S204R mutations located in close vicinity to the basic finger. These data highlight short- and long-range effects of the mutations on the protein flexibility, with a large enhancement in BA5 only missing the P151S mutation as compared to BA4 that may increase replication in the former variant.

Associated with these changes in the cavity, we measured the protein flexibility of the three proteins, estimated by the root mean square fluctuations (RMSF). The model of wt N FL suggested that the α-helix close to the basic finger in the NTD was quite mobile (when the protein was not engaged in interaction with ligands), with flexibility of some linkers as part of the serine-rich domain and the C-tail (see also Supplementary Figure S1).

The R203K, G204R mutations in both variants tilted the edge of the basic finger that binds RNA towards the linker that carries these two positively charged mutated residues. Interestingly, the flexibility of the α-helix edge (R209) in close proximity to the R203K, G204R mutations was enhanced in the BA5 variant as compared to BA4, which may help to grasp the RNA more quickly and/or more efficiently. The data suggested subtle, long-range effects of the mutations (here the mutation status of residue 151 influencing the mobility of R209) that may modify replication. The N variant carrying the R203K, G204R mutations increased the infectivity, fitness and virulence of SARS-CoV-2 [30]. Altogether, the variants presented a modified cavity as compared to wt N, which nevertheless remained quite large and very flexible.

2.2. *Characterization of Recombinant N, N-NTD and Its Mutants and N Full Length N-FL*

The recombinant NTD protein was expressed as previously described [17]. NTD purified as a monomer deduced from SEC-MALS analysis [4]. The full-length protein (FL) was expressed and purified according to a similar protocol than that for NTD. However, the purified FL found in the soluble fraction remained contaminated by bacterial nucleic acids as attested by an absorbance ratio 260 nm/280 nm ranging between 0.9 to 1.3, despite the use of benzonase and RNase [26]. Instead, we used the insoluble fraction that was denatured with 6 M urea and then renatured in Tris-NaCl buffer. In these conditions, the absorbance ratio 260 nm/280 nm decreased to 0.47, consistent with no significant contamination by bacterial nucleic acids. The FL was then tested by DLS for its ability to bind RNA (TAR-polyA). The size of the native protein with bacterial nucleic acids contamination was 14.5 ± 1.0 nm in volume (data not shown). As shown in Figure 1D, the size of the protein depleted from nucleic acids contaminants was 12.0 ± 0.4 nm in volume (14.0 ± 0.3 nm in intensity) that increased to 15.1 ± 0.3 nm (20.3 ± 0.3 nm in intensity) upon addition of RNA (Figure 1D), showing that the protein is functional. By comparison, the size of the NTD monomeric protein was 6.0 ± 0.3 nm and increased to 15.7 ± 0.5 nm in intensity in the presence of nucleic acids. The size of NTD could not be fitted by a spherical model to obtain its value in volume. To further characterize FL protein, the apparent melting temperature (Tm) of the protein alone was determined by DLS as a function of temperature. FL presented a first transition at 43 ± 1 °C followed by a denaturation at about 54 °C, (Supplementary Figure S2), consistent with the hypothesis that the protein could be (at least partly) dimeric at 25 °C, with a transition to random coil upon heating at 54 °C, a transition also determined by circular dichroism [31]. Preliminary CD experiments showed that the spectrum of FL is dependent upon the protein concentration as previously reported.

2.3. Virtual Drug Screening on NTD and FL

We performed virtual screening to identify ligands binding to the large C-terminal cavity of the full-length protein (Figure 2A) using the data base of the Sigma-Aldrich catalog. We also screened for ligands binding to NTD. We first used the Libdock software of Discovery studio. Then, docking of the interesting hits was further repeated using CDocker. Finally, MD simulations were performed on the best hits. Tables 1 and 2 summarize the mean-full hits both in terms of docking and biological function.

Table 1. NTD and FL ligands identified by virtual screening and their reported biological functions.

NTD Ligands	FL Ligands	Biological Function
DHF	DHF	Metabolism
THF	THF	Metabolism
AICAR (AMPK agonist)	Lauroyl coA	Metabolism
	Hemin	Metabolism
Naproxen	Prostaglandin E2 and F2	COX pathway/NSAID
Acetamine	and	COX pathway/NSAID
Indomethacin	*Other eicosanoids*	COX pathway/NSAID
	D Ala_2-Leu_5-Enkephalin YAGFL-OH (DADLE)	Neuropeptide involved in pain reduction, agonist of the μ− opioid receptor
	Substance P (1-7) RPKPQQF-OH	Neuropeptide mediator of inflammation, pain, and vasodilation, agonist of neuro- -kinin-1 receptor.
	Cholecystokinin DYMGWMDF (CCK8)	Intestinal hormone peptide binding to a receptor on nerve fibers of the vagus nerve

Table 2. Summary of the ligand-N interactions obtained by modeling compared to the measured complex by surface plasmon resonance with wt or mutant NTD or wt FL.

Ligand		Protein	-CDocker Interaction Energy Kcal/mol	Critical Residues	Mutations SPR (% Inhibi- tion/NTD wt)	K_D
DHF	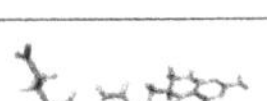	NTD monomer	58.0	G44, L45, N77, H145, R149, N150, N153	H145 (96%), R149 (83%)	1 μM

Table 2. *Cont.*

Ligand		Protein	-CDocker Interaction Energy Kcal/mol	Critical Residues	Mutations SPR (% Inhibition/NTD wt)	K_D
THF	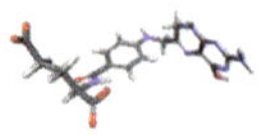	NTD monomer	61.7			
Naproxen	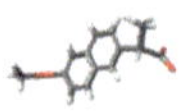	NTD monomer	38.0	W52, I146, R149 (site 1) [17] A50, R88, R92, R93 (site 2)	R88 (55%), R149 (44%)	4.4 ± 1.4 μM [17]
Acetamine	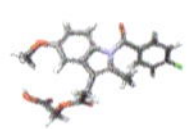	FL	39.7	F286, I304, S318, T334, G335, I337		
DHF		FL	57.9	R189, K233, K237, N239		
Enkephalin	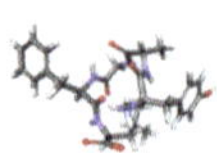	FL	64.4	A155, K261 [a], F274, F286, G287, Y298, I304, S318, Y333, G335, A336		
Lauroyl coA		FL	79.5	S187, S188, R189, K261 [a], R262 [a],		
Substance P (1-7)	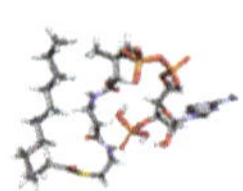	FL	82.4	S188, R189, R259 [a],T263 [a], A264 [a], R277, F286, L291, G295, Y298, Y333, T334, H356		
Cholecystokinin	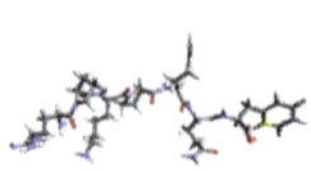	FL	82.0	A55, R107, Y109, V158, A264 [a], V270, L291, W301, P302, H356		
Hemin	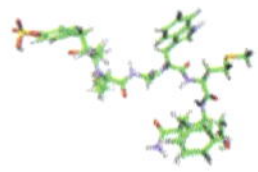	FL	47	A264 [a], V270,F274, L291, I304, A305, I337		
ZnTPPS	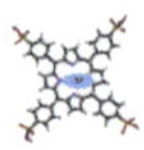	FL				0.4 ± 0.1 μM

[a]: Residue belonging to a putative nuclear localization signal NLS [32,33].

- *Ligands involved in metabolism*

Dihydrofolate (DHF) and tetrahydrofolate (THF) are, respectively, the substrate and product of the enzyme dihydrofolate reductase (DHFR).

DHF binding to NTD: DHF interacted with NTD in its monomeric and dimeric forms (Figure 3A,B, respectively). It was anchored by arginine R149 (or R107 in NTD dimer) forming a π-cation complex with one of the aromatic rings of DHF, a $\pi-\pi$ complex with H145 and further stabilized by a number of H bonds or electrostatic interactions with G44, N77, N150 (NTD monomer) and R92, S105 (NTD dimer). DHF binding on one face of the protein induced a compaction of the basic finger binding RNA on the other face, suggesting long range or allosteric effects. This compaction involved, in particular, R88 stabilized by salt bridges with D98 and E118. As R88, together with R92 and R107, have been all involved in RNA binding to NTD by NMR and further confirmed by mutation studies [2], it is likely that DHF binding strongly reduced RNA binding to N NTD.

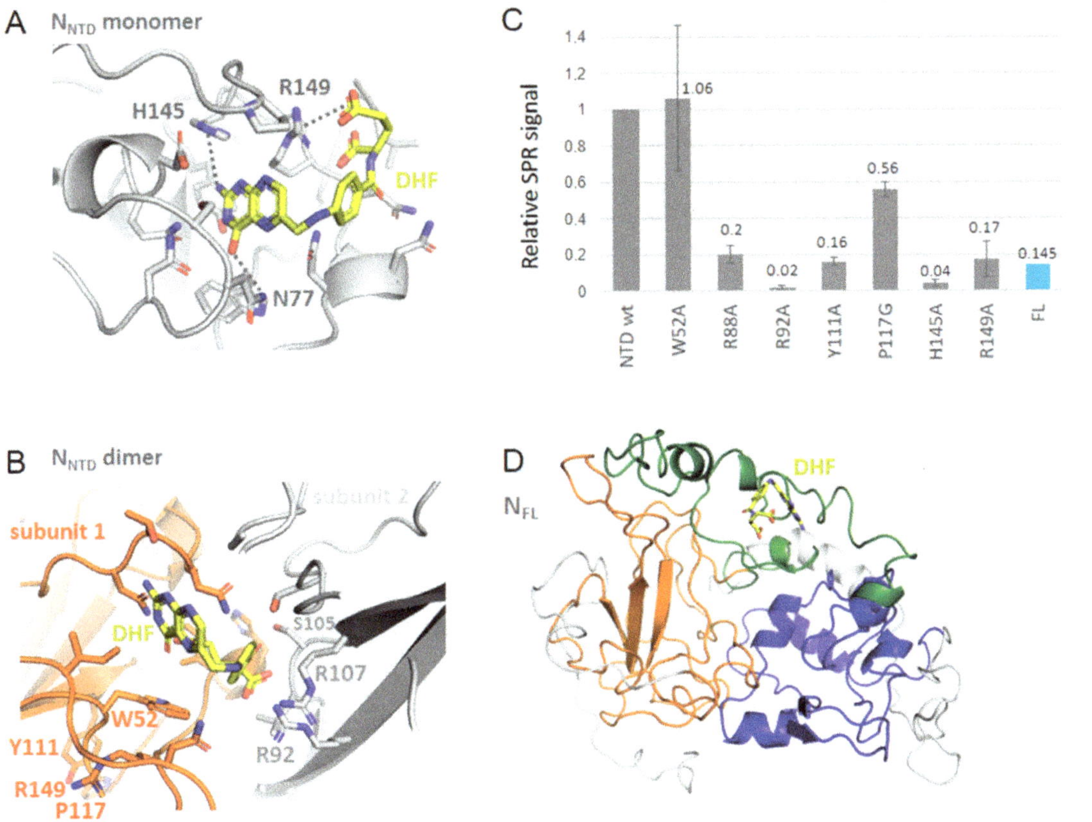

Figure 3. Interactions of N with dihydrofolate DHF. (**A**) Complex of monomeric NTD with DHF shown in yellow based on PDB 7ACT; (**B**) Complex of dimeric NTD with DHF based on PDB 6WYO; note the presence of cation–π interactions in both complexes; (**C**) SPR signal of wt NTD or mutants or of wt FL with DHF (10 µM) in 20 mM HEPES buffer: note the large decrease of the signal with the H145A and R149A mutants as predicted in A; the decrease with the R88A, R92A, and Y111A mutants with a small decrease in the P117G mutant, are in agreement with the structure in (**B**); (**D**) shows the modeled complex of DHF with FL, also observed in (**C**).

DHF in FL (Table 2) Although DHF formed stronger polar interactions in FL than DHF in NTD, it did not form multiple hydrophobic and π–π interactions as in NTD.

AICAR (Table 1) AICAR is an agonist of AMPK; it bound through electrostatic interactions at a site at the basic finger where RNA binds, involving K61, R89, R93, R95, K102 and R107 via its charged and polar groups.

Lauroyl-CoA in FL (Tables 1 and 2) The long aliphatic chain of lauroyl CoA was recognized by hydrophobic interactions with a series of five leucines and isoleucines residues and H356, F286 while its polar moiety forms H-bonds with S187 and S188 and K261 and R259 residues.

- *Ligands involved in the COX-prostaglandins pathway:*

The NSAIDs as naproxen bound the NTD [18]. They also can bind the C-terminal cavity although their size was not fitted to the large cavity. Table 2 describes the binding of acetamine at the C-terminal. Additionally, prostaglandins A2 and F2 and other eicosanoids bound FL with low affinity, further indicating the importance of this pathway in the host response to viral infection.

- *Neuropeptides:*

Substance P (1-7) in FL bound at a similar site than observed with other neuropeptides, enkephalin, cholescystokinin and lauroyl coenzyme A (Figures 4–6 and Table 2). Substance P (1-7) formed marked hydrophobic interactions with F286, Y298, I304 and Y333 as observed for enkephalin, and also binds S188 and R189 (Figure 4). This binding also resulted in a compaction of the RNA-binding finger, stabilized through interactions involving R95, D98, K100 and T205 although no direct interaction with the NTD was observed, including long-range effects between different sub-domains of the protein. Other fragments of substance P (4–11) also bound FL at a similar site and interaction energy than the P (1-7) fragment.

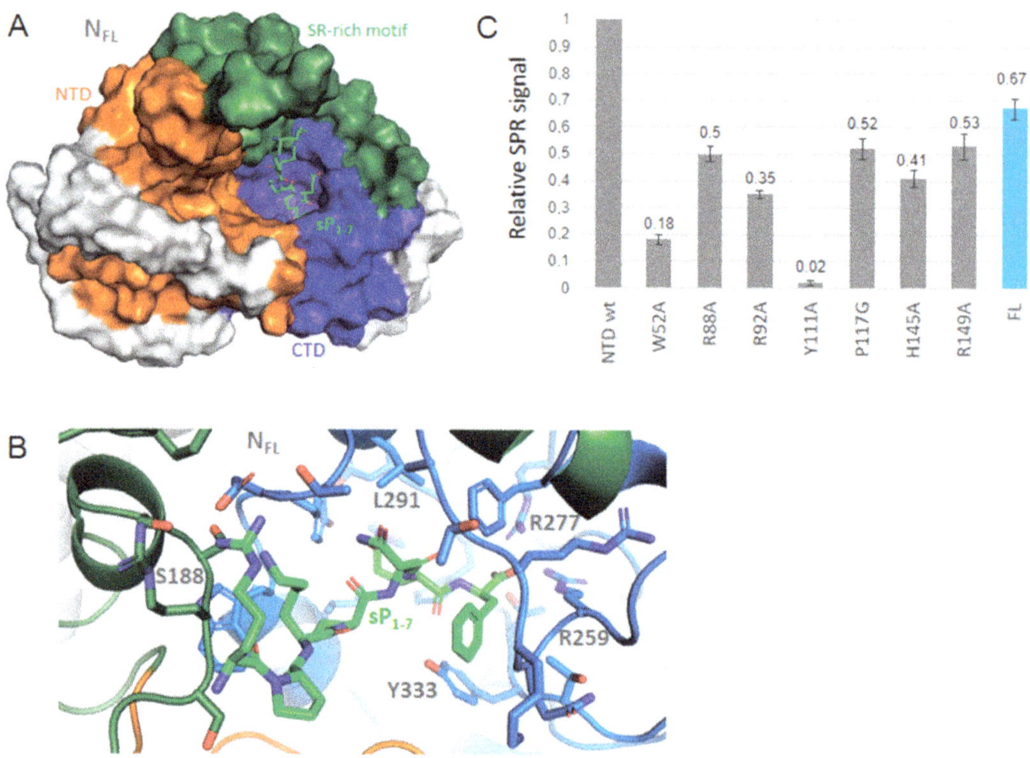

Figure 4. Interactions of wt N with substance P (1-7). (**A**) shows that the binding site of substance P (1-7) in a large binding site at FL C-terminal; (**B**) zoom on the complex of N with substance P (1-7) shown in green that formed multiple stacking and hydrophobic interactions, further stabilized by electrostatic and polar interactions; (**C**) consequently, the effect of the NTD mutations on the SPR signal was mainly observed with Y111A and W52A, with little effect seen on the other mutations. Substance P (1-7) also bound the N FL protein as measured by SPR.

To further test whether substance P (1-7) may bind the BA4 and BA5 variants of N, we introduced the relevant mutations and deletion in the wt FL-substance P (1-7) model. Figure 5 shows that substance P (1-7) tightly bound to both wt and BA5 variant of N with similar interactions. Figure 5C also highlights the decrease of flexibility of N due to peptide binding in the structural element involved in RNA binding, in agreement with the competition between the ligand and RNA for binding to N (see below Table 3).

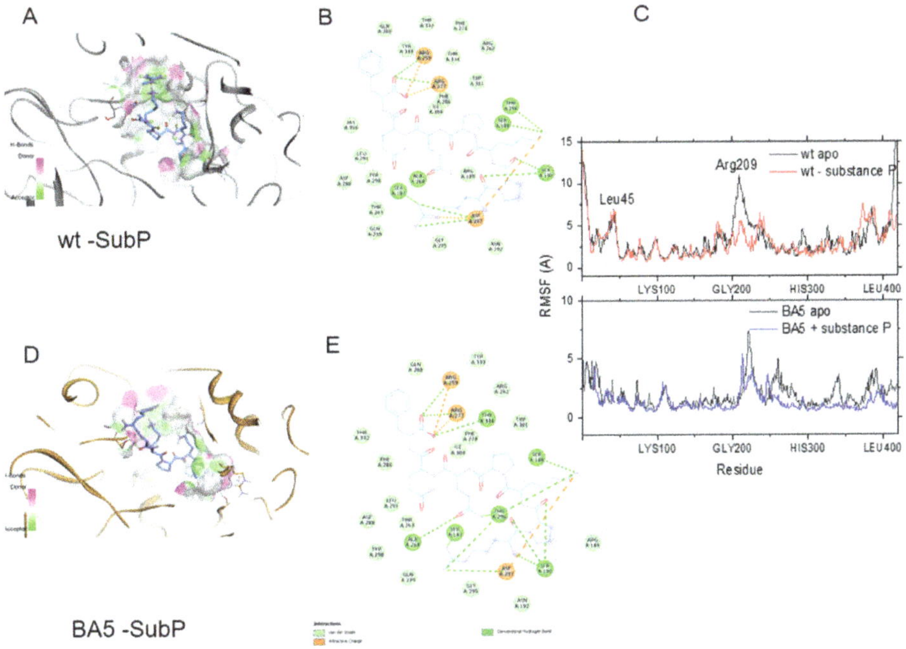

Figure 5. Binding of substance P (1-7) to the wt and the BA5 variant of N. (**A**) Binding site of substance P (1-7) in the wt N colored according to H-bond donor (pink) or acceptor (green) of the cavity; (**B**) 2D plots of the interactions of substance P (1-7) with wt N. Hydrogen bond and ionic interactions are shown as dashed green and orange colored lines, respectively; (**C**) upper: comparison of the flexibility of the wt apo protein (black line) with the wt-substance P (1-7) complex (red line); lower: comparison of the RMSF of BA5 apo (black line) and BA5- substance P (1-7) (complex (blue line); note the large decrease in the movement of the α-helix (R209) due to peptide binding in both complexes; (**D**) Binding site of substance P (1-7) in BA5; (**E**) 2D plots of the interactions of substance P (1-7) with BA5.

Table 3. Size of N and its complexes with the studied ligands with or without nucleic acids determined by DLS.

Ligand	Ligand Concentration (μM)	Size (nm) Intensity	Size (nm) Volume	N NTD/FL
none		5.3 ± 0.3 (80%)	Does not fit	NTD (60 μM)
none		13.2 ± 1.0 (85%)	12.0 ± 1.0 (96%)	FL 2 μM
Naproxen	2 6	12.2 ± 0.5 (40%) 8.7 ± 0.3 (60%)	11.3 ± 0.4 (99%) 8.9 ± 0.3 (96%)	FL 2 μM
Indomethacin	2 6	11.0 ± 0.4 (30%) 6.7 ± 0.3 (60%)	10.3 ± 0.5 (100%) 6.3 ± 0.4 (96%)	FL 2 μM

Table 3. *Cont.*

Ligand	Ligand Concentration (μM)	Size (nm) Intensity	Size (nm) Volume	N NTD/FL
Enkephalin	2 6	12.6 ± 0.6 (31%) 8.7 ± 0.2 (40%)	11.3 ± 0.6 (96%) 8.0 ± 0.3 (100%)	FL 2 μM
Substance P	2 6	10.3 ± 0.6 (50%) 8.0 ± 1.0 (76%)	9.5 ± 0.6 (98%) 7.2 ± 1.1 (99%)	FL 2 μM
ZnTPPS	2	7.6 ± 0.6 (25%)	7.3 ± 0.6 (99%)	FL 2 μM
RNA (TAR-polyA)	4	21.5 ± 2.0 (75%)	15.1 ± 1.1 (98%)	FL 4 μM
48m-DNA	4	25.3 ± 3.1 (56%)	15.7 ± 1.2 (96%)	FL 4 μM
RNA + subP	4	16.1 ± 0.6 (75%)	11.3 ± 0.6 (97%)	FL 4 μM
DNA + subP	4	13.8 ± 0.6 (70%)	10.1 ± 0.6 (91%)	FL 4 μM
DNA + enkephalin	4	10.5 ± 1.2 (89%)	9.0 ± 1.7 (99%)	FL 4 μM

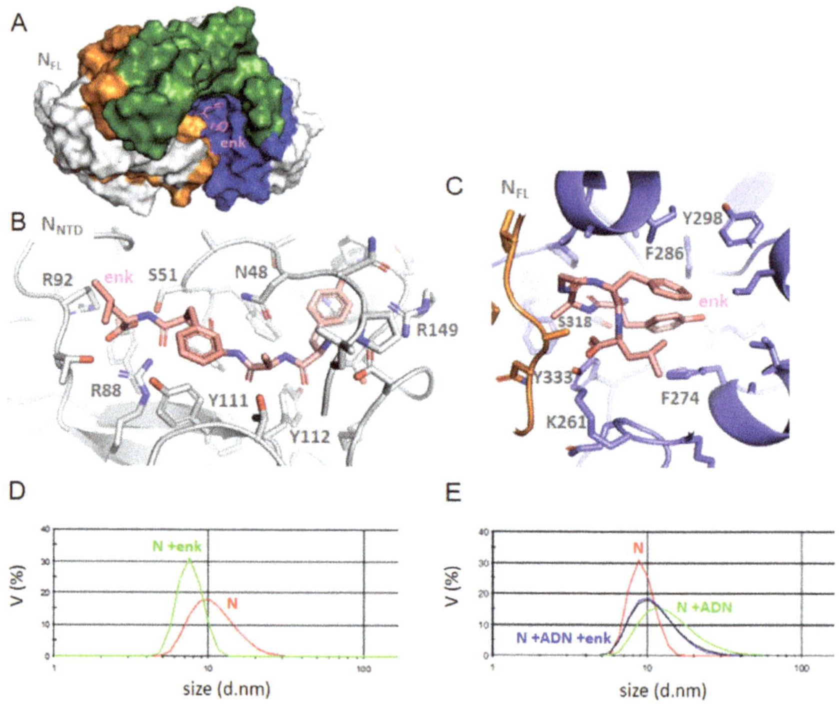

Figure 6. Interactions of Leu-enkephalin with NTD or FL. (**A**) enkephalin binding in CTD site in N FLwt, (**B**) the zoom of NTD-enkephalin complex, (**C**) FL-enkephalin complex; note the multiple hydrophobic interactions of this ligand in FL with P274, F286, Y298 and Y333, while the aromatic residues of the NTD β-sheet are involved in ligand binding in the NTD; (**D**) size distribution of N and its complex with enkephalin, which decreased its size as compared to the size of N alone; (**E**) the size of the N-DNA complex was reduced upon addition of enkephalin, in agreement with a competition of the ligand with nucleic acid binding to N FL.

Enkephalin in NTD (Figure 6): Enkephalin bound at a close-by site from that of DHF via electrostatic interactions with R88 and R92. In addition, enkephalin extended toward the central β−sheet forming hydrophobic and polar interactions with Y111 and Y112.

This binding induced some structuration of the linker carrying H145 and R149 in a short α−helix.

Enkephalin in FL (Figure 6A,C): Enkephalin interacted with FL through hydrophobic and π–π interactions with F274, F286, F298 and Y333 surrounding the ligand, stabilized by K261 and S318. Even though enkephalin bound FL through multiple hydrophobic interactions in the C-terminal cavity, it "closed" the basic finger on the N-terminal by H-bonds involving R95 with G97, G99 and T205, again suggesting long-range effects within the N structure. This hypothesis is confirmed by competition studies of enkephalin with RNA binding to N monitored by DLS (Paragraph 2–5).

Although we did not test directly the binding of enkephalin in the BA4 and BA5 variants of N FL, we anticipate that enkephalin would accommodate even more easily the variants than substance P as it can adopt an extended or folded conformation.

Cholescystokinin similarly bound FL through extensive hydrophobic interactions in the C-terminal cavity and also extended towards the N-terminal β-sheet via R107 and Y109. This binding mode was similar to that observed for substance P (Table 2).

- Hemin in FL:

Hemin bound FL through hydrophobic interactions with V270, L291, I304, I337 and π-π stacking with F274, in agreement with [32]. The stability of the complex was lower than that observed with the neuro/vasopeptides (Table 2).

2.4. SPR Study of DHF and Substance P (1-7) Interactions with NTD WT and Mutants, and FL

To comparatively assess the binding of DHF and substance P (1-7) to N, we generated mutants of the NTD recombinant protein by targeting some of the residues involved in ligand binding as detailed in experimental Section 4.1.3. The effect of such mutations on DHF binding to NTD are shown in Figure 3. Mutations H145A and R149A strongly reduced the signal of DHF as compared to DHF binding to NTD (monomer) wt as expected from the modeling. In addition to these mutations, we also observed strong effects of the mutations R92A, Y111A and R88A. These residues corresponded to those involved in DHF binding in NTD dimer. As compared to DHF signal in NTD, binding of DHF to the full-length protein was smaller. The relative signal ratio (DHF in NTD)/(DHF in FL) was dependent on the buffer used. It is possible that the oligomeric status of FL was different in HEPES buffer supplemented with 0.05% Tween, compared to Tris buffer. FL was reported to be either tetrameric or a mixture of monomer and dimer [2,26]. Monomer or dimer form of N FL, in particular in the C-terminal provided more accessible site(s) for ligand binding than the tetrameric form did. DHF K_D for binding to NTD was about 1 μM, deduced from the variation of the signal as a function of DHF concentration in the range 0.3–10.0 μM.

Substance P (1-7) binding to NTD and FL was also observed as shown in Figure 4C. This large ligand bound NTD and exhibited relatively small effects of the mutations with notable exceptions of W52A and Y111A. This suggested that substance P (1-7) bound on the β-sheet at the conserved sequence of the five aromatic residues $W^{108}YFYY^{112}$. Substance P (1-7) binding to N therefore seemed driven by hydrophobic interactions, consistent with the sequence of the peptide. It was also in line with both the binding of the peptide to FL seen by SPR and the multiple hydrophobic contacts of this peptide in the C-terminal cavity of the FL protein suggested by modeling.

2.5. FL Interactions with the Ligands by DLS; Competition with Single-Stranded DNA or RNA Binding

To further address whether the ligands may modify the oligomeric status of the FL protein, we performed dynamic light scattering experiments. FL (2 μM) presented a main peak at 12.0 $\pm$ 0.6 nm in volume (14 nm in intensity); the size of the complex decreased to 10.3 and 7.8 $\pm$ 0.3 nm upon addition of 2 μM and 6 μM enkephalin, respectively (Figure 6D). As shown in Table 3, the same trend was observed upon addition of naproxen, indomethacin, substance P (1-7) or a porphyrin ZnTPPS with a larger decrease observed with substance P (1-7), ZnTPPS and indomethacin. This suggested that ligand binding decreased the oligomerization state of the protein. We then tested whether ligand binding to FL could be competitive with nucleic acids binding. Table 3 indeed shows that the substance P (1-7) or enkephalin (Figure 6E) are competitive with nucleic acids binding to FL as the size of the RNA-FL or DNA-FL complex about 14 nm observed without ligand always decreased by addition of the ligands.

2.6. FL Interactions with ZnTPPS

In our in silico screening, we identified hemin as a potential ligand binding to FL C-terminal site as recently proposed; hemin binding to N was shown to reduce viral replication [32]. Figure 7A shows that the binding site of hemin colored in red. Here, we chose a water-soluble, negatively charged porphyrin with properties of a photosensitizer, able to produce singlet oxygen and ROS in the perspective of a potential antiviral which could sensitize infected tissues containing N. ZnTPPS has a Soret band at 420 nm and two Q bands, being a metalated porphyrin. Upon addition of FL, the Soret band decreased and red-shifted to 431 nm with the presence of an isosbestic point at 426 nm, while the first Q band shifted from 557 to 561 nm. The data were repeated at various concentrations of ZnTPPS, yielding a $K_D = 0.40 \pm 0.10$ μM (Figure 7B,C).

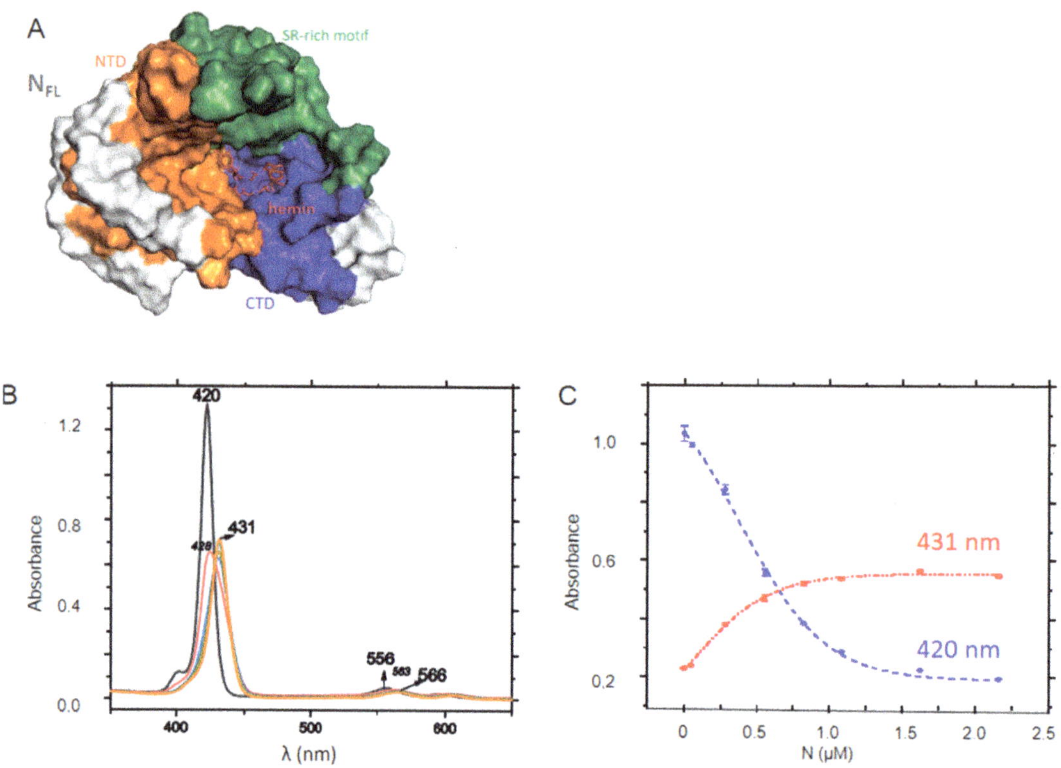

Figure 7. Interactions of FL with ZnTPPS. (**A**) the porphyrin shown in red binds at the C-terminal cavity of FL (See also Table 2). (**B**) Absorption changes of ZnTPPS upon titration with FL; (**C**) Absorbance at Soret maxima of the free (420 nm) or FL-bound (431 nm) porphyrin; the dashed lines correspond to fits of the experimental data according to a dose-response function, yielding $K_D = 0.40 \pm 0.10$ μM.

3. Discussion

In this work, we identified new ligands of N in the perspective of drug repurposing, with perspectives in basic understanding of N functions.

The model of FL, the full-length nucleoprotein of the wt sequence we built, was generated by Alphafold, with subsequent refinements using molecular dynamics simulations. The model clearly showed that the protein is very flexible, nevertheless the simulations successfully produced a folded model, presenting a large cavity at the C-terminal of the protein able to bind ligands with a broad distribution of molecular weights, similar to the cavity found in the X-ray structure of the isolated C-terminal protein. By introducing the mutations and deletion found in the Omicron BA4 and BA5 variants of N, we also detected a large cavity at the C-terminal, although modified as compared to the wt one. It is interesting to note the increased flexibility of the BA5 variant as compared to that of the BA4 variant, in particular in a linker carrying the 203–204 mutations just opposite the basic finger that bound RNA. This suggested a better fitness to replication in the BA5 variant because of a better/faster adaptation to the viral RNA as compared to that of the BA4 variant [30].

The ligands we identified by in silico screening are likely to decrease viral replication as: (i) the ligands competed with RNA binding; (ii) the ligands reduced FL (oligomer) size. Since N is known to oligomerize upon binding to RNA, both mechanisms are expected to interfere with viral replication. Four ligands we identified in our screening were indeed recently shown to decrease viral replication: naproxen, indomethacin, DHF and hemin [17,18,34,35]. The DLS data were consistent with a competition of DHF, naproxen, indomethacin, enkephalin, substance P (1-7) and ZnTPPS with RNA binding to FL.

Besides its roles in viral transcription and replication, N is involved in immunity via its control in host interferon release, in cytoskeleton rearrangement. The literature further suggests additional roles of N that our data may further highlight, as detailed below.

3.1. N as a Mediator of Inflammation and Its Effect in Long-Haul COVID-19

A sustained inflammation that extended well beyond clearance of the primary infection was observed in long-hauler patients and also found in a hamster model of SARS-CoV-2 infection [15,36]. The expression of prostaglandin receptors was increased in patients with severe COVID-19, as part of the cytokine storm developed by these patients [37]. Accordingly, our virtual screening identified prostaglandin E2 and other eicosanoids as N FL binders, these metabolites belonging to the COX–arachidonic -eicosanoids (including prostaglandin) pathways of inflammation and metabolism as an (early) host response triggered by the viral infection, in agreement with recent reports [16,38]. Treatment of patients hospitalized for mild and moderate COVID-19 with the COX inhibitor indomethacin helped the patients to recover [39]. Naproxen inhibited replication and reduced inflammation in a model of reconstituted lung epithelium [17]. In silico studies suggested that naproxen and acetaminophen may not only bind the N protein but also SARS-CoV-2 main protease, the RBD domain of spike and the polymerase [40].

The cytokine storm could also be considered as a possible driving factor for the expansion of neuropathies after severe COVID-19 infection, contributing to the chronic pain that appeared after acute infection recovery. The cytokine storm was at least partly mediated by (lung-resident and brain-penetrating) macrophages blocked in an M1 state that released pro-inflammatory cytokines as IL6 [41–43]. The neuropeptides identified in this work were known to have a role in inflammation: substance P is a mediator of inflammation and pain [41,44,45], while enkephalin had the opposite role to release pain [46,47]. In dopaminergic neurons, subpicomolar levels of substance P activated NADPH Oxydase-2 (NOX2) to increase ROS concentrations and subsequently synergistically activated the MAPK and NF-κB pathways, contributing to a potentiated pro-inflammatory cytokine production [47,48]. In addition, substance P and cholestocystokinin were vasoactive peptides [47]. We hypothesize that the sequestration by N of the identified ligands of N could have consequences for the host in terms of endothelial dysfunction, micro-thrombus in the brain and in other microvessels, with increased risk of stroke. Indeed, in the brain of SARS-CoV-2 infected non-human primates, the occurrence of a neuronal injury, brain micro-hemorrhages and hypoxia and rare viruses in brain-associated endothelium where N co-localized with the von Willebrand factor were observed [21]. It was also proposed that substance P and bradykinin, were likely to drive microvascular permeability, and be responsible for a phenomenon called «vasoactive peptide storm» as part of the development of COVID-19 pathology [48]. An agonist of cholecystokinin A (CCK-A) showed benefit in reduction of inflammation, among those hospitalized with moderate COVID-19 [49].

Therefore, we propose the hypothesis that the identified neuropeptides (and/or their fragments) could contribute to long-term inflammation in the CNS via N and possibly other mechanisms.

3.2. *N Implication in Immunity, with Possible Long-Term Neurological Effects and Putative Viral Latency*

N implication in immunity of the host against SARS-CoV-2 was expressed at multiple levels.

(i) *T cells*: Patients with persistent symptoms over 4 months following COVID-19 onset presented a lower frequency of $CD8^+$ T cells expressing CD107a, a marker of degranulation, in response to Nucleocapsid (N) peptide pool stimulation, and a more rapid decline in the frequency of N-specific interferon-γ-producing $CD8^+$ T cells [10].

(ii) *Opioid peptides*: Immune system and neuronal system cross-talk; this signaling is mediated by various molecules such as opioid peptides such as enkephalin. Enkephalins can impact lymphocytes proliferation, antibody synthesis. Enkephalin can enhance the release of pro-inflammatory cytokines like IL6 [50]. The cross-talk can take place between cytokines as CCL2 and opioid peptides and alter nociceptive synaptic transmission [51]. Thus, more work is required to test whether the enkephalin peptide identified here as a ligand of N may increase pain and immune disorders in the context of COVID-19, despite its physiological pain release function.

(iii) *Vasoactive peptides*: Among the possible routes through which SARS-CoV-2 can invade the CNS, SARS-CoV-2 can directly invade the vagus nerve and retrograde into the CNS, or indirectly stimulate the enteric nervous system through immune pathways. Cholecystokinin could participate in this process. In addition, peripheral nerves may spread SARS-CoV-2 into the brain through the retro-neural route, including the olfactory nerve, trigeminal nerve, glossopharyngeal nerve and vagus nerve. Substance P is the main neuropeptide, neuromodulator and neuro-hormone of the trigeminal ganglion (TG), associated with nociception and inflammation through its receptor, the neurokinin-1. As observed with other viruses such as herpes or HIV, it was hypothesized that SARS-CoV-2 virus infection might become latent if it is acting through TG [41,45], involving substance P action.

3.3. *N as a Mediator of Perturbed Metabolism via Its Ligands, with Possible Long-Term Effects*

DHF and THF linked N to DHFR, an important enzyme that participates to DNA synthesis, being coupled to methionine metabolism. Additionally, folate exerted a protective role in the cardiovascular system [52,53]. Moreover, folate levels were usually low in patients with viral infection [54,55]. The reduction of viral replication by DHF/THF came at the expense of large perturbations of the host purine metabolism, nicely described in [35]; it is likely that the agonist of AMPK, AICAR would also have a dual effect on replication and lipid metabolism. The screening also identified Lauroyl-coA as a ligand of N, potentially linking N to beta oxidation of fatty acids and to triacylglycerol biosynthesis.

3.4. *Zntpps as a Prototype of a Photoactive N Ligand for Photodynamic Therapy (PDT)*

The binding site of hemin was deduced from docking [32], with a putative NLS signal (amino acids 258–268) at the edge of this site [33]. This site shared a number of predicted residues involved in the binding of the neuropeptides and DHF. Binding of

ZnTPPS was demonstrated by spectroscopic changes of the porphyrin. ZnTPPS is one of the sensitizers that can release singlet oxygen and ROS upon light/ultrasound activation. Such compounds have been useful in the treatment of some cancers and proposed as antiviral treatment to inactivate SARS-CoV-2 [56].

3.5. *Heme Sequestering by N and Its Potential Effect on NO and ROS Signaling*

Dysfunctional, low hemoglobin levels, observed in patients with COVID-19 were possibly linked to well-characterized brain hypoxia [22,57]. Here, we further suggested that folate binding to N could also be responsible for an altered iron metabolism via perturbation in folate levels [35]. Moreover, we speculate that heme binding to FL could increase disruption of labile hemes from proteins such as guanylate cyclase, with potential effects in NO signaling, reduced oxygen/hypoxia sensing and/or impediment of their gaseous ligand binding such as O_2, NO and CO, with all the known consequences of altered NO/O_2 signaling in the cardiovascular system (eNOS), in neurotransmission, control of serotonin levels (nNOS) and in immunity (iNOS).

4. Materials and Methods

4.1. *Models of N, in Silico Screening, Docking and Molecular Dynamics Simulations*

4.1.1. Models of wt N

The following X-ray structures extracted from the Protein Data Bank (PDB) have been used: N-terminal domain of SARS-CoV-2: 6VYO [58], 7ACT [2].

A computational model of the full-length nucleoprotein of SARS-CoV-2 was generated with the ColabFold [59] version of AlphaFold 2 [27] using an API hosted at the Södinglab based on the MMseqs2 server [60] for the multiple sequence alignment creation. 300 sequences were aligned on uniprot entry A0A6C0T6Z7 as query sequence and no template (419 residues). Predicted lDDT is highest in the NTD and CTD regions that were previously determined by experimental methods (PDB entries 7act, 6wzo).

MD simulations were performed based on Alphafold and initial minimization and equilibration steps for 21 ns in implicit solvent. The wt protein was prepared with Discovery Studio 2022 (Dassault Systèmes BIOVIA, Discovery Studio Modeling Environment, Release 2022, (San Diego, CA, USA): Dassault Systèmes 2022) the models were first protonated at pH 7.4, typed with CHARMM36 force field and the protein was placed in an orthorhombic box solvated with a pre-equilibrated solvent containing TIP3 waters, with a 15Å minimum distance from boundary. Two trajectories were generated with the NAMD 2.13 engine and CUDA acceleration as implemented in Discovery Studio 2022, by varying the Random Seed Number. The MD simulations were run for 100 ns each under constant pressure (NPT ensemble) at 300 K using a Langevin thermostat and piston, for temperature and pressure control, respectively, and long-range electrostatic interactions were computed using Particle Mesh Ewald (PME) algorithm [61]. The trajectory was analyzed with the Analyze Trajectory Protocol to compute the RMSD (Root Mean Square Deviation) and RMSF (Root Mean Square Fluctuation).

4.1.2. Model of N Variants

This wt model was then used to generate the BA4 and BA5 variants of N with the MODELER algorithm [62] as implemented in Discovery Studio version 2022 (Dassault Systèmes BIOVIA, Discovery Studio Modeling Environment, Release 2022, San Diego: Dassault Systèmes 2022). For each of the two mutants BA4 and BA5, a 100 ns trajectory was also generated with the same protocol as described for the WT protein.

All the cavities were obtained with the Define Site from Receptor Cavities of Discovery Studio 2022 as shown in Figure 2.

4.1.3. Ligand Screening and Docking to wt N

In silico screening of ligands that bind the protein structures was performed using the Sigma catalog data base and assessed in a two-step protocol. For this screening, we first used the docking program Libdock of Discovery Studio version 2021, with the exclusion of ligands with molecular weights above 2000 and according to the binding site(s) defined by cavity detection. A site-docking was then performed using Cdocker with energy minimization (Discovery Studio version 2021) to more precisely identify the ligand binding mode. For each protein, the most representative pose of the ligand was selected. The resulting protein–ligand complexes were finally refined by a molecular dynamics simulation using the CHARMm force field [63] and the standard dynamic cascade protocol of Discovery studio version 2021. This protocol started with a first minimization of 1000 steps using the Steepest Descent algorithm and a RMSD gradient of 1 Å, followed by a second minimization of 2000 steps using the Adopted Basis Newton–Raphson algorithm and an rmsd gradient of 0.1 Å. The third step involved heating from 50 K to 300 K, with a fourth step of equilibration during 1 ns and a fifth step, production. The time of the production step was initially set at 10 ns, but extension to 20 ns was applied if the ligand was not stable. Three replicas were carried out for each complex. For each trajectory, the displacement of the ligands was studied by RMSD calculation. The representative structure (i.e., with the smallest average rmsd from all other structures of the cluster) of the largest cluster of each complex was selected.

4.1.4. Substance P (1-7) Bound to BA5 Variant

An homology model of BA5 in complex with substance P (1-7) was built with MODELER, using the wt-substance P (1-7) as a template. For each of the two mutants BA4 and BA5 in complex with the substance P (1-7), a 100 ns trajectory was also generated with the same protocol as described for the wt protein.

4.2. *Chemicals and Oligonucleotides*

We bought from Sigma Aldrich, St Quentin Fallavier, France: dihydrofolate, Leu-enkephalin, 5-Aminoimidazole-4-carboxamide ribonucleotide (AICAR), naproxen, indomethacin and Tris, NaCl, the oligonucleotides with or without TEG biotin tags were purchased with HPLC purification. The sequence of the 48 mer DNA was: 5′ATA TAT ATC TAT GTC CAT ATA TAT ATA AAA CAC AGC GTG TGT GTG TAA 3′. The sequence of TAR-polyA was: 5′ $(A)_{21}$ GAA AGG AGC CUG GGA GCU CC 3′. Substance P (1-7) was synthesized by Genecust, Boynes, France. The plasmids for NTD, mutants and the full-length N were obtained from Genecust, Boynes, France.

4.3. *Proteins Expression and Purification*

The N-terminal domain of the nucleoprotein of SARS-CoV-2 (residues 50–173) with an N-terminal His6 tag was cloned in pET-28a vector (pET28a-His6-NP-NTD). Heterologous expression at 15 °C for 16 h in *E. coli* BL21 bacterial strain (DE3 (NEB, Evry, France). The recombinant protein (15 kDa) found in the soluble fraction was purified on a Ni^{2+}-NTA affinity column and SP sepharose ion exchange chromatography, and presented a single band revealed by SDS PAGE. The yield was 1.5 mg of pure protein per liter of culture.

The expression of the FL protein was performed as described above. The insoluble fraction of the pellet was resuspended in lysis buffer containing 6 M urea for an hour at room temperature, the insoluble part was removed by centrifugation over 10 min. The purification involved an affinity Ni^{2+} column, with a first washing at 3 M urea, followed by washings in a 20 mM Tris-HCl pH = 7.9 and 1 M NaCl, and then increasing concentrations of imidazole (10, 20 and 50 mM). Elution was performed with 500 mM imidazole. Imidazole was then removed from the solution by dilution-ultrafiltration (centricon molecular weight cutoff 10,000, Millipore). The protein was then left at 4 °C for three days for renaturation in a 20 mM Tris-HCl buffer at pH = 7.9, 100 mM NaCl.

4.4. *SPR Experiments*

The recombinant FL protein was tagged with six His for purposes of identification and affinity purification, and the same principle was exploited for capture of the recombinant proteins on the sensor surface. Poly-histidine is a commonly used tag that can chelate with Ni^{2+} ions in complex with nitrilotriacetic acid (NTA, Biacore, Uppsala, Sweden), providing a convenient approach for capturing tagged constructs on Sensor Chip NTA. The FL protein was injected at a flow rate of 30 µL/minute; the ligand was injected after the protein capture; then, the protein or protein–ligand complex was removed from the surface by applying 350 mM EDTA, with a very good reproducibility. The experiments were usually performed in 20 mM HEPES buffer (Biacore), 100 mM NaCl and 0.05% Tween. The FL protein was captured at a concentration of 0.1 µM as the signal decreased at higher FL concentrations, presumably because the protein adopts higher oligomeric state(s). The NTD wt and mutant proteins were captured at a concentration of 1 µM for shorter times. In both FL and NTD proteins, signals of 5000 to 8000 RU were captured on the surface. The NTD proteins were always used as freshly prepared solutions. The ligands concentration range was usually in the range 0.3–10 µM.

4.5. Dynamic Light Scattering

The experiments were performed on a Malvern nanosizer apparatus. The temperature was set at 20 °C, and 10 scans with a duration of 10 s each were acquired in duplicate for each time and sample. The size distribution in the intensity of the scattered light was obtained using the Cumulants method from the instrumental software, yielding the hydrodynamic diameter. The N NTD concentration was in the range of 40–60 µM, the FL concentration was in the range of 1 to 10 µM in 20 mM Tris buffer at pH = 7.9 containing 100 mM NaCl. The melting experiments were performed at a heating rate of 1 °C/min over the range 30 °C to 65 °C in sealed disposable cuvettes. The apparent melting temperature, Tm, was determined by the first derivative of the melting curve.

5. Conclusions

The abundant N protein potentially could disrupt many signaling pathways, through sequestration of signaling molecules as the neuropeptides identified here or other useful metabolites at its large C-terminal cavity. An increased flexibility of N seemed to emerge in the BA5 N variant as compared to the BA4 one, which may both speed up replication and enhance ligand binding, with possible cooperative or long-range effects of the mutations/deletion. Although these neuropeptides usually signal in the pico-to nanomolar concentration range in physiological conditions, their concentrations in the blood, in the CNS and in TG can increase locally, in particular in COVID-19 patients [64–70], in line with the hypothesis suggested by this study. N may affect the cell metabolism via folate and AMPK and neurotransmission via neuro/vasoactive peptides. The physiological function of these ligands or their fragments would probably be altered consecutive to their association with N.

Altogether, the host likely senses the N protein as a "danger hub"; any possible mean seems to be undertaken to neutralize N. Therefore, instead of letting valuable neuropeptides be sequestered by N, blocking N with antivirals may decrease symptoms associated with long COVID-19, which accounts for 5–20% of the patients [22,71]. Based on the biological functions of the neuropeptides, we speculate that N could be involved in pain, neurological brain fog and inflammation and possibly immune imbalance and latency. Current NSAIDs as naproxen and indomethacin are readily available and bind to N; extended ligands could be also adequate to fill the C-terminal cavity of N. The hypothesis raised in this discussion offers a comprehensive picture that may link multiple long-haul COVID-19 symptoms through N, which is yet to be demonstrated by clinical data.

Supplementary Materials: The following supporting information can be downloaded at: https://www.mdpi.com/article/10.3390/molecules27228094/s1, Figure S1: Analysis of the models of N; Figure S2: Melting curve of recombinant FL determined by DLS.

Author Contributions: Conceptualization, A.S.S., C.D.P. and J.H.; methodology, A.S.S., L.M. and J.H.; software, A.G.-L. and S.D.; validation, L.M. and K.M.; formal analysis, A.G.-L. and L.M.; resources,

A.G.-L., C.D.P. and J.H.; writing—original draft preparation, A.S.S.; writing—review and editing, A.S.S., C.D.P. and J.H.; visualization, A.S.S.; funding acquisition, A.S.S. and J.H. All authors have read and agreed to the published version of the manuscript.

Funding: This research was funded by The CovNucleovir from the faculty of Medicine of Sorbonne University and The Sorbonne University Foundation, grant CoVNucleovir on Flash COVID funding.

Institutional Review Board Statement: Not applicable.

Informed Consent Statement: Not applicable.

Data Availability Statement: Modeled structures can be available upon request to A.S.S.

Acknowledgments: The authors are grateful to HO Bertrand from Biovia for the gift of a temporary license for Discovery Studio (Dassault Systemes). We thank N. Griffete and C. Menager from the Phenix Laboratory for the use of their Malvern DLS apparatus. We are grateful to the structural biology facility (UMS 3033/US001) of the Institut Européen de Chimie et Biologie (Pessac, France) for using the Biacore T200 instrument in SPR experiments. We thank the following colleagues for their support in the initial stages of the CoVNucleovir project: Bruno Fève, Michèle Sabbah, Vincent Maréchal and Francis Berenbaum, from CRSA, Sorbonne Université, Olivier Terrier, Rosa Calavatra, B. Lina, Virpath, Lyon University and the team of Evzen Boura, the Czech Republic.

Conflicts of Interest: The authors declare no conflict of interest. The funders had no role in the design of the study; in the collection, analyses, or interpretation of data; in the writing of the manuscript, or in the decision to publish the project results.

Sample Availability: Samples of the compounds are not available from the authors.

Dedication: This work is dedicated to the so many long-hauler COVID-19 patients, with love and hope that this work will open new and useful insight to helping them.

References

1. Cubuk, J.; Alston, J.J.; Incicco, J.J.; Singh, S.; Stuchell-Brereton, M.D.; Ward, M.D.; Zimmerman, M.I.; Vithani, N.; Griffith, D.; Wagoner, J.A.; et al. The SARS-CoV-2 nucleocapsid protein is dynamic, disordered, and phase separates with RNA. *Nat. Commun.* **2021**, *12*, 1936. [CrossRef]
2. Dinesh, D.C.; Chalupska, D.; Silhan, J.; Koutna, E.; Nencka, R.; Veverka, V.; Boura, E. Structural basis of RNA recognition by the SARS-CoV-2 nucleocapsid phosphoprotein. *PLoS Pathog.* **2020**, *16*, e1009100. [CrossRef] [PubMed]
3. Kang, S.; Yang, M.; Hong, Z.; Zhang, L.; Huang, Z.; Chen, X.; He, S.; Zhou, Z.; Zhou, Z.; Chen, Q.; et al. Crystal structure of SARS-CoV-2 nucleocapsid protein RNA binding domain reveals potential unique drug targeting sites. *Acta Pharm. Sin. B* **2020**, *10*, 1228–1238. [CrossRef] [PubMed]
4. Ye, Q.; West, A.M.V.; Silletti, S.; Corbett, K.D. Architecture and self-assembly of the SARS-CoV-2 nucleocapsid protein. *Protein Sci.* **2020**, *29*, 1890–1901. [CrossRef] [PubMed]
5. Gao, Y.; Yan, L.; Huang, Y.; Liu, F.; Zhao, Y.; Cao, L.; Wang, T.; Sun, Q.; Ming, Z.; Zhang, L.; et al. Structure of the RNA-dependent RNA polymerase from COVID-19 virus. *Science* **2020**, *368*, 779–782. [CrossRef]
6. Klein, S.; Cortese, M.; Winter, S.L.; Wachsmuth-Melm, M.; Neufeldt, C.J.; Cerikan, B.; Stanifer, M.L.; Boulant, S.; Bartenschlager, R.; Chlanda, P. SARS-CoV-2 structure and replication characterized by in situ cryo-electron tomography. *Nat. Commun.* **2020**, *11*, 5885. [CrossRef]
7. Chen, K.; Xiao, F.; Hu, D.; Ge, W.; Tian, M.; Wang, W.; Pan, P.; Wu, K.; Wu, J. SARS-CoV-2 Nucleocapsid Protein Interacts with RIG-I and Represses RIG-Mediated IFN-β Production. *Viruses* **2021**, *13*, 47. [CrossRef]
8. Mu, J.; Xu, J.; Zhang, L.; Shu, T.; Wu, D.; Huang, M.; Ren, Y.; Li, X.; Geng, Q.; Xu, Y.; et al. SARS-CoV-2-encoded nucleocapsid protein acts as a viral suppressor of RNA interference in cells. *Sci. China Life Sci.* **2020**, *63*, 1413–1416. [CrossRef]
9. Lu, S.; Ye, Q.; Singh, D.; Cao, Y.; Diedrich, J.K.; Yates, J.R., 3rd; Villa, E.; Cleveland, D.W.; Corbett, K.D. The SARS-CoV-2 nucleocapsid phosphoprotein forms mutually exclusive condensates with RNA and the membrane-associated M protein. *Nat. Commun.* **2021**, *12*, 502. [CrossRef]
10. Peluso, M.J.; Deitchman, A.N.; Torres, L.; Iyer, N.S.; Munter, S.E.; Nixon, C.C.; Donatelli, J.; Thanh, C.; Takahashi, S.; Hakim, J.; et al. Long-term SARS-CoV-2-specific immune and inflammatory responses in individuals recovering from COVID-19 with and without post-acute symptoms. *Cell Rep.* **2021**, *36*, 109518. [CrossRef]
11. Qian, Y.; Lei, T.; Patel, P.S.; Lee, C.H.; Monaghan-Nichols, P.; Xin, H.-B.; Qiu, J.; Fu, M. Direct Activation of Endothelial Cells by SARS-CoV-2 Nucleocapsid Protein Is Blocked by Simvastatin. *J. Virol.* **2021**, *95*, e01396-21. [CrossRef] [PubMed]
12. Bai, Z.; Cao, Y.; Liu, W.; Li, J. The SARS-CoV-2 Nucleocapsid Protein and Its Role in Viral Structure, Biological Functions, and a Potential Target for Drug or Vaccine Mitigation. *Viruses* **2021**, *13*, 1115. [CrossRef] [PubMed]

13. Chen, J.S.; Alfajaro, M.M.; Chow, R.D.; Wei, J.; Filler, R.B.; Eisenbarth, S.C.; Wilen, C.B. Nonsteroidal Anti-inflammatory Drugs Dampen the Cytokine and Antibody Response to SARS-CoV-2 Infection. *J. Virol.* **2021**, *95*, e00014-21. [CrossRef] [PubMed]
14. Yan, X.; Hao, Q.; Mu, Y.; Timani, K.A.; Ye, L.; Zhu, Y.; Wu, J. Nucleocapsid protein of SARS-CoV activates the expression of cyclooxygenase-2 by binding directly to regulatory elements for nuclear factor-kappa B and CCAAT/enhancer binding protein. *Int. J. Biochem. Cell Biol.* **2006**, *38*, 1417–1428. [CrossRef] [PubMed]
15. Xia, J.; Tang, W.; Wang, J.; Lai, D.; Xu, Q.; Huang, R.; Hu, Y.; Gong, X.; Fan, J.; Shu, Q.; et al. SARS-CoV-2 N Protein Induces Acute Lung Injury in Mice via NF-kB Activation. *Front. Immunol.* **2021**, *12*, 791753. [CrossRef]
16. Perico, N.; Cortinovis, M.; Suter, F.; Remuzzi, G. Home as the new frontier for the treatment of COVID-19: The case for anti-inflammatory agents. *Lancet Infect. Dis.* **2022**. [CrossRef]
17. Terrier, O.; Dilly, S.; Pizzorno, A.; Chalupska, D.; Humpolickova, J.; Bouřa, E.; Berenbaum, F.; Quideau, S.; Lina, B.; Fève, B.; et al. Antiviral Properties of the NSAID Drug Naproxen Targeting the Nucleoprotein of SARS-CoV-2 Coronavirus. *Molecules* **2021**, *26*, 2593. [CrossRef]
18. Terrier, O.; Dilly, S.; Pizzorno, A.; Henri, J.; Berenbaum, F.; Lina, B.; Fève, B.; Adnet, F.; Sabbah, M.; Rosa-Calatrava, M.; et al. Broad-spectrum antiviral activity of naproxen: From Influenza A to SARS-CoV-2 Coronavirus. *bioRxiv* **2020**. [CrossRef]
19. Dilly, S.; Fotso Fotso, A.; Lejal, N.; Zedda, G.; Chebbo, M.; Rahman, F.; Companys, S.; Bertrand, H.C.; Vidic, J.; Noiray, M.; et al. From Naproxen Repurposing to Naproxen Analogues and Their Antiviral Activity against Influenza A Virus. *J. Med. Chem.* **2018**, *61*, 7202–7217. [CrossRef]
20. Lejal, N.; Tarus, B.; Bouguyon, E.; Chenavas, S.; Bertho, N.; Delmas, B.; Ruigrok, R.W.; Di Primo, C.; Slama-Schwok, A. Structure-based discovery of the novel antiviral properties of naproxen against the nucleoprotein of influenza A virus. *Antimicrob. Agents Chemother.* **2013**, *57*, 2231–2242. [CrossRef]
21. Rutkai, I.; Mayer, M.G.; Hellmers, L.M.; Ning, B.; Huang, Z.; Monjure, C.J.; Coyne, C.; Silvestri, R.; Golden, N.; Hensley, K.; et al. Neuropathology and virus in brain of SARS-CoV-2 infected non-human primates. *Nat. Commun.* **2022**, *13*, 1745. [CrossRef] [PubMed]
22. Thompson, E.J.; Williams, D.M.; Walker, A.J.; Mitchell, R.E.; Niedzwiedz, C.L.; Yang, T.C.; Huggins, C.F.; Kwong, A.S.F.; Silverwood, R.J.; Di Gessa, G.; et al. Long COVID burden and risk factors in 10 UK longitudinal studies and electronic health records. *Nat. Commun.* **2022**, *13*, 3528. [CrossRef] [PubMed]
23. Guseva, S.; Perez, L.M.; Camacho-Zarco, A.; Bessa, L.M.; Salvi, N.; Malki, A.; Maurin, D.; Blackledge, M. (1)H, (13)C and (15)N Backbone chemical shift assignments of the n-terminal and central intrinsically disordered domains of SARS-CoV-2 nucleoprotein. *Biomol. NMR Assign.* **2021**, *15*, 255–260. [CrossRef] [PubMed]
24. Khan, W.H.; Khan, N.; Mishra, A.; Gupta, S.; Bansode, V.; Mehta, D.; Bhambure, R.; Ansari, M.A.; Das, S.; Rathore, A.S. Dimerization of SARS-CoV-2 nucleocapsid protein affects sensitivity of ELISA based diagnostics of COVID-19. *Int. J. Biol. Macromol.* **2022**, *200*, 428–437. [CrossRef] [PubMed]
25. Peng, Y.; Du, N.; Lei, Y.; Dorje, S.; Qi, J.; Luo, T.; Gao, G.F.; Song, H. Structures of the SARS-CoV-2 nucleocapsid and their perspectives for drug design. *EMBO J.* **2020**, *39*, e105938. [CrossRef] [PubMed]
26. Di, D.; Dileepan, M.; Ahmed, S.; Liang, Y.; Ly, H. Recombinant SARS-CoV-2 Nucleocapsid Protein: Expression, Purification, and Its Biochemical Characterization and Utility in Serological Assay Development to Assess Immunological Responses to SARS-CoV-2 Infection. *Pathogens* **2021**, *10*, 1039. [CrossRef]
27. Jumper, J.; Evans, R.; Pritzel, A.; Green, T.; Figurnov, M.; Ronneberger, O.; Tunyasuvunakool, K.; Bates, R.; Zidek, A.; Potapenko, A.; et al. Highly accurate protein structure prediction with AlphaFold. *Nature* **2021**, *596*, 583–589. [CrossRef]
28. Yang, M.; He, S.; Chen, X.; Huang, Z.; Zhou, Z.; Zhou, Z.; Chen, Q.; Chen, S.; Kang, S. Structural Insight Into the SARS-CoV-2 Nucleocapsid Protein C-Terminal Domain Reveals a Novel Recognition Mechanism for Viral Transcriptional Regulatory Sequences. *Front. Chem.* **2020**, *8*, 624765. [CrossRef]
29. Peronace, C.; Tallerico, R.; Colosimo, M.; Fazio, M.; Pasceri, F.; Talotta, I.; Panduri, G.; Pintomalli, L.; Oteri, R.; Calantoni, V.; et al. The First Identification in Italy of SARS-CoV-2 Omicron BA.4 Harboring KSF141_del: A Genomic Comparison with Omicron Sub-Variants. *Biomedicines* **2022**, *10*, 1839. [CrossRef]
30. Wu, H.; Xing, N.; Meng, K.; Fu, B.; Xue, W.; Dong, P.; Tang, W.; Xiao, Y.; Liu, G.; Luo, H.; et al. Nucleocapsid mutations R203K/G204R increase the infectivity, fitness, and virulence of SARS-CoV-2. *Cell Host Microbe* **2021**, *29*, 1788–1801.e6. [CrossRef] [PubMed]
31. Zeng, W.; Liu, G.; Ma, H.; Zhao, D.; Yang, Y.; Liu, M.; Mohammed, A.; Zhao, C.; Yang, Y.; Xie, J.; et al. Biochemical characterization of SARS-CoV-2 nucleocapsid protein. *Biochem. Biophys. Res. Commun.* **2020**, *527*, 618–623. [CrossRef] [PubMed]
32. Lechuga, G.C.; Souza-Silva, F.; Sacramento, C.Q.; Trugilho, M.R.O.; Valente, R.H.; Napoleao-Pego, P.; Dias, S.S.G.; Fintelman-Rodrigues, N.; Temerozo, J.R.; Carels, N.; et al. SARS-CoV-2 Proteins Bind to Hemoglobin and Its Metabolites. *Int. J. Mol. Sci.* **2021**, *22*, 9035. [CrossRef] [PubMed]
33. Panagiotopoulos, A.; Tseliou, M.; Karakasiliotis, I.; Kotzampasi, D.M.; Daskalakis, V.; Kesesidis, N.; Notas, G.; Lionis, C.; Kampa, M.; Pirintsos, S.; et al. p-cymene impairs SARS-CoV-2 and Influenza A (H1N1) viral replication: In Silico predicted interaction with SARS-CoV-2 nucleocapsid protein and H1N1 nucleoprotein. *Pharm. Res. Perspect.* **2021**, *9*, e00798. [CrossRef] [PubMed]
34. Amici, C.; Di Caro, A.; Ciucci, A.; Chiappa, L.; Castilletti, C.; Martella, V.; Decaro, N.; Buonavoglia, C.; Capobianchi, M.R.; Santoro, M.G. Indomethacin has a potent antiviral activity against SARS coronavirus. *Antivir. Ther.* **2006**, *11*, 1021–1030. [CrossRef] [PubMed]

35. Zhang, Y. SARS-CoV-2 hijacks folate and one-carbon metabolism for viral replication. *Nat. Commun.* **2021**, *12*, 1676. [CrossRef] [PubMed]
36. Frere, J.J.; Serafini, R.A.; Pryce, K.D.; Zazhytska, M.; Oishi, K.; Golynker, I.; Panis, M.; Zimering, J.; Horiuchi, S.; Hoagland, D.A.; et al. SARS-CoV-2 infection in hamsters and humans results in lasting and unique systemic perturbations post recovery. *Sci. Transl. Med.* **2022**, *14*, eabq3059. [CrossRef]
37. Ricke-Hoch, M.; Stelling, E.; Lasswitz, L.; Gunesch, A.P.; Kasten, M.; Zapatero-Belinchon, F.J.; Brogden, G.; Gerold, G.; Pietschmann, T.; Montiel, V.; et al. Impaired immune response mediated by prostaglandin E2 promotes severe COVID-19 disease. *PLoS ONE* **2021**, *16*, e0255335. [CrossRef]
38. Kishk, A.; Pacheco, M.P.; Sauter, T. DCcov: Repositioning of drugs and drug combinations for SARS-CoV-2 infected lung through constraint-based modeling. *iScience* **2021**, *24*, 103331. [CrossRef]
39. Ravichandran, R.; Mohan, S.K.; Sukumaran, S.K.; Kamaraj, D.; Daivasuga, S.S.; Ravi, S.; Vijayaraghavalu, S.; Kumar, R.K. Author Correction: An open label randomized clinical trial of Indomethacin for mild and moderate hospitalised COVID-19 patients. *Sci. Rep.* **2022**, *12*, 10389, Correction in *Sci. Rep.* **2022**, *12*, 6413. [CrossRef] [PubMed]
40. Shah, F.H.; Lim, K.H.; Kim, S.J. Do fever-relieving medicines have anti-COVID activity: An in silico insight. *Future Virol.* **2021**, *16*, 293–300. [CrossRef]
41. Mehboob, R.; Kurdi, M.; Bamaga, A.; Aldardeir, N.; Nasief, H.; Moshref, L.H.; Alsinani, T.; Rayes, A.O.; Jabbad, R.H. Substance P/Neurokinin-1 Receptor, Trigeminal Ganglion, Latency, and Coronavirus Infection-Is There Any Link? *Front. Med.* **2021**, *8*, 727593. [CrossRef] [PubMed]
42. Salina, A.C.G.; Dos-Santos, D.; Rodrigues, T.S.; Fortes-Rocha, M.; Freitas-Filho, E.G.; Alzamora-Terrel, D.L.; Castro, I.M.S.; Fraga da Silva, T.F.C.; de Lima, M.H.F.; Nascimento, D.C.; et al. Efferocytosis of SARS-CoV-2-infected dying cells impairs macrophage anti-inflammatory functions and clearance of apoptotic cells. *eLife* **2022**, *11*, e74443. [CrossRef] [PubMed]
43. Chau, C.W.; Sugimura, R. COVID-19: Locked in a pro-inflammatory state. *eLife* **2022**, *11*, e80699. [CrossRef] [PubMed]
44. Harris, J.A.; Faust, B.; Gondin, A.B.; Damgen, M.A.; Suomivuori, C.M.; Veldhuis, N.A.; Cheng, Y.; Dror, R.O.; Thal, D.M.; Manglik, A. Selective G protein signaling driven by substance P-neurokinin receptor dynamics. *Nat. Chem. Biol.* **2022**, *18*, 109–115. [CrossRef] [PubMed]
45. Mehboob, R.; Lavezzi, A.M. Neuropathological explanation of minimal COVID-19 infection rate in newborns, infants and children—A mystery so far. New insight into the role of Substance P. *J. Neurol. Sci.* **2021**, *420*, 117276. [CrossRef] [PubMed]
46. Corder, G.; Castro, D.C.; Bruchas, M.R.; Scherrer, G. Endogenous and Exogenous Opioids in Pain. *Annu. Rev. Neurosci.* **2018**, *41*, 453–473. [CrossRef]
47. Wang, Q.; Chu, C.H.; Qian, L.; Chen, S.H.; Wilson, B.; Oyarzabal, E.; Jiang, L.; Ali, S.; Robinson, B.; Kim, H.C.; et al. Substance P exacerbates dopaminergic neurodegeneration through neurokinin-1 receptor-independent activation of microglial NADPH oxidase. *J. Neurosci. Off. J. Soc. Neurosci.* **2014**, *34*, 12490–12503. [CrossRef]
48. Karamyan, V.T. Between two storms, vasoactive peptides or bradykinin underlie severity of COVID-19? *Physiol. Rep.* **2021**, *9*, e14796. [CrossRef]
49. Lattmann, E.; Bhalerao, P.; ShashiBhushan, B.L.; Nargundkar, N.; Lattmann, P.; Pillai, K.S.; Balaram, P.N. Randomized, Comparative, Clinical Trial to Evaluate Efficacy and Safety of PNB001 in Moderate COVID-19 Patients. *medRxiv* **2021**. [CrossRef]
50. Singh, V.; Allawadhi, P.; Khurana, A.; Banothu, A.K.; Bharani, K.K. Critical neurological features of COVID-19: Role of imaging methods and biosensors for effective diagnosis. *Sens. Int.* **2021**, *2*, 100098. [CrossRef]
51. Heles, M.; Mrozkova, P.; Sulcova, D.; Adamek, P.; Spicarova, D.; Palecek, J. Chemokine CCL2 prevents opioid-induced inhibition of nociceptive synaptic transmission in spinal cord dorsal horn. *J. Neuroinflammation* **2021**, *18*, 279. [CrossRef] [PubMed]
52. Gao, L.; Chalupsky, K.; Stefani, E.; Cai, H. Mechanistic insights into folic acid-dependent vascular protection: Dihydrofolate reductase (DHFR)-mediated reduction in oxidant stress in endothelial cells and angiotensin II-infused mice: A novel HPLC-based fluorescent assay for DHFR activity. *J. Mol. Cell. Cardiol.* **2009**, *47*, 752–760. [CrossRef] [PubMed]
53. Oak, J.H.; Cai, H. Attenuation of angiotensin II signaling recouples eNOS and inhibits nonendothelial NOX activity in diabetic mice. *Diabetes* **2007**, *56*, 118–126. [CrossRef] [PubMed]
54. Hara, M.; Tanaka, K.; Hirota, Y. Immune response to influenza vaccine in healthy adults and the elderly: Association with nutritional status. *Vaccine* **2005**, *23*, 1457–1463. [CrossRef] [PubMed]
55. Xiao, S.; Tang, Y.S.; Khan, R.A.; Zhang, Y.; Kusumanchi, P.; Stabler, S.P.; Jayaram, H.N.; Antony, A.C. Influence of physiologic folate deficiency on human papillomavirus type 16 (HPV16)-harboring human keratinocytes in vitro and in vivo. *J. Biol. Chem.* **2012**, *287*, 12559–12577. [CrossRef] [PubMed]
56. Ziganshyna, S.; Szczepankiewicz, G.; Kuehnert, M.; Schulze, A.; Liebert, U.G.; Pietsch, C.; Eulenburg, V.; Werdehausen, R. Photodynamic Inactivation of SARS-CoV-2 Infectivity and Antiviral Treatment Effects In Vitro. *Viruses* **2022**, *14*, 1301. [CrossRef] [PubMed]
57. Courrol, L.C.; de Oliveira Silva, F.R.; Masilamani, V. SARS-CoV-2, hemoglobin and protoporphyrin IX: Interactions and perspectives. *Photodiagnosis Photodyn. Ther.* **2021**, *34*, 102324. [CrossRef]
58. Chang, C.; Michalska, K.; Jedrzejczak, R.; Maltseva, N.; Endres, M.; Godzik, A.; Kim, Y.; Joachimiak, A. Crystal structure of RNA binding domain of nucleocapsid phosphoprotein from SARS coronavirus 2. *Protein Data Bank* **2020**. [CrossRef]
59. Mirdita, M.; Schütze, K.; Moriwaki, Y.; Heo, L.; Ovchinnikov, S.; Steinegger, M. ColabFold: Making protein folding accessible to all. *Nat. Methods* **2022**, *19*, 679–682. [CrossRef]

60. Mirdita, M.; Steinegger, M.; Soding, J. MMseqs2 desktop and local web server app for fast, interactive sequence searches. *Bioinformatics* **2019**, *35*, 2856–2858. [CrossRef]
61. Darden, T.; York, D.; Pedersen, L. Particle mesh Ewald: An N·log(N) method for Ewald sums in large systems. *J. Chem. Phys.* **1993**, *98*, 10089–10092. [CrossRef]
62. Šali, A.; Blundell, T.L. Comparative protein modeling by satisfaction of spatial restraints. *J. Mol. Biol.* **1993**, *234*, 779–815. [CrossRef] [PubMed]
63. Brooks, B.R.; Bruccoleri, R.E.; Olafson, B.D.; States, D.J.; Swaminathan, S.; Karplus, M. CHARMM: A program for macromolec ular energy, minimization, and dynamics calculations. *J. Comput. Chem.* **1983**, *4*, 187–217. [CrossRef]
64. Dagistan, Y.; Kilinc, E.; Balci, C.N. Cervical sympathectomy modulates the neurogenic inflammatory neuropeptides following experimental subarachnoid hemorrhage in rats. *Brain Res.* **2019**, *1722*, 146366. [CrossRef] [PubMed]
65. Lorente, L.; Martín, M.M.; Almeida, T.; Hernández, M.; Ramos, L.; Argueso, M.; Cáceres, J.J.; Solé-Violán, J.; Jiménez, A. Serum substance P levels are associated with severity and mortality in patients with severe traumatic brain injury. *Crit. Care* **2015**, *19*, 192. [CrossRef] [PubMed]
66. Martinez, A.N.; Philipp, M.T. Substance P and Antagonists of the Neurokinin-1 Receptor in Neuroinflammation Associated with Infectious and Neurodegenerative Diseases of the Central Nervous System. *J. Neurol. Neuromedicine* **2016**, *1*, 29–36. [CrossRef]
67. Niemela, V.; Landtblom, A.M.; Nyholm, D.; Kneider, M.; Constantinescu, R.; Paucar, M.; Svenningsson, P.; Abujrais, S.; Burman, J.; Shevchenko, G.; et al. Proenkephalin Decreases in Cerebrospinal Fluid with Symptom Progression of Huntington's Disease. *Mov. Disord. Off. J. Mov. Disord. Soc.* **2021**, *36*, 481–491. [CrossRef]
68. Otsuka, K.; Niimi, A.; Matsumoto, H.; Ito, I.; Yamaguchi, M.; Matsuoka, H.; Jinnai, M.; Oguma, T.; Takeda, T.; Nakaji, H.; et al. Plasma substance P levels in patients with persistent cough. *Respiration* **2011**, *82*, 431–438. [CrossRef]
69. Smieszek, S. Late Breaking Abstract—Increased Substance P levels in COVID-19 hospitalized patients. *Eur. Respir. J.* **2021**, *58*, OA4114. [CrossRef]
70. Tyagi, A.; Daliri, E.B.-M.; Kwami Ofosu, F.; Yeon, S.-J.; Oh, D.-H. Food-Derived Opioid Peptides in Human Health: A Review. *Int. J. Mol. Sci.* **2020**, *21*, 8825. [CrossRef]
71. Boaventura, P.; Macedo, S.; Ribeiro, F.; Jaconiano, S.; Soares, P. Post-COVID-19 Condition: Where Are We Now? *Life* **2022**, *12*, 517. [CrossRef] [PubMed]

molecules

MDPI

Communication

Effects of a Shift of the Signal Peptide Cleavage Site in Signal Peptide Variant on the Synthesis and Secretion of SARS-CoV-2 Spike Protein

Zhikai Zhang [1,†], Xuan Wan [2,†], Xinyue Li [1] and Chengsong Wan [1,*]

1 BSL-3 Laboratory (Guangdong), Guangdong Provincial Key Laboratory of Tropical Disease Research, School of Public Health, Southern Medical University, Guangzhou 510515, China
2 Chronic Airways Diseases Laboratory, Department of Respiratory and Critical Care Medicine, Nanfang Hospital, Southern Medical University, Guangzhou 510515, China
* Correspondence: gzwcs@smu.edu.cn
† These authors have contributed equally to this work.

Abstract: The COVID-19 pandemic is caused by SARS-CoV-2; the spike protein is a key structural protein that mediates infection of the host by SARS-CoV-2. In this study, we aimed to evaluate the effects of signal peptide on the secretion and release of SARS-CoV-2 spike protein. Therefore, we constructed a signal peptide deletion mutant and three signal peptide site-directed mutants. The (H) region and (C) region in the signal peptide of L5F-S13I mutant have changed significantly, compared with wild type, L5F and S13I. We demonstrated the effects of signal peptide on the secretion and synthesis of RBD protein, finding that mutation of S13 to I13 on the signal peptide is more conducive to the secretion of RBD protein, which was mainly due to the shift of the signal peptide cleavage site in the mutant S13I. Here, we not only investigated the structure of the N-terminal signal peptide of the SARS-CoV-2 spike protein but also considered possible secretory pathways. We suggest that the development of drugs that target the signal peptide of the SARS-CoV-2 spike protein may have potential to treat COVID-19 in the future.

Keywords: signal peptide; RBD protein; secretion; mutant; SARS-CoV-2

Citation: Zhang, Z.; Wan, X.; Li, X.; Wan, C. Effects of a Shift of the Signal Peptide Cleavage Site in Signal Peptide Variant on the Synthesis and Secretion of SARS-CoV-2 Spike Protein. *Molecules* **2022**, *27*, 6688. https://doi.org/10.3390/molecules27196688

Academic Editors: Giovanni N. Roviello and Caterina Vicidomini

Received: 2 September 2022
Accepted: 5 October 2022
Published: 8 October 2022

1. Introduction

At the beginning of 2020, a serious outbreak of pneumonia was attributed to a novel coronavirus [1,2]. Six human coronaviruses (HCoVs), HCoV-229E, HCoV-OC43, HCoV-NL63, HCoV-HKU1, SARS-CoV and MERS-CoV, were known to infect humans [3]. The first four viruses usually cause relatively mild cold-like symptoms in those with a healthy immune system, whereas the latter two viruses are zoonotic viruses and can cause serious respiratory disease and death. The beta-coronavirus SARS-CoV-2 has now become the seventh discrete coronavirus species that is capable of causing human disease [4]. SARS-CoV-2 is easily transmitted and highly pathogenic [5] and COVID-19 was declared a pandemic by the WHO (https://www.who.int/, accessed on 27 July 2022).

The total ssRNA genome of a SARS-CoV-2 strain, isolated from a patient with novel coronavirus pneumonia, was found to contain 29,903 base pairs (GenBank: MN908947.3). Coronavirus spike proteins are involved in binding and fusion of the virus with the host cell membrane, and initial studies showed that the spike protein of SARS-CoV-2 is very similar to that of SARS-CoV. This suggested that angiotensin converting enzyme 2 (ACE2), which was known to be a receptor for SARS-CoV spike protein, may also be an important receptor for the SARS-CoV-2 spike protein [6,7]. Further studies showed that the SARS-CoV-2 spike protein has a higher binding affinity than SARS-CoV spike protein for ACE2, indicating that SARS-CoV-2 may be more invasive than SARS-CoV [8]. The SARS-CoV-2 spike protein has two main domains, S1, which is responsible for binding to the receptor, and S2, which is responsible for fusion with the host cell membrane [9].

A trend towards variation has been observed in SARS-CoV-2. Multiple SARS-CoV-2 variants have been reported, such as the RBD mutation that appears in lineages B.1.1.7 (Alpha), B.1.351 (Beta), P.1 (Gamma), B.1.429 (Epsilon), B.1.617.1 (Kappa), B.1.351 (Beta), P.2 (Zeta), B.1.526 (Iota), B.1.617.2 (Delta) and B.1.1.529 (Omicron) [10–13]. These studies indicate that genetic variations in the virus lead to changes in the interactions between virus and host and will affect the selection of drugs and treatment regimens in clinical practice.

Many studies have shown that the spike protein is key to viral infection and could potentially be a therapeutic target. Research on the SARS-CoV-2 spike protein, mainly focused on receptor identification [14,15], structural analysis [16], regulation of spike protein binding to its receptor [17,18] and natural products as inhibitors to affect the function of SARS-CoV-2 spike protein [19]. In particular, the receptor binding site of the spike protein had become an important target for the development of SARS-CoV-2 therapeutic antibodies and vaccine design [14,20]. However, there are few reports describing regulation of spike protein expression or processing modifications and release of the protein. In eukaryotic cells, translated proteins must undergo a series of processing modifications and be secreted to become functional proteins with biological activity, and these steps are typically governed by an N-terminal signal peptide [21]. In this study, we investigated the potential effects of signal peptides on the expression and secretion of SARS-CoV-2 spike protein to find novel drug targets for the treatment of COVID-19.

2. Results

2.1. Bioinformatics Analysis Signal Peptide of SARS-CoV-2 Spike(S) Protein

Sequences analysis of the signal peptide of SARS-CoV-2 spike protein was shown (Figure 1A). SignalP (Department of Health Technology, Technical University of Denmark, Kgs Lyngby, Denmark) is a freely available web-based tool that uses a deep neural network-based approach to predict the presence and cleavage sites of signal peptides in amino acid sequences from different organisms [22]. The signal peptide structure of SARS-CoV-2 spike protein was analysis using bioinformatics online analysis tools SignalP version 3.0 (http://www.cbs.dtu.dk/services/SignalP-3.0/, accessed on 18 May 2022). The signal peptide of SARS-CoV-2 spike protein can be divided into three regions, an N-terminal (N) region, a central hydrophobic (H) region and a C-terminal (C) region (Figure 1B). The positive charge on the N region of the signal peptide has been shown to contribute to efficient post-translational translocation of small prerequisite proteins [23]. The H region of the signal peptide sequence is mainly responsible for recognition and binding by the signal recognition particle [24]. Comparing with the structures of wild type signal peptide, we found no significant change in the single-point mutant L5F, but significant changes in the (C) region of single point mutant S13I (Figure 1B). It is interesting that the (H) region and (C) region of L5F-S13I mutant have changed significantly, compared with wild type, L5F and S13I (Figure 1B). In addition, after the mutation of S13 to I13, the cleavage position at Q14 and V16 of the region (C) of mutants S13I and L5F-S13I also shifted compared with wild-type and mutant L5F, respectively (Figure 1B) The results indicated that the mutation of signal peptide may affect the expression, secretion and modification of SARS-CoV-2 spike protein.

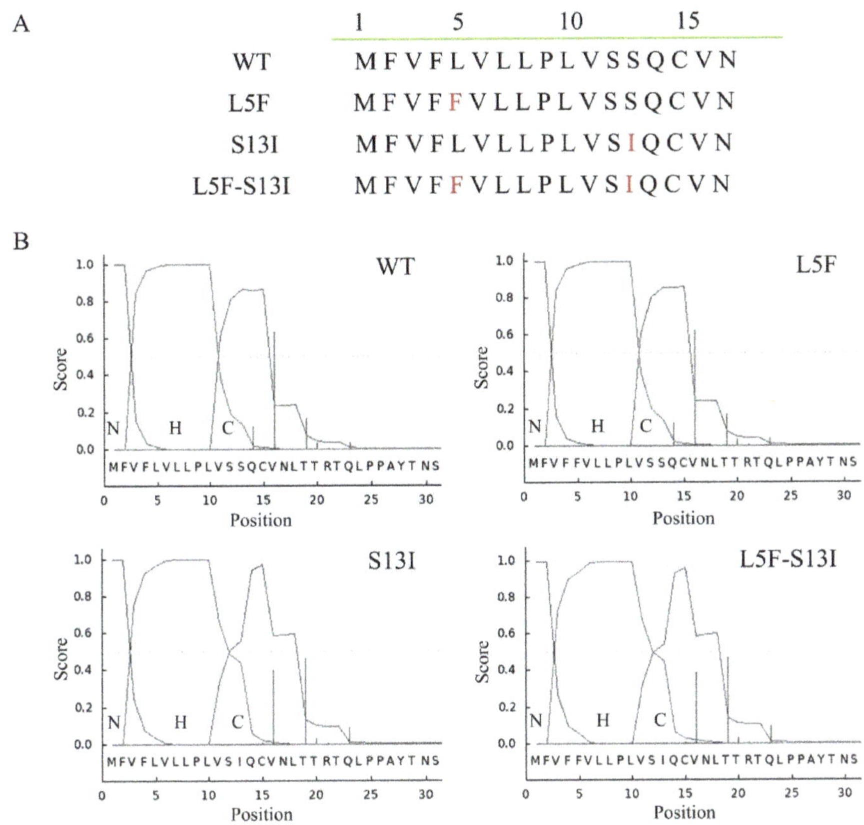

Figure 1. Signal peptide comparison of SARS-CoV-2 spike protein. (**A**) Signal peptide sequences of SARS-CoV-2 spike protein. (**B**) Signal peptide structural analysis by SignalP 3.0. WT means wild-type signal peptide of S protein; The mutants L5F and S13I were present in Iota (B.1.526) and Epsilon (B.1.429) variants; L5F-S13I was a newly constructed double mutant.

2.2. Signal Peptide Targets SARS-CoV-2 RBD Protein to the Endoplasmic Reticulum

In order to determine the effect of signal peptides on the expression of the RBD protein and the localization of the endoplasmic reticulum, recombinant plasmids pEGFP-ΔRBD, pEGFP-RBD, pEGFP-RBD-L5F, pEGFP-RBD-S13I and pEGFP-RBD-L5F-S13I were transiently transfected into HEK293T cells, respectively, and then cell fixation, DAPI, red fluorescent probe staining of endoplasmic reticulum and fluorescence microscopy were performed. As shown in Figure 2, EGFP protein and all fusion proteins could be expressed normally, indicating that the signal peptide had no significant effect on the expression of RBD protein. However, after the deletion of signal peptides, EGFP protein and ΔRBD-EGFP fusion protein were mainly distributed in the nucleus, and only a small amount is distributed in the cytoplasm. The RBD-EGFP protein and mutant RBD-EGFP fusion protein containing signal peptides are mainly distributed in the cytoplasm, and a small amount is distributed in the nucleus (Figure 2). In addition, there are many more proteins in the cytoplasm of mutations S13I and L5F-S13I, compared to wild type and mutant L5F, which may be mainly due to the mutation at the S13 site in the signal peptide. These results suggest that the (C) region amino acid sites in signal peptide have an important influence on the localization of the endoplasmic reticulum of the SARS-CoV-2 spike protein, which means that the (C) region sites may become important candidate targets for drug design.

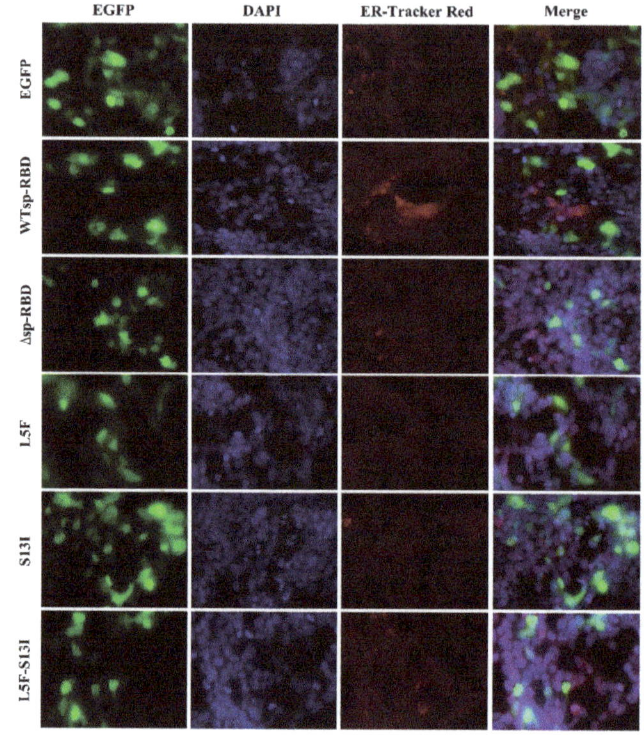

Figure 2. Signal peptides mediate the localization of SARS-CoV-2 RBD protein. WT means wild-type signal peptide of S protein; The mutants L5F and S13I were present in Iota (B.1.526) and Epsilon (B.1.429) variants; L5F-S13I was a newly constructed double mutant. The fluorescence for EGFP (green), DAPI (blue), ER-Tracker-Red (red) and the merge of the three channels are displayed.

2.3. Signal Peptide Promotes the Secretion Levels of SARS-CoV-2 RBD Protein

To further evaluate the effects of signal peptide on the secretion and release of SARS-CoV-2 spike protein, the green fluorescence intensity in cell culture supernatants were detected to determine the secretion of the fusion protein. After the HEK293T, EGFP, Δsp-RBD, WTsp-RBD, L5F, S13I and L5F-S13I cell lines were normally cultured for 48 h, the culture supernatant was collected by low speed centrifugation, and then the green fluorescence intensity in the supernatant was detected by a multifunctional microplate reader. The results showed that the relative fluorescence intensity in the supernatant of EGFP and Δsp-RBD cells were lower, which was significantly different from that in the supernatant of mutant L5F, S13I and L5F-S13I cells (Figure 3). This implies that EGFP and Δsp-RBD proteins could not be secreted to the extracellular without the signal peptide, while WTsp-RBD, L5F, S13I and L5F-S13I proteins could be secreted extracellular after modification mediated by signal peptide. The relative fluorescence value of mutant S13I and L5F-S13I was significantly, compared to the mutant Δsp-RBD, which was also suggested that mutation of S13 to I13 on the signal peptide is more conducive to the secretion of RBD protein.

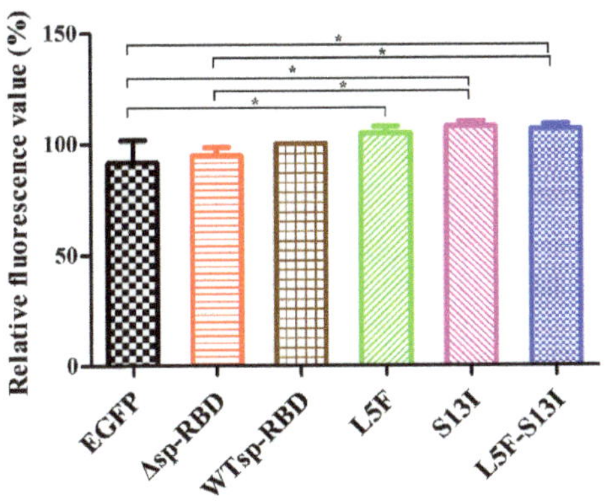

Figure 3. Analysis of RBD protein release regulated by signal peptide. WT means wild-type signal peptide of S protein; the mutants L5F and S13I were present in Iota (B.1.526) and Epsilon (B.1.429) variants; L5F-S13I was a newly constructed double mutant. Statistical significance was defined as $p < 0.05$, and * $p < 0.05$.

3. Discussion

Recent progress in biology and medicine has enabled the development of many new drugs for the treatment and prevention of diseases. It is interesting that many membrane-bound and secretory proteins in the conventional secretory pathway have been identified as potential targets for drug design [25]. In the conventional secretory pathway, signal peptides located at the N-terminus of precursor proteins guide these proteins into the ER to complete post-translational processing and modification [21,26]. Some research groups have, therefore, chosen to focus on identifying small molecules that can specifically bind to the signal peptide, and thus inhibit the normal expression of the target protein [25,27,28]. Such studies have clearly demonstrated that signal peptides can be used to design drugs for specific target proteins.

It has been shown that increasing the hydrophobicity of the (H) region of a signal peptide can enhance the production of full length monoclonal antibodies [29]. The effects of a variety of different signal peptides on antibody yield were evaluated in an attempt to increase production of VRC01, a broadly neutralizing antibody against HIV [30]. As another example, the insertion of a cleavable leucine-rich signal peptide into olfactory receptors has been shown to increase expression of the receptors on the surface of HEK293T cells [31]. In our research, the mutants L5F and S13I were present in SARS-CoV-2 of Iota (B.1.526) and Epsilon (B.1.429) variants. However, although the mutant of L5F and L5F-S13I on the signal peptide of RBD protein, the expression of proteins could not significantly increase (Figures 2 and 3). Studies had shown that the positive charge on the (N) region of the signal peptide has been shown to contribute to binding by the signal recognition particle and efficient post-translational translocation of small prerequisite proteins [32,33]. It was suggested the (N) region of the signal peptide plays a key role in protein expression. Because the mutant L5F was in the (N) region of the signal peptide, we hypothesized that the L5 sites of the signal peptide may not have significant effects on the expression of SARS-CoV-2 spike protein.

Protein glycosylation is one of the most important forms of post-translational modification and 50–70% of cellular proteins may be glycosylated [34]. Glycosylation plays a role in regulating the localization, function and activity of proteins in tissues and cells [35–37]. In eukaryotes, glycosylation of most cellular proteins takes place along the secretory pathway, which begins in the ER and is completed in the Golgi apparatus [38]. The N-terminus of the SARS-CoV-2 spike protein contains a signal peptide (Figure 1), which indicates that synthesis, processing and release of the spike protein in the ER are mediated by the signal peptide. So far, 22 N-linked glycosylation sites and 17 O-linked glycosylation sites have been identified in the SARS-CoV-2 spike protein [39]. In addition, Veesler et al. though mass spectrometry and structural studies, revealing that the S13I mutation resulted in total loss of neutralization for 10 of 10 NTD-specific mAbs because the NTD antigenic supersite was remodeled by a shift of the signal peptide cleavage site and the formation of a new disulfide bond [40]. Therefore, signal peptide plays an important role in the expression, secretion and modification of SARS-CoV-2 spike protein. In our study, mutants L5F and S13I occur on the signal peptide, and S13I, in particular, could significantly increase the secretory capacity of the RBD protein, which suggests that S13 was a key site on the signal peptide. In addition, the glycosylation of viral structural proteins is closely associated with viral replication, infectivity and the host immune response [41–43]. The shielding of receptor binding sites by glycosylation is a common feature of viral glycoproteins and can be observed with SARS-CoV spike protein, HIV-1 envelope protein, influenza virus hemagglutinin and Lassa virus glycoprotein precursor [44]. HIV-1 envelope glycoprotein gp160 is directed to the ER by its signal peptide and the premature cleavage of the mutated gp160, compared with wild-type gp160, results in a virus with significantly reduced adaptability [45].

Based on the above analyses, we are convinced that the synthesis and secretion of SARS-CoV-2 spike proteins take place through the conventional secretion pathway (Figure 4). We speculate, therefore, that the signal peptides may play an important role in the synthesis, processing modification and secretion of SARS-CoV-2 spike protein, and thus represent a new target for drug design. Drugs that interfere with the co-translational translocation of new polypeptide chains have the potential to reduce the expression of many cell surface receptors and secretory proteins that are important therapeutic targets [46,47]. For example, the small molecule cyclotriazadisulfonamide (CADA) selectively downregulates the expression of CD4, the primary receptor for human immunodeficiency virus (HIV) and simian immunodeficiency virus, and thus inhibits viral replication and reduces pathogenicity [27,48]. Another small molecule, CAM741, which is an analog of the cycloheptadepsipeptide fungal metabolite HUN-7293, is a signal peptide-selective inhibitor of protein co-translation translocation [28].

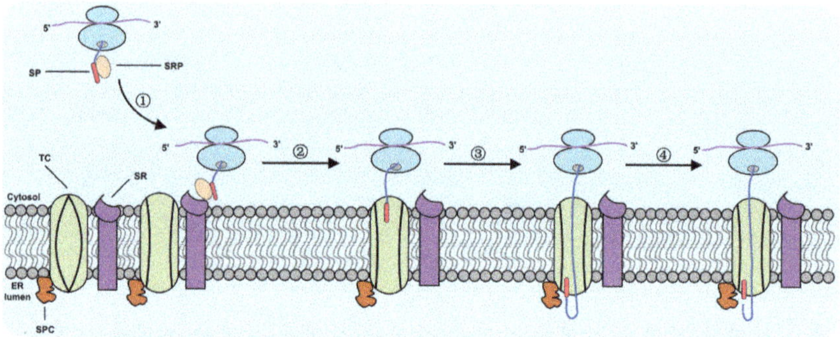

Figure 4. Proposed secretion pathway of SARS-CoV-2 spike protein. ① New polypeptide-SRP-ribosomal complex binding to SR on ER membrane. ② SRP separates from its receptors and promotes the tight binding of ribosomes and ER membranes to protein translocon channel. ③ The extended polypeptide passes through the membrane structure into the ER cavity as translation continues. ④ The signal peptidase inside the ER membrane cavity cleaves the signal peptide after recognizing the signal peptide cleavage site of the polypeptide, and the remaining polypeptides continue to undergo co-translational translocation through the ER membrane. SP: signal peptide. SRP: signal recognition particle. SR: signal receptor. TC: translocon channel. ER: endoplasmic reticulum. SPC: signal peptidase cleavage.

4. Materials and Methods

4.1. Bacterial, Plasmids Construct and Cells Culture Conditions

The *E. coli* DH5α was used to cloning host strain. All bacterial strains in our experiment were cultured in LB medium at 37 °C and kanamycin was added as needed. HEK293T cells were cultured in DMEM (Gibco, Waltham, MA, USA) broth containing 6% fetal bovine serum (FBS) and 1% at 5% CO_2 and 37 °C. SARS-CoV-2 RBD gene fragment was derived from our previous experiment. The plasmid fragment of pEGFP-N1 was obtained by PCR to construct recombinant expression vector. The signal peptide mutants of SARS-CoV-2 spike protein were constructed by reverse PCR. The primers were listed in Table 1.

Table 1. Primers used in this study.

Primers Name	Sequence (5′–3′)
pEGFP-F	CATCATCACCATCACCATGGATCCACCGGTCGCCACCATGGTG
pEGFP-R	GGTGGCGAATTCGAAGCTTGAGCTC
ΔRBD-F	GAGCTCAAGCTTCGAATTCGCCACCATGAATATTACAAACTTGTGCCCTTTTG
ΔRBD-R	TGGATCCATGGTGATGGTGATGATGCTCAAGTGTCTGTGGATCACGGAC
RBD-F	CTTGTTTTATTGCCACTAGTCTCTAGTCAGTGTGTTAATATTACAAACTTGTGCCCTTTTG
RBD-R	CTAGAGACTAGTGGCAATAAAACAAGAAAAACAAACATGGTGGCGAATTCGAAGCTTGAGCTC
L5F-F	TTGTTTTTTTTGTTTTATTGCCACTAGTCTCTAGTC
L5F-R	CAATAAAACAAAAAAAACAAACATGGTGGCGAATTCG
S13I-F	CTAGTCTCTATTCAGTGTGTTAATATTACAAACTTGT
S13I-R	ACACACTGAATAGAGACTAGTGGCAATAAAACAAGA

4.2. RBD Proteins Expressed and Intracellular Fluorescence Assay in HEK293T Cell

The SARS-CoV-2 RBD proteins were expressed by transiently transfecting HEK293T cells with recombinant plasmid pEGFP-ΔRBD (None signal peptide), pEGFP-RBD, pEGFP-RBD-L5F, pEGFP-RBD-S13I and pEGFP-RBD-L5F-S13I. HEK293T cells were placed in a 35 mm confocal dish with 1×10^4 per plate in advance and cultured in 5% CO_2 cell incubator at 37 °C for 24 h. After the cells grew well, recombinant plasmids were transfected according to the method of PEI transfection reagent. Cell culture medium was discarded after 6 h and a new serum-free medium was used for further culture. After 24 h of culture, the medium were removed and cells were rinsed with PBS buffer twice. Next, the cells were fixed with 4% paraformaldehyde at room temperature for 15 min. Paraformaldehyde was removed and cells were rinsed with PBS 3 times, DAPI dye (Cat. No.: C1005, Beyotime, Shanghai, China) or endoplasmic reticulum red fluorescent probe (Cat. No.: C1041, Beyotime, Shanghai, China) diluted 1000 times were added, and the cells were incubated at room temperature without light for 5 min or 20 min. The cells were rinsed with PBS 3 times, and 1–2 drops of anti-fluorescence quenching agent were added. Photos were taken under the fluorescence microscope after the slides were covered.

4.3. RBD Protein Secretion Detection

Recombinant plasmids were transiently transfected HEK293T cells; SMM 293-TI medium (Sino Biological, Beijing, China) was used to replace DMEM medium before transfection. After 48 h culture, cell fragments were removed and supernatant was collected by centrifugation at 12,000× g for 3 min at room. The 200 μL supernatant sample was added to a 96-well black plate, and the fluorescence absorption value was read by the microplate instrument. The detection conditional excitation wavelength was 485 mm, and the transmission wavelength was 515 mm. The relative fluorescence value is expressed by the following Equation (1):

$$Rf = \frac{S - N}{P - N} \times 100\% \tag{1}$$

Rf: the relative fluorescence value;
N: fluorescence value of negative control (HEK293T);

P: fluorescence value of positive control (pEGFP-RBD);
S: fluorescence value of target sample.

4.4. Statistical Analysis

For all experiments, unless stated otherwise, three independent were performed. Statistical analysis were drawn using GraphPad Prism software (version 5.0, GraphPad Software, San Diego, CA, USA). Significant differences were determined by one-way analysis of variance followed by the Tukey's multiple comparison test. Statistical significance was defined as $p < 0.05$.

5. Conclusions

Signal peptides have huge potential as drug design targets and are also very important in vaccine production. This is the first study to report about the signal peptide of SARS-CoV-2 spike proteins. We analyzed the signal peptide structure of SARS-CoV-2 spike protein and found that the mutants L5F and S13I could change regions. In particular, the mutant S13I could shift the signal peptide cleavage site in the (C) region on the signal peptide. Compared with wild type and mutant L5F, the mutants S13I and L5F-S13I can promote the secretion of SARS-CoV-2 RBD protein. In order to further clarify the role of L5F and S13I mutations on signal peptides, it was found that S13 is a key site on signal peptides that may play an important role in the expression, modification and secretion of SARS-CoV-2 spike protein. Moreover, we propose that signal peptides may be effective targets for the design of drugs to treat SARS-CoV-2 infections.

Author Contributions: Conceptualization, C.W.; investigation, Z.Z. and X.W.; resources, X.W. and X.L.; software and data curation, X.W. and X.L.; writing—original draft preparation, Z.Z.; writing—review and editing, C.W.; funding acquisition, Z.Z. and C.W. All authors have read and agreed to the published version of the manuscript.

Funding: This research was funded by the China Postdoctoral Science Foundation (2021M701594), Guangdong Science and Technology Program Key Projects (No.2021B1212030014).

Institutional Review Board Statement: Not applicable.

Informed Consent Statement: Not applicable.

Data Availability Statement: Not applicable.

Acknowledgments: We thank the bioinformatics online analysis tools SignalP version 3.0 (http://www.cbs.dtu.dk/services/SignalP-3.0/, accessed on 18 May 2022).

Conflicts of Interest: The authors declare no conflict of interest.

References

1. Perlman, S. Another Decade, Another Coronavirus. *N. Engl. J. Med.* **2020**, *382*, 760–762. [CrossRef] [PubMed]
2. Zhu, N.; Zhang, D.; Wang, D.; Wang, W.; Li, X.; Yang, B.; Song, J.; Zhao, X.; Huang, B.; Shi, W.; et al. A Novel Coronavirus from Patients with Pneumonia in China, 2019. *N. Engl. J. Med.* **2020**, *382*, 727–733. [CrossRef] [PubMed]
3. Andersen, K.G.; Rambaut, A.; Lipkin, W.I.; Holmes, E.C.; Garry, R.F. The proximal origin of SARS-CoV-2. *Nat. Med.* **2020**, *26*, 450–452. [CrossRef] [PubMed]
4. Lu, R.; Zhao, X.; Li, J.; Niu, P.; Yang, B.; Wu, H.; Wang, W.; Song, H.; Huang, B.; Zhu, N.; et al. Genomic characterisation and epidemiology of 2019 novel coronavirus: Implications for virus origins and receptor binding. *Lancet* **2020**, *395*, 565–574. [CrossRef]
5. Steven, S.; Yen, T.L.; Chonggang, X.; Ethan, R.; Nick, H.; Ruian, K. High Contagiousness and Rapid Spread of Severe Acute Respiratory Syndrome Coronavirus 2. *Emerg. Infect. Dis. J.* **2020**, *26*, 1470–1477.
6. Coutard, B.; Valle, C.; de Lamballerie, X.; Canard, B.; Seidah, N.G.; Decroly, E. The spike glycoprotein of the new coronavirus 2019-nCoV contains a furin-like cleavage site absent in CoV of the same clade. *Antivir. Res.* **2020**, *176*, 104742. [CrossRef]
7. Tang, X.; Wu, C.; Li, X.; Song, Y.; Yao, X.; Wu, X.; Duan, Y.; Zhang, H.; Wang, Y.; Qian, Z.; et al. On the origin and continuing evolution of SARS-CoV-2. *Natl. Sci. Rev.* **2020**, *7*, 1012–1023. [CrossRef]
8. Wang, Q.; Zhang, Y.; Wu, L.; Niu, S.; Song, C.; Zhang, Z.; Lu, G.; Qiao, C.; Hu, Y.; Yuen, K.; et al. Structural and Functional Basis of SARS-CoV-2 Entry by Using Human ACE2. *Cell* **2020**, *181*, 894–904. [CrossRef]
9. Walls, A.C.; Park, Y.; Tortorici, M.A.; Wall, A.; McGuire, A.T.; Veesler, D. Structure, Function, and Antigenicity of the SARS-CoV-2 Spike Glycoprotein. *Cell* **2020**, *181*, 281–292. [CrossRef]

10. Kumar, V.; Singh, J.; Hasnain, S.E.; Sundar, D. Possible Link between Higher Transmissibility of Alpha, Kappa and Delta Variants of SARS-CoV-2 and Increased Structural Stability of Its Spike Protein and hACE2 Affinity. *Int. J. Mol. Sci.* **2021**, *22*, 9131. [CrossRef] [PubMed]
11. Escalera, A.; Gonzalez-Reiche, A.S.; Aslam, S.; Mena, I.; Laporte, M.; Pearl, R.L.; Fossati, A.; Rathnasinghe, R.; Alshammary, H.; van de Guchte, A.; et al. Mutations in SARS-CoV-2 variants of concern link to increased spike cleavage and virus transmission. *Cell Host Microbe* **2022**, *30*, 373–387. [CrossRef]
12. Biswas, S.; Dey, S.; Chatterjee, S.; Nandy, A. Combatting future variants of SARS-CoV-2 using an in-silico peptide vaccine approach by targeting the spike protein. *Med. Hypotheses* **2022**, *161*, 110810. [CrossRef]
13. Kannan, S.R.; Spratt, A.N.; Sharma, K.; Chand, H.S.; Byrareddy, S.N.; Singh, K. Omicron SARS-CoV-2 variant: Unique features and their impact on pre-existing antibodies. *J. Autoimmun.* **2022**, *126*, 102779. [CrossRef] [PubMed]
14. Lan, J.; Ge, J.; Yu, J.; Shan, S.; Zhou, H.; Fan, S.; Zhang, Q.; Shi, X.; Wang, Q.; Zhang, L.; et al. Structure of the SARS-CoV-2 spike receptor-binding domain bound to the ACE2 receptor. *Nature* **2020**, *81*, 215–220. [CrossRef]
15. Vankadari, N.; Wilce, J.A. Emerging WuHan (COVID-19) coronavirus: Glycan shield and structure prediction of spike glycoprotein and its interaction with human CD26. *Emerg. Microbes Infect.* **2020**, *9*, 601–604. [CrossRef]
16. Yan, R.; Zhang, Y.; Li, Y.; Xia, L.; Guo, Y.; Zhou, Q. Structural basis for the recognition of SARS-CoV-2 by full-length human ACE2. *Science* **2020**, *367*, 1444–1448. [CrossRef]
17. Hoffmann, M.; Kleine-Weber, H.; Schroeder, S.; Krüger, N.; Herrler, T.; Erichsen, S.; Schiergens, T.S.; Herrler, G.; Wu, N.; Nitsche, A.; et al. SARS-CoV-2 Cell Entry Depends on ACE2 and TMPRSS2 and Is Blocked by a Clinically Proven Protease Inhibitor. *Cell* **2020**, *181*, 271–280. [CrossRef] [PubMed]
18. Song, H.; Seddighzadeh, B.; Cooperberg, M.R.; Huang, F.W. Expression of ACE2, the SARS-CoV-2 Receptor, and TMPRSS2 in Prostate Epithelial Cells. *Eur. Urol.* **2020**, *78*, 296–298. [CrossRef]
19. Aatif, M.; Muteeb, G.; Alsultan, A.; Alshoaibi, A.; Khelif, B.Y. Dieckol and Its Derivatives as Potential Inhibitors of SARS-CoV-2 Spike Protein (UK Strain: VUI 202012/01): A Computational Study. *Mar. Drugs* **2021**, *19*, 242. [CrossRef]
20. Wrapp, D.; Wang, N.; Corbett, K.S.; Goldsmith, J.A.; Hsieh, C.; Abiona, O.; Graham, B.S.; McLellan, J.S. Cryo-EM structure of the 2019-nCoV spike in the prefusion conformation. *Science* **2020**, *367*, 1260–1263. [CrossRef] [PubMed]
21. Nothwehr, S.F.; Gordon, J.I. Targeting of proteins into the eukaryotic secretory pathway: Signal peptide structure/function relationships. *Bioessays* **1990**, *12*, 479–484. [CrossRef]
22. Almagro Armenteros, J.J.; Tsirigos, K.D.; Sønderby, C.K.; Petersen, T.N.; Winther, O.; Brunak, S.; von Heijne, G.; Nielsen, H. SignalP 5.0 improves signal peptide predictions using deep neural networks. *Nat. Biotechnol.* **2019**, *37*, 420–423. [CrossRef] [PubMed]
23. Guo, H.; Sun, J.; Li, X.; Xiong, Y.; Wang, H.; Shu, H.; Zhu, R.; Liu, Q.; Huang, Y.; Madley, R.; et al. Positive charge in the n-region of the signal peptide contributes to efficient post-translational translocation of small secretory preproteins. *J. Biol. Chem.* **2018**, *293*, 1899–1907. [CrossRef] [PubMed]
24. Nilsson, I.; Lara, P.; Hessa, T.; Johnson, A.E.; von Heijne, G.; Karamyshev, A.L. The Code for Directing Proteins for Translocation across ER Membrane: SRP Cotranslationally Recognizes Specific Features of a Signal Sequence. *J. Mol. Biol.* **2015**, *427*, 1191–1201. [CrossRef] [PubMed]
25. Lumangtad, L.A.; Bell, T.W. The signal peptide as a new target for drug design. *Bioorganic. Med. Chem. Lett.* **2020**, *30*, 127115. [CrossRef] [PubMed]
26. Zhang, M.; Liu, L.; Lin, X.; Wang, Y.; Li, Y.; Guo, Q.; Li, S.; Sun, Y.; Tao, X.; Zhang, D.; et al. A Translocation Pathway for Vesicle-Mediated Unconventional Protein Secretion. *Cell* **2020**, *181*, 637–652. [CrossRef] [PubMed]
27. Van Puyenbroeck, V.; Pauwels, E.; Provinciael, B.; Bell, T.W.; Schols, D.; Kalies, K.U.; Hartmann, E.; Vermeire, K. Preprotein signature for full susceptibility to the co-translational translocation inhibitor cyclotriazadisulfonamide. *Traffic* **2019**, *21*, 250–264. [CrossRef]
28. Van Puyenbroeck, V.; Vermeire, K. Inhibitors of protein translocation across membranes of the secretory pathway: Novel antimicrobial and anticancer agents. *Cell Mol. Life Sci.* **2018**, *75*, 1541–1558. [CrossRef]
29. Zhou, Y.; Liu, P.; Gan, Y.; Sandoval, W.; Katakam, A.K.; Reichelt, M.; Rangell, L.; Reilly, D. Enhancing full-length antibody production by signal peptide engineering. *Microb. Cell Factories* **2016**, *15*, 47. [CrossRef]
30. Aw, R.; McKay, P.F.; Shattock, R.J.; Polizzi, K.M. A systematic analysis of the expression of the anti-HIV VRC01 antibody in Pichia pastoris through signal peptide optimization. *Protein Expr. Purif.* **2018**, *149*, 43–50. [CrossRef]
31. Shepard, B.D.; Natarajan, N.; Protzko, R.J.; Acres, O.W.; Pluznick, J.L. A cleavable N-terminal signal peptide promotes widespread olfactory receptor surface expression in HEK293T cells. *PLoS ONE* **2013**, *8*, e68758. [CrossRef]
32. Yuan, M.; Wu, N.C.; Zhu, X.; Lee, C.D.; So, R.T.Y.; Lv, H.; Mok, C.K.P.; Wilson, I.A. A highly conserved cryptic epitope in the receptor-binding domains of SARS-CoV-2 and SARS-CoV. *Science* **2020**, *368*, 630–633. [CrossRef] [PubMed]
33. Song, Z.; Xu, Y.; Bao, L.; Zhang, L.; Yu, P.; Qu, Y.; Zhu, H.; Zhao, W.; Han, Y.; Qin, C. From SARS to MERS, Thrusting Coronaviruses into the Spotlight. *Viruses* **2019**, *11*, 59. [CrossRef]
34. Steentoft, C.; Vakhrushev, S.Y.; Joshi, H.J.; Kong, Y.; Vester-Christensen, M.B.; Schjoldager, K.T.; Lavrsen, K.; Dabelsteen, S.; Pedersen, N.B.; Marcos-Silva, L.; et al. Precision mapping of the human O-GalNAc glycoproteome through SimpleCell technology. *EMBO J.* **2013**, *32*, 1478–1488. [CrossRef]

35. Hargett, A.A.; Renfrow, M.B. Glycosylation of viral surface proteins probed by mass spectrometry. *Curr. Opin. Virol.* **2019**, *36*, 56–66. [CrossRef]
36. Breiman, A.; Smith, G.L. Vaccinia virus B5 protein affects the glycosylation, localization and stability of the A34 protein. *J. Gen. Virol.* **2010**, *91*, 1823–1827. [CrossRef]
37. Walski, T.; De Schutter, K.; Van Damme, E.J.M.; Smagghe, G. Diversity and functions of protein glycosylation in insects. *Insect Biochem. Mol. Biol.* **2017**, *83*, 21–34. [CrossRef]
38. Eichler, J. Protein glycosylation. *Curr. Biol.* **2019**, *29*, R229–R231. [CrossRef]
39. Tian, W.; Li, D.; Zhang, N.; Bai, G.; Yuan, K.; Xiao, H.; Gao, F.; Chen, Y.; Wong, C.; Gao, G.F. O-glycosylation pattern of the SARS-CoV-2 spike protein reveals an "O-Follow-N" rule. *Cell Res.* **2021**, *31*, 1123–1125. [CrossRef]
40. McCallum, M.; Bassi, J.; De Marco, A.; Chen, A.; Walls, A.C.; Di Iulio, J.; Tortorici, M.A.; Navarro, M.J.; Silacci-Fregni, C.; Saliba, C.; et al. SARS-CoV-2 immune evasion by the B.1.427/B.1.429 variant of concern. *Science* **2021**, *373*, 648–654. [CrossRef]
41. Zheng, J.; Yamada, Y.; Fung, T.S.; Huang, M.; Chia, R.; Liu, D.X. Identification of N-linked glycosylation sites in the spike protein and their functional impact on the replication and infectivity of coronavirus infectious bronchitis virus in cell culture. *Virology* **2018**, *513*, 65–74. [CrossRef]
42. Tan, L.; Chen, Z.; Ma, X.; Huang, Q.; Sun, H.; Zhang, F.; Yang, S.; Xu, C.; Cui, N. Glycosylation of the hemagglutinin protein of H9N2 subtype avian influenza virus influences its replication and virulence in mice. *J. Integr. Agric.* **2019**, *18*, 1443–1450. [CrossRef]
43. Kotwal, G.J. Influence of glycosylation and oligomerization of vaccinia virus complement control protein on level and pattern of functional activity and immunogenicity. *Protein Cell* **2010**, *1*, 1084–1092. [CrossRef]
44. Watanabe, Y.; Allen, J.D.; Wrapp, D.; McLellan, J.S.; Crispin, M. Site-specific glycan analysis of the SARS-CoV-2 spike. *Science* **2020**, *369*, b9983. [CrossRef]
45. Snapp, E.L.; McCaul, N.; Quandte, M.; Cabartova, Z.; Bontjer, I.; Källgren, C.; Nilsson, I.; Land, A.; von Heijne, G.; Sanders, R.W.; et al. Structure and topology around the cleavage site regulate post-translational cleavage of the HIV-1 gp160 signal peptide. *eLife* **2017**, *6*, e26067. [CrossRef]
46. Goh, E.; Lin, Z.; Ahn, B.Y.; Lopes-Rodrigues, V.; Dang, N.H.; Salim, S.; Berger, B.; Dymock, B.; Senger, D.L.; Ibanez, C.F. A Small Molecule Targeting the Transmembrane Domain of Death Receptor p75(NTR) Induces Melanoma Cell Death and Reduces Tumor Growth. *Cell Chem. Biol.* **2018**, *25*, 1485–1494. [CrossRef]
47. Asada, H.; Inoue, A.; Ngako Kadji, F.M.; Hirata, K.; Shiimura, Y.; Im, D.; Shimamura, T.; Nomura, N.; Iwanari, H.; Hamakubo, T.; et al. The Crystal Structure of Angiotensin II Type 2 Receptor with Endogenous Peptide Hormone. *Structure* **2020**, *28*, 418–425. [CrossRef]
48. Van Puyenbroeck, V.; Claeys, E.; Schols, D.; Bell, T.W.; Vermeire, K. A Proteomic Survey Indicates Sortilin as a Secondary Substrate of the ER Translocation Inhibitor Cyclotriazadisulfonamide (CADA). *Mol. Cell. Proteom.* **2017**, *16*, 157–167. [CrossRef]

molecules

MDPI

Article

In Silico Study towards Repositioning of FDA-Approved Drug Candidates for Anticoronaviral Therapy: Molecular Docking, Molecular Dynamics and Binding Free Energy Calculations

Wesam S. Qayed [1,*], Rafaela S. Ferreira [2] and José Rogério A. Silva [2,*]

1 Medicinal Chemistry Department, Faculty of Pharmacy, Assiut University, Assiut 71526, Egypt
2 Laboratório de Planejamento e Desenvolvimento de Fármacos, Instituto de Ciências Exatas e Naturais, Universidade Federal do Pará, Belém 66075-110, Brazil
* Correspondence: wesam.qayed@aun.edu.eg (W.S.Q.); rogerio@ufpa.br (J.R.A.S.)

Abstract: The SARS-CoV-2 targets were evaluated for a set of FDA-approved drugs using a combination of drug repositioning and rigorous computational modeling methodologies such as molecular docking and molecular dynamics (MD) simulations followed by binding free energy calculations. Six FDA-approved drugs including, Ouabain, Digitoxin, Digoxin, Proscillaridin, Salinomycin and Niclosamide with promising anti-SARS-CoV-2 activity were screened in silico against four SARS-CoV-2 proteins—papain-like protease (PLpro), RNA-dependent RNA polymerase (RdRp), SARS-CoV-2 main protease (Mpro), and adaptor-associated kinase 1 (AAK1)—in an attempt to define their promising targets. The applied computational techniques suggest that all the tested drugs exhibited excellent binding patterns with higher scores and stable complexes compared to the native protein cocrystallized inhibitors. Ouabain was suggested to act as a dual inhibitor for both PLpro and Mpro enzymes, while Digitoxin bonded perfectly to RdRp. In addition, Salinomycin targeted PLpro. Particularly, Niclosamide was found to target AAK1 with greater affinity compared to the reference drug. Our study provides comprehensive molecular-level insights for identifying or designing novel anti-COVID-19 drugs.

Keywords: anti-COVID-19; drug repositioning; molecular docking; molecular dynamic simulations; binding free energy

Citation: Qayed, W.S.; Ferreira, R.S.; Silva, J.R.A. In Silico Study towards Repositioning of FDA-Approved Drug Candidates for Anticoronaviral Therapy: Molecular Docking, Molecular Dynamics and Binding Free Energy Calculations. *Molecules* **2022**, *27*, 5988. https://doi.org/10.3390/molecules27185988

Academic Editors: Giovanni N. Roviello and Caterina Vicidomini

Received: 29 July 2022
Accepted: 6 September 2022
Published: 14 September 2022

Publisher's Note: MDPI stays neutral with regard to jurisdictional claims in published maps and institutional affiliations.

1. Introduction

Severe acute respiratory syndrome coronavirus 2 (SARS-CoV-2) caused the well-known coronavirus disease of 2019, COVID-19 [1]. The World Health Organization (WHO) has declared a public health emergency of worldwide importance [2]. The fundamental difficulty is that COVID-19 is a very contagious disease with a high fatality rate [3]. It has exhibited strong transmissibility and has caused a substantial number of deaths worldwide since its inception. It is still creating havoc all around the world [4]. This signifies that appropriate treatment is required as soon as an epidemic begins in order to prevent the disease from spreading [2,5].

The global research and clinical communities have been under tremendous pressure to produce treatments to battle the ever-growing pandemic since the outbreak of the unique COVID-19 disease. However, drug development can take several years and there is no certainty that it will be completed successfully. Furthermore, with the existing drug production or application method, this cannot be done, and it takes several years for newly approved drugs to reach the market. Instead, SARS-CoV-2 infection could be treated with existing drugs. When it comes to developing new drugs/vaccines or repurposing approved ones, detailed information on the possible target(s) is essential [6].

Since the start of the pandemic, scientists have been working on developing vaccinations against SARS-CoV-2. As of July 2021, the WHO had approved 14 vaccinations under the Emergency Use Listing (EUL), with many more on the waiting list. Although all the licensed vaccinations have proven to be rather effective up to this point, the emergence of new strains due to rapid virus mutations poses a significant imminent danger to vaccine efficacy. Various routes are being targeted to limit viral replication inside human cells, and studies are being conducted to understand the complexities of how the virus affects the human body [7].

Potential anti-coronavirus drugs can be divided into two categories based on the target: One targets the human immune system or human cells, while the other targets the coronavirus itself. The innate immune system response is critical in suppressing coronavirus replication and infection in the human immune system, and interferon is believed to improve the immune response [8]. Furthermore, inhibiting viral endocytosis is pivotal in COVID-19 therapy [9]. The therapies that target the coronavirus include blocking the virus's binding to human cell receptors or inhibiting the virus self-assembly process by acting on some structural proteins, as well as preventing the synthesis of viral RNA by acting on the virus genetic material or inhibiting virus replication by acting on critical viral enzymes [10]. The spike protein, membrane protein, envelop protein, nucleocapsid protein, protease, helicase, and hemagglutinin esterase are the seven key structural protein targets for SARS-CoV-2 [11], while replicase proteins, polyprotein 1a (PP1a) and polyprotein 1ab (PP1ab) represent the other 16 nonstructural proteins (NSPs) formed from the virus [12]. All NSPs have a unique role in replication and transcription. On the other hand, human targets, including many kinases, e.g., adaptor-associated kinase 1 (AAK1), cathepsin L, and the transmembrane serine protease 2 (TMPRSS2), are involved in the advancement of symptoms associated with SARS-CoV-2 infections, such as pneumonia, inflammation, and fibrosis [13]. Kinases react to and control the synthesis of potentially damaging cytokines (IL-6, IL-10, and TNF) and proteins linked to inflammation and the induction of pulmonary fibrosis (such as the pro-inflammatory cytokine TGF-). Thus, kinase inhibitors can have antiviral, anti-inflammatory, cytokine-suppressive, and antifibrotic properties, all of which could be beneficial in the fight against respiratory viral infections. Several kinase inhibitors are now being investigated in COVID-19 clinical studies [13,14].

Because there is a pressing need to treat the disease, most research is focusing on repurposing already approved drugs rather than classical drug discovery. Remdesivir was the first of all the COVID-19 drugs to be approved for hospitalized patients by the US Food and Drug Administration (FDA). The concept of drug repositioning, also known as drug repurposing, drug reprofiling, and therapeutic switching, has gained considerable attention from pharmaceutical companies and researchers due to its potential to discover new uses for approved drugs [15]. Compared to traditional de novo methods, the approach is more efficient in terms of time and cost savings. The shorter development time, in particular, corresponds to an example of the general scenario, in this case, COVID-19. Furthermore, contemporary and powerful artificial intelligence (AI) technologies have substantially improved the throughput and accuracy of drug repurposing [16]. A multi-target molecular modeling protocol was employed to identify druggable targets associated with viral replication, in addition to a molecular dynamic simulation against the SARS-CoV-2 viral genome, as mutations are one of the most challenging obstacles to overcome with antiviral therapeutics [17,18].

Since the outbreak started, a flood of computational articles identifying potential antiviral drug repurposing candidates has appeared in peer-reviewed journals and preprint services, but most of them lack experimental confirmation [10,19–23]. The purpose of this research was to look at the primary SARS-CoV-2 proteins as potential targets for active FDA-approved drugs with a more enhanced antiviral profile with the aid of computational tools, docking, and molecular dynamic simulations. As the selected listed drugs have been previously experimentally screened against the SARS-CoV-2 virus, we tried to elucidate their activity to find their viral targets from structural and nonstructural protein targets, which are crucial for viral replication and translation, as well as one of the promising human kinases. These drugs can be used in future research as a potential pharmacophore scaffold for the development of promising anti-COVID-19 therapies.

2. Results and Discussion

2.1. Rationale of Drug and Proteins Selection

A public health emergency with regard to COVID-19 in all countries and efforts to develop an effective vaccine or drug for prevention or treatment are critical. New compound approval as a drug necessitates a lengthy investigation period and a considerable financial investment. So, the best starting point is repurposing already approved drugs when developing novel COVID-19 therapeutics. We virtually reprofiled drugs against multiple SARS-CoV-2 targets, including SARS-CoV-2 Mpro, PLpro, and RdRp, in addition to AAK1 as a human target.

Previously, Meehyun Ko et al. tested a total of 24 FDA-approved drugs against SARS-CoV-2 in Vero cells and Calu3 cells—a well-known human lung cell line—as potential antiviral candidates [24,25]. For deep computation, we created a panel of the most active (IC_{50} less than 2.5 μM) FDA-approved drugs that showed antiviral activity compared to the clinically applied drugs Chloroquine, Remdesivir, and Lopinavir (IC_{50} in Vero cells 11.4, 7.28 and 9.12 μM, IC_{50} Calu cells, respectively). The current study tried to find the potential target for each drug against SARS-CoV-2 through virtual screening. The included drugs are currently on the market and some are in clinical trials against SARS-CoV-2 infections. The drugs that are currently in clinical studies were given special consideration. The drugs' generic name, clinical use, and reported activity against SARS-CoV-2 (IC_{50} μM) are presented in Table 1.

Table 1. FDA-approved drugs are considered tested ligands that were selected for molecular docking. Drug generic names, clinical use, and reported activity against SARS-CoV-2.

Drug Generic Name	Clinical Use	(IC_{50}, μM) in Vero Cell	(IC_{50}, μM) in Calu Cell
Ouabain (OUB)	Cardiovascular agents	<0.097	0.1
Digoxin (DGX)	Cardiovascular agents	0.19	0.72
Digitoxin (DIG)	Cardiovascular agents	0.23	0.16
Salinomycin (SLM)	Antibacterial agents	0.24	0.5
Niclosamide (NIS)	Antiparasitic agents	0.28	0.84
Proscillaridin (PRO)	Cardiovascular agents	2.04	5.95

2.2. Molecular Docking

The structures of four SARS-CoV-2 proteins (macromolecules) and antiviral drugs (ligands) were retrieved and optimized for the docking simulation to investigate the binding preference between the protein binding pocket residues and ligands. The molecular docking resulted in ten different docked poses for each drug. Each drug's optimal pose was chosen based on how closely its binding pattern in the active pocket matched that of the original ligand, taking into consideration the binding energy. Accordingly, the drugs under investigation were scored, with the top scorers being those with the lowest binding energy. For each ligand, the complex with the best-predicted affinity energy (AE) was used for the MD simulations.

In this research, the docking efficacy of six FDA-approved drugs previously tested against SARS-CoV-2 (Scheme 1) against a host and viral SARS-CoV-2 proteins was studied. Four SARS-CoV-2 proteins were investigated. Among them, three are viral SARS-CoV-2 proteins: Mpro, PLpro, and RdRp. In addition to human protein AAK1, docking studies were conducted using MOE software. The MOE software's output was further evaluated and displayed. These docking studies predicted the AE for the compounds against the studied SARS-CoV-2 systems listed in Table 2.

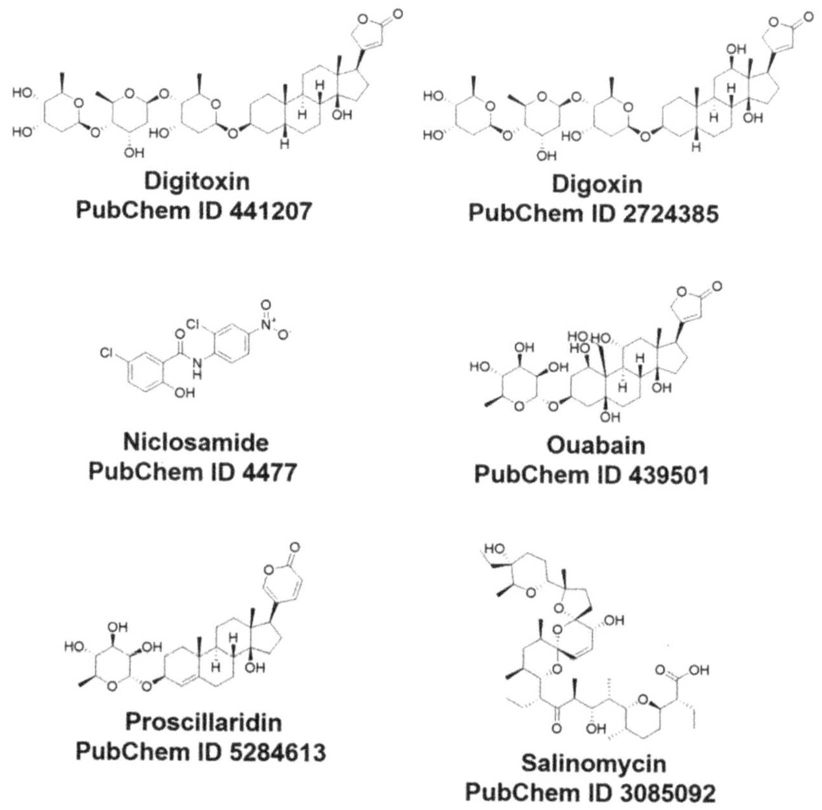

Scheme 1. FDA-approved drugs that were included in the virtual study.

Table 2. Affinity energy (AE) values (Kcal/mol) of binding for FDA-approved repositioned drugs to selected SARS-CoV-2 target proteins.

Drug Name	PLpro	Mpro	RdRp	AAK1
Ouabain (OUB)	−9.52	−8.21	−6.35	−8.17
Digoxin (DGX)	−7.88	−7.64	−7.63	−9.08
Digitoxin (DIG)	−8.01	−6.56	−7.69	−9.9
Salinomycin (SLM)	−9.02	−4.97	−5.19	−9.05
Niclosamide (NIS)	−6.43	−5.87	−5.66	−11.84
Proscillaridin (PRO)	−6.22	−7.28	−7.11	−8.25
Cocrystallized ligand GRM	−8.46			
Cocrystallized ligand X77		−8.05		
Cocrystallized ligand RDV			−7.8	
Cocrystallized ligand LKB				−11.72

The repositioned drug with the highest score for each investigated target was selected. Ouabain (OUB) showed the greatest affinity for SARS-CoV-2 Mpro and PLpro (−8.21 and −9.52 Kcal/mol, respectively) contrasted to the enzyme native ligand X77 and GRM (−8.05 and −8.46 Kcal/mol), respectively. DGX and DIG showed AEs comparable to RdRp with binding energy (−7.63 and −7.69 Kcal/mol, respectively) compared to cocrystallized ligand Remdesivir (RDV) which revealed AE equal to −7.8 Kcal/mol. Salinomycin (SLM) showed binding affinity for PLpro (AE = −9.05 Kcal/mol) compared to cocrystallized ligand GRM (AE = −8.46 Kcal/mol). Niclosamide (NIS) showed excellent interaction and binding affinity (AE = −11.84 Kcal/mol) for the AAK1 system in comparison to its cocrystallized ligand LKB (AE = −11.72 Kcal/mol). Finally, Proscillaridin (PRO) showed optimum binding affinity for RdRp (AE = −7.11 Kcal/mol).

2.3. *Analysis of the Intermolecular Binding Pattern of Repurposed Systems by Docking and MD Simulations*

Molecular docking is a powerful approach for evaluating binding affinity and investigating the binding pattern of ligands that bind to the active region of target proteins [26]. The most likely interaction of the ligand with the protein receptor was identified and visualized using molecular docking studies performed using MOE software. Both the binding modalities and affinity of docked ligands at the active pocket of the protein were predicted using the docking score and hydrogen bonds generated between amino acids and the interacted atoms. A specific ligand binds and interacts at the active site residues of a target protein with a certain affinity and strength that is referred to as the binding energy. A docking investigation was conducted for FDA-approved drugs: OUB, DGX, DIG, SLM, NIS, and PRO using MOE software, as discussed above.

Although molecular docking may shed light on repurposing drug protocols [27], their usual fast and approximate algorithms lack protein flexibility, which may be related to the recognition and binding involved in ligand–protein complexes [28]. In this sense, molecular dynamic (MD) simulation techniques, which are more computationally expensive but more accurate than docking protocols, may provide a better complementary result [29]. In summary, MD simulations can be used to study the macromolecule characteristics and provide a suitable ensemble for thermodynamics analysis.

Here, MD simulation calculations were conducted to understand the structural stability of some favorable docking results. We can summarize it as (i) PLpro–GRM, PLpro–OUB, and PLpro–SLM systems; (ii) Mpro–X77 and Mpro–OUB systems; (iii) RdRp–RDV and RdRp–DIG systems; (iv) AAK1–LKB and AAK1–NIS systems. Initially, The RMSD analysis from the 200 ns of MD trajectories were explored to examine flexibility across the simulation periods.

2.3.1. PLpro Systems

PLpro is an essential aspect of the replicase–transcriptase complex as a nonstructural protein 3 (NSP3). By releasing NSP1–3 from the viral polyprotein, which is required for viral replication, human proteases with a comparable cleavage selectivity can be known, and inhibitors of that enzyme are unlikely to be harmful [30]. The comprehensive examination of the Mpro catalytic mechanism makes it a promising target for anti-COVID-19 medication development [30]. PLpro acts as a negative regulator of the immune response to viral infection [31]. All of these factors combine to make it an attractive target for antiviral drugs. PLpro was represented with the cocrystallized lead compound GRM [32]. Of the six selected drugs on the active pocket of PLpro, Ouabain (OUB) was approved by the FDA to treat heart conditions of patients for over 10 decades. OUB showed the highest docking score (−9.52 Kcal/mol) among the tested compounds, which was higher than that of the native ligand GRM (-8.46 Kcal/mol). GRM hydrogen-bonded with the carboxylate of Asp165 and with the backbone carbonyl oxygen of Tyr269. OUB rings similarly showed binding to the PLpro active pocket Asp165 and Tyr269. In addition, it showed an extra hydrogen-bond interaction with Leu163 and Glu168 (Figure 1). This outcome is consistent with in vitro screening as it has the lowest IC_{50} values of all the tested drugs.

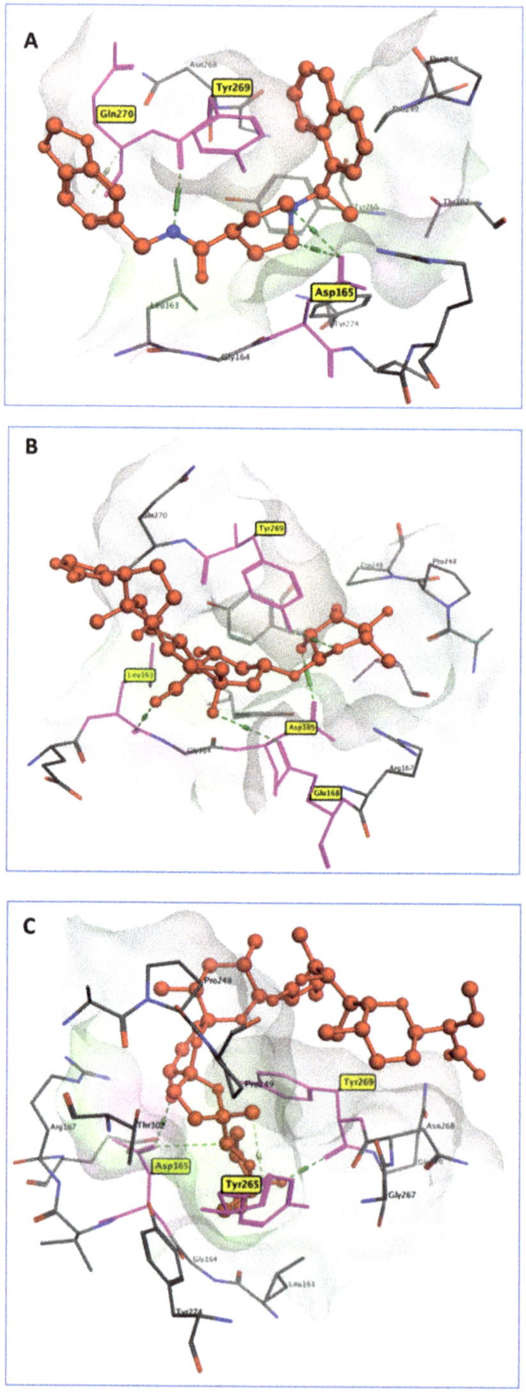

Figure 1. 3D model of the interactions between cocrystallized ligand GRM (**A**), OUB (**B**), and SLM (**C**) with PLpro active site residues.

Streptomyces albus produces Salinomycin (SLM), a carboxylic polyether ionophore. Ionophores have a wide range of bioactivity, including antibacterial, antifungal, antiparasitic, antiviral, and, more recently, anti-tumor properties [33]. In a trial to reposition SLM for the treatment of COVID-19, it showed antiviral activity, with a lower IC_{50} compared to the control drugs. This activity may be attributed to the SLM binding affinity for PLpro, as manifested from the docking simulation. SLM is connected to key PLpro residues through three hydrogen bonds (Figure 1). It bonded to the Asp165, Tyr265, and Tyr269 amino acids with a docking score of −9.02 Kcal/mol, which is more than that of the crystallized ligand (8.46 Kcal/mol).

A suitable structural stabilization for PLpro systems by computing RMSD/RMSF can be observed in Figure 2A,B, where all systems converged to their respective average structures during the total MD simulation scale. In addition, all the systems showed very small RMSDs (Figure 2A), which changed from 1.41 ± 0.30 Å (PLpro–GRM) to 1.33 ± 0.34 Å (PLpro–OUB) and 1.34 ± 0.24 Å (PLpro–SLM), respectively. From the RMSF analysis (Figure 2B), the values changed from 1.24±0.80 Å (PLpro–GRM) to 1.17 ± 0.74 Å (PLpro–OUB) and 1.18 ± 0.70 Å (PLpro–SLM), respectively. All these results suggest that the crystal (PLpro–GRM) and repurposed (PLpro–OUB and PLpro–SLM) systems were in a very similar conformation during the MD simulation.

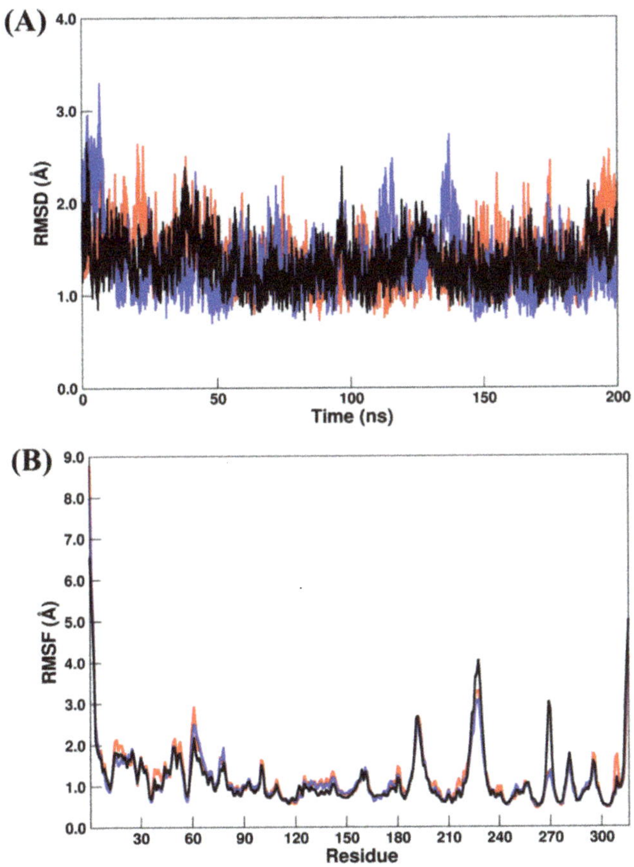

Figure 2. (**A**) RMSD and (**B**) RMSF plots for PLpro–GRM (red line), PLpro–OUB (blue line), and PLpro–SLM (black line). All values are reported in Å.

2.3.2. Mpro Systems

The comprehensive examination of the Mpro catalytic mechanism identified it as a promising target for anti-COVID-19 medication development. For Mpro systems, it should be highlighted that previous experimental studies show that in the biological environment, Mpro acts as a dimer instead of a monomer [34]. However, for computational analysis and computer-aided drug design strategies, only the monomeric form is necessary [35–37]. For the comparison of the docking data of the studied compounds, the cocrystallized ligand X77 (N-(4-tert-butylphenyl)-N-[(1R)−2-(cyclohexylamino)−2-oxo−1-(pyridine−3-yl)ethyl]−1H-imidaz-ole−4-carboxamide) was employed as a reference ligand. It interacted with Asn142, Gly143, His163, and Glu166, through four hydrogen bonds [38], while OUB created six hydrogen bonds with Met49, Asn142, Gly143, Met165, and Glu166 (Figure 3). It exhibited a binding score of −8.21 Kcal/mol, similar to that of the reference ligand (−8.05 Kcal/mol).

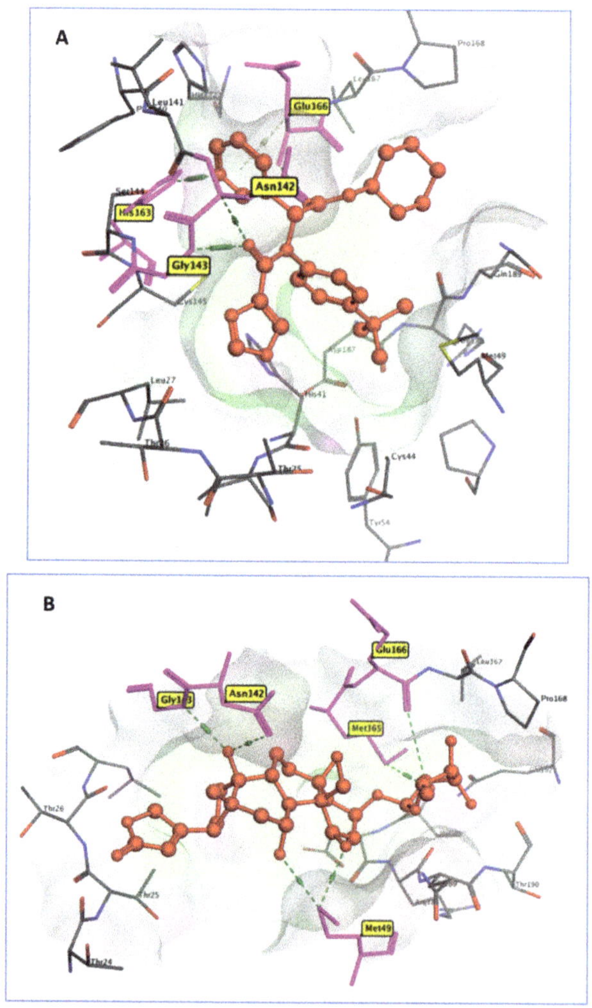

Figure 3. 3D model of the interactions of cocrystallized ligand X77 (**A**) and OUB (**B**) with Mpro active site residues.

Here, Mpro–X77 (crystal reference) and Mpro–OUB (repurposed system) were submitted to 200 ns production MD simulations. For the RMSD analysis (Figure 4A), the values for Mpro–X77 and Mpro–OUB changed from 1.37 ± 0.48 Å to 1.27 ± 0.25 Å, respectively. A similar trend was found for the RMSF analysis (Figure 4B), where the values changed from 1.37 ± 0.54 Å to 1.10 ± 0.71 Å. Both results suggest excellent structural stabilization during all the MD production scales. Particularly, Mpro–X77 system showed an increase of about 1 Å on the last 20 ns MD simulation. However, it did not affect the binding of the crystal inhibitor (X77). Our findings are in good concordance with a previous computational study involving a drug repurposing analysis for the Mpro system [35]. Consequently, OUB can be considered as a dual inhibitor of SARS-CoV-2 through in silico screening against two SARS-CoV-2 targets: Mpro and PLpro.

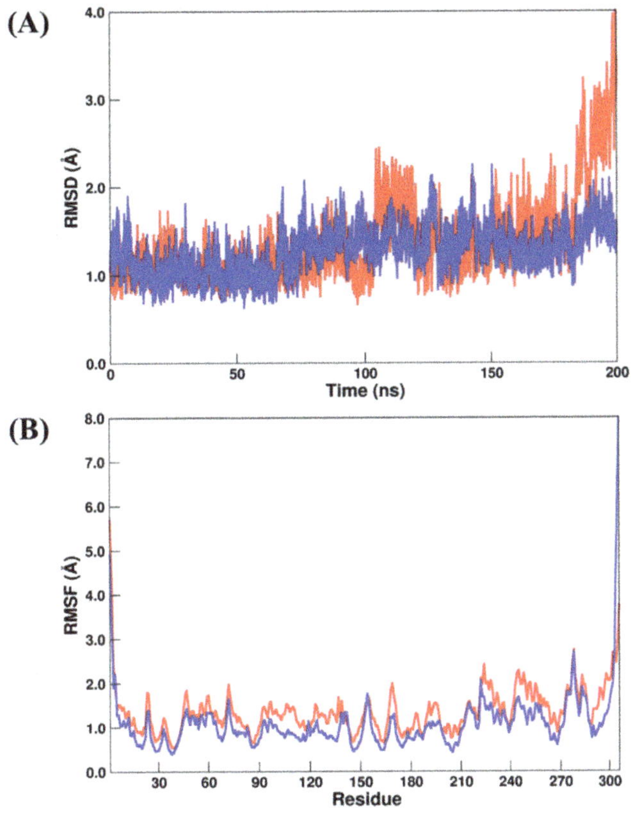

Figure 4. (**A**) RMSD and (**B**) RMSF plots for Mpro–X77 (red line) and Mpro–OUB (blue line). All values are reported in Å.

2.3.3. RdRp Systems

Natural cardiac glycosides, such as Digoxin, Digitoxin, and Proscillaridin, can be utilized to treat congestive heart failure and cardiac arrhythmia. The virtual screening of our selected drugs indicated the affinity of these drugs for SARS-CoV-2 RNA-dependent RNA polymerase (RdRp) (PDB code: 7BV2). The RdRp enzyme allows the viral genome to be transcribed into new RNA copies using the host cell's machinery [38]. RdRp is required for the RNA virus life cycle but has no homolog in the host cell. This allows for the development of antiviral drugs and minimizes the risk of a protein in human cells being impacted [39]. Remdesivir, a repositioned drug, was taken as a standard. The in silico docking of RDV on (RdRp) protein hydrogen bonded to the active site residues Asp730,

Arg588, and Arg523 with a docking score of −7.8 Kcal/mol. DIG had a binding pattern to RdRp similar to RDV as it was connected to the active site residues Arg553, Arg555, Lys621, and Asp623 through hydrogen bonds with an almost similar binding score of −7.69 Kcal/mol, compared to the RDV docking score of −7.8 Kcal/mol, as illustrated in Figure 5.

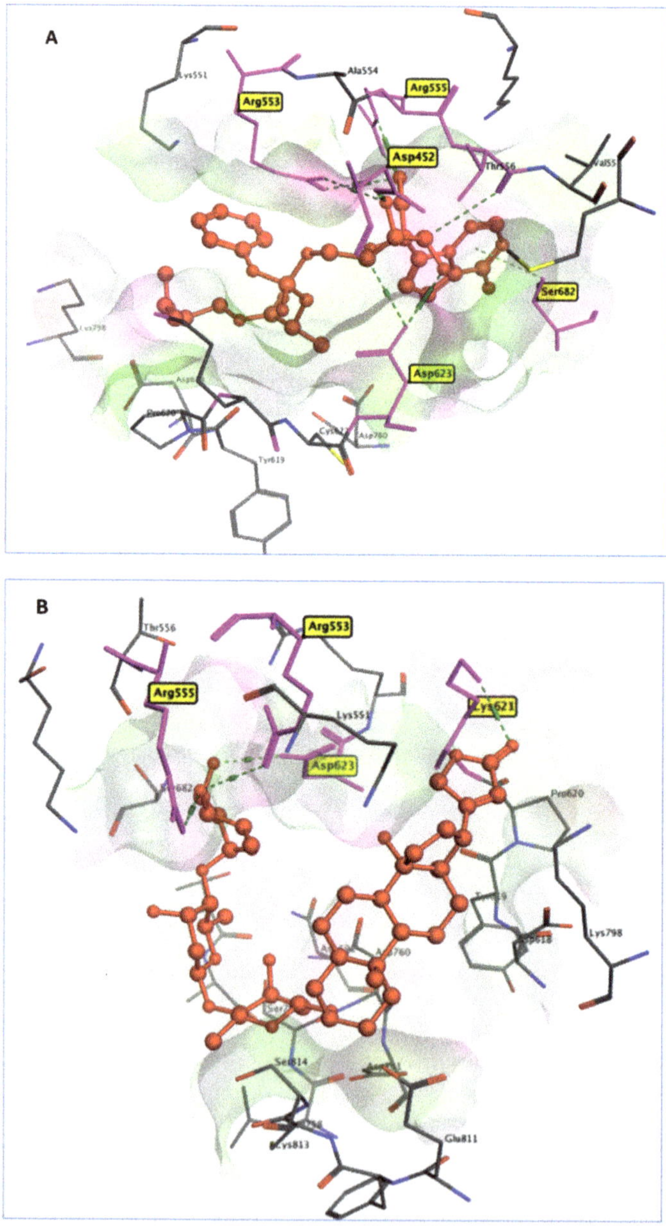

Figure 5. 3D interactions between RDV (**A**) and DIG (**B**) with 7BV2 active site residues.

Recently, MD simulation studies for RdRp have been carried out using RDV as a bound inhibitor [40,41]. Particularly, Srivastava et al. [40] used different RDV models (named monophosphate and triphosphate). On the other hand, Koulgi et al. [41] used RDV before any metabolic change during 100 ns of MD simulations. Here, we chose to follow Koulgi's proposal, but we performed 200 ns of MD simulations. According to our RMSD and RMSF analysis, the RdRp–RDV (crystal reference) and RdRp–DIG (repurposed) systems had similar structural stabilization. The RMSD values (Figure 6A) for RdRp–RDV and RdRp–DIG changed from 1.40 ± 0.17 Å to 1.48 ± 0.22 Å, respectively, while the RMSF values (Figure 6B) changed from 1.08 ± 0.75 Å to 1.23 ± 0.80 Å, respectively.

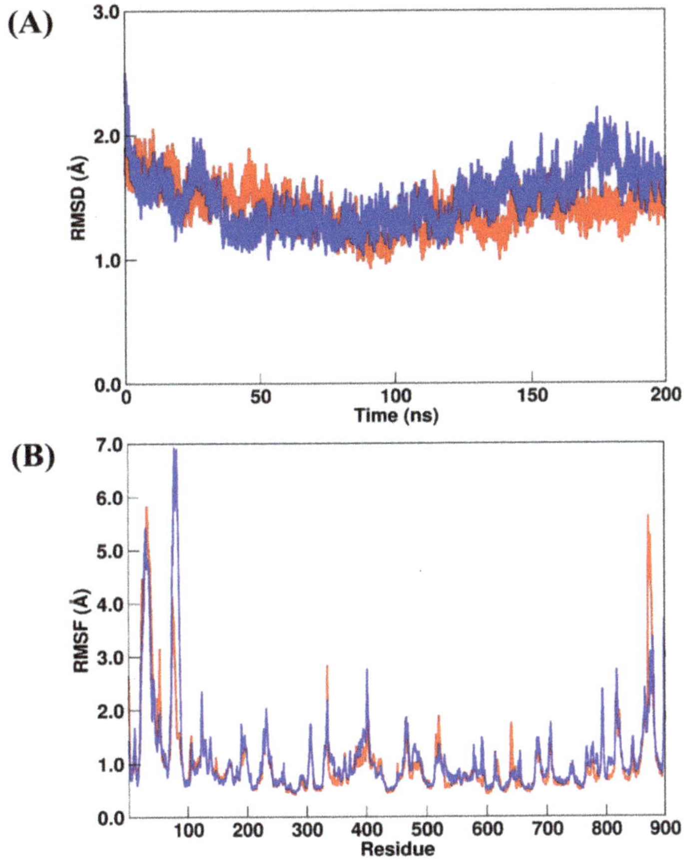

Figure 6. (**A**) RMSD and (**B**) RMSF plots for RdRp–RDV (red line) and RdRp–DIG (blue line). All values are reported in Å.

2.3.4. AAK1 Systems

Niclosamide (NIS) is an antiparasitic drug that prevents the tapeworm from absorbing glucose, as well as preventing oxidative phosphorylation and anaerobic metabolism [42]. In comparison to the cocrystallized inhibitor (AE = −11.72 Kcal/mol), it was able to robustly bind to the AAK1 active site (AE = −11.84 Kcal/mol). AKK1 is a key kinase of receptor-mediated endocytosis found in the host cell, which regulates viral entrance and processing. Previous research has found that inhibiting AKK1 selectively can be helpful in SARS-CoV-2 treatment, either by blocking viral endocytosis or lowering the presence of pro-inflammatory molecules (IFN-γ and IL-6) [43]. Due to its affinity for the AP2-associated protein AAK1, Baricitinib, an efficient AAK1 and GAK inhibitor, has recently been discovered to have antiviral effects, reducing SARS-CoV-2 endocytosis [44,45]. From the docking results, NIS interacted with the AKK1 active site through four hydrogen bonds between carbonyl, chloride, phenolic hydroxyl, and nitro moieties and the active site residues Asp133, Lys127, Cys129, and Asp194, respectively. Further, it formed three π interactions with the amino acids Leu52 and Val60 (Figure 7), while the native ligand LKB formed four hydrogen bonds with Lys74, Asp127, Cys129 and Asn136 in addition to a Pi–H interaction with the amino acids Leu52 and Val60.

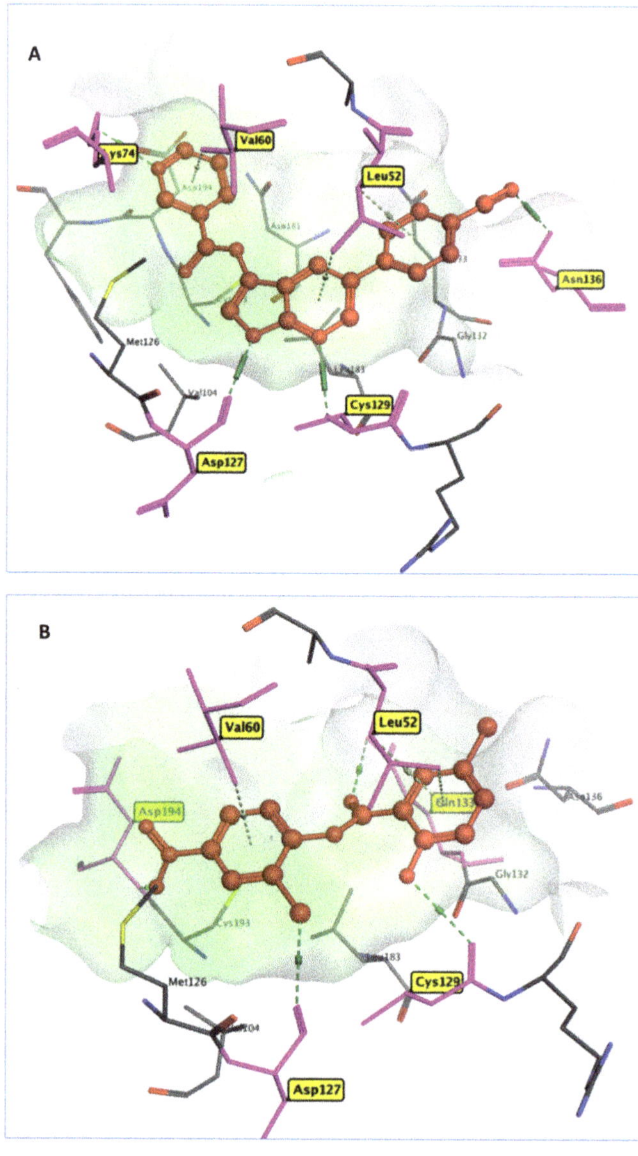

Figure 7. 3D model of the interactions between crystallized ligand LKB (**A**) and NIS (**B**) in 5L4Q active site residues.

The structural analysis for the AAK1 systems by considering the RMSD and RMSF calculations for AAK1–LKB (crystal reference) and AAK1–NIS (repurposed) systems suggested that they were more structurally stable than the previous simulated systems. The RMSD values (Figure 8A) for AAK1–LKB and AAK1–NIS changed from 0.96 ± 0.17 Å to 1.29 ± 0.51 Å, respectively, while the RMSF values (Figure 8B) changed from 0.86 ± 0.51 Å to 0.83 ± 0.43 Å, respectively. However, by comparing the RMSF plots of AAK1–LKB and AAK1–NIS, we can observe a suitable divergence in the region that comprises residues from Trp214 to Tyr260. For the previous simulated systems, no similar differences were found.

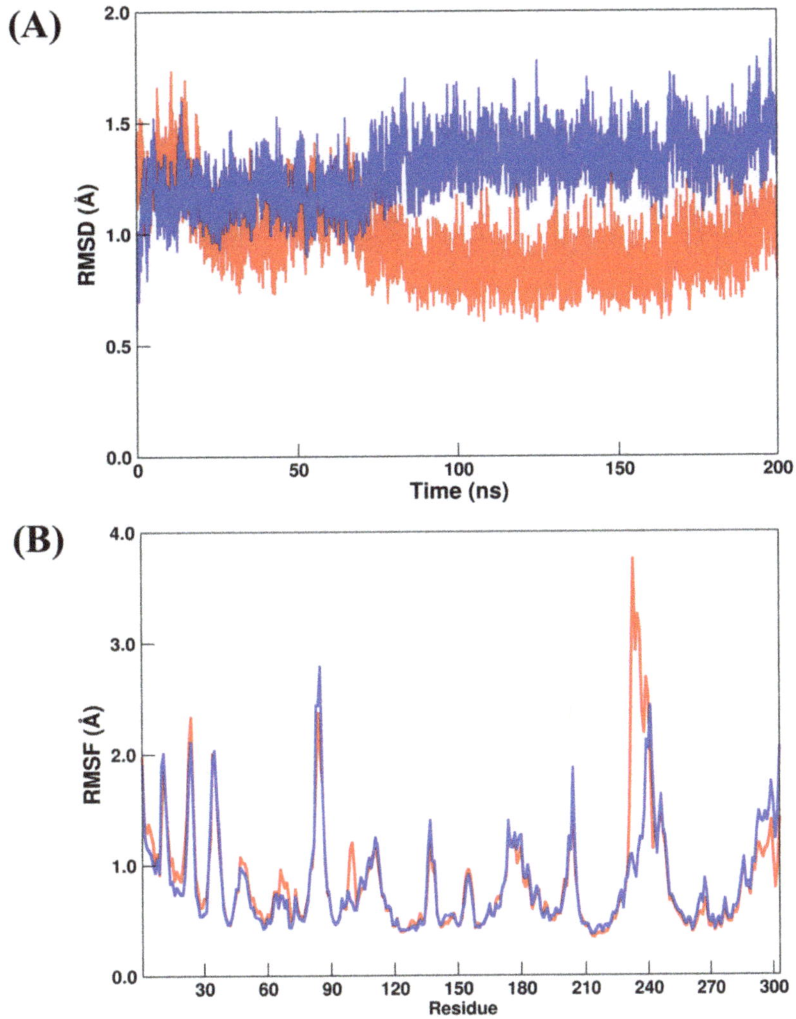

Figure 8. (**A**) RMSD and (**B**) RMSF plots for AAK1–LKB (red line) and AAK1–NIS (blue line). All values are reported in Å.

2.4. Binding Free Energy Calculations

As described in the Material and Methods section, relative binding free energies for all simulated systems were computed using the MM/GBSA approach [46–48] as implemented in *MMPBSA.py* [49] and are presented in Table 3. A single MD trajectory of the bound complexes was used to calculate the relative binding free energy (ΔG_{bind}); the ΔE_{int} term (in Equation (3)) was canceled once the changing energy between complex systems and their components were computed using the same MD ensemble [47,48].

Table 3. Binding free energy (ΔG_{bind}) values and their components for the crystal and repurposed systems by MM/GBSA. The errors were computed by using the single-trajectory protocol [48]. All values are reported in Kcal/mol.

System	ΔE_{vdW}	ΔE_{elec}	ΔG_{polar}	$\Delta G_{non-polar}$	ΔG_{bind}
GRM–PLpro	−39.7 (0.1)	0.7 (0.1)	13.9 (0.1)	−4.6 (0.1)	−29.7 (0.1)
OUB–PLpro	−38.2 (0.2)	−55.4 (0.3)	61.9 (0.2)	−5.7 (0.1)	−37.4 (0.2)
PLpro–SLM	−36.7 (0.1)	52.9 (0.2)	−41.4 (0.2)	−4.6 (0.1)	−29.8 (0.1)
Mpro–X77	−52.9 (0.1)	−22.8 (0.2)	39.1 (0.1)	−6.2 (0.1)	−42.8 (0.1)
Mpro–OUB	−41.9 (0.1)	−29.3 (0.2)	47.7 (0.1)	−5.2 (0.1)	−28.7 (0.1)
RdRp–RDV	−36.9 (0.1)	−219.8 (0.8)	250.2 (0.7)	−6.5 (0.1)	−13.0 (0.1)
RdRp–DIG	−50.6 (0.1)	−34.6 (0.3)	76.8 (0.3)	−6.3 (0.1)	−14.7 (0.1)
AAK1–LKB	−44.7 (0.1)	−31.3 (0.1)	40.3 (0.1)	−5.7 (0.1)	−41.4 (0.1)
AAK1–NIS	−41.0 (0.1)	−22.0 (0.1)	29.1 (0.1)	−5.3 (0.1)	−39.2 (0.1)

All the binding free energy terms are also listed in Table 3. For the PLpro systems, we can see that in PLpro–GRM, the vdW term (ΔE_{vdW}) was the most important component for binding free energy followed by the polar term (ΔG_{polar}). The PLpro–SLM showed similar values of ΔE_{vdW} and ΔG_{bind} when compared with PLpro–GRM. However, a significant difference can be observed in electrostatic (ΔE_{elec}) and polar terms, where the first term increased from 0.70 Kcal/mol to 52.9 Kcal/mol for PLpro–GRM and PLpro–SLM, respectively, while the second term (ΔG_{polar}) decreased from 13.9 Kcal/mol to −41.4 Kcal/mol, respectively. The most interesting result can be observed for PLpro–OUB, where the ΔE_{elec} and ΔG_{polar} appeared to be the most important terms for OUB binding with the catalytic site of the PLpro enzyme. As can be seen in Table 3, there was a decrease in the ΔE_{elec} value from 0.70 Kcal/mol to −41.4 Kcal/mol for PLpro–GRM and PLpro–OUB, respectively. In addition, the ΔG_{polar} term increased from 13.9 Kcal/mol to 61.9 Kcal/mol. These changes resulted in a large decrease in ΔG_{bind} from −29.7 Kcal/mol to −37.4 Kcal/mol, respectively. Our findings suggest that OUB is a more favorable inhibitor than the crystal inhibitor (GRM). This agrees with its reported activity against SARS-CoV-2, as well as the IC_{50} values in Vero and Calu cells (Table 1).

In the Mpro systems, Mpro–X77 showed the lowest ΔG_{bind} value (−42.8 Kcal/mol) in comparison with the Mpro–OUB (−28.7 Kcal/mol) systems. Particularly, the increase in the ΔE_{vdW} and ΔG_{polar} terms from −52.9 Kcal/mol to −41.9 Kcal/mol and from 39.1 Kcal/mol to 47.7 Kcal/mol, respectively, for Mpro–X77 and Mpro–OUB, explain the large increase in ΔG_{bind} for the Mpro systems.

Interestingly, for the RdRp systems, the RDV and DIG inhibitors showed similar ΔG_{bind} values. However, the free energy components were very different. As can be observed in Table 3, the ΔE_{elec} and ΔG_{polar} showed significant changes between these compounds. The ΔE_{elec} increased from −219.8 Kcal/mol to −34.6 Kcal/mol for RDV and DIG, respectively. This occurred due to the highest electronic density of RDV, mainly due to the presence of its phosphate group. On the other hand, the ΔG_{polar} decreased from 250.2 Kcal/mol to 76.8 Kcal/mol for RDV and DIG, respectively. Probably, it resulted due to compensation for the electrostatic effects of the electronic density of the RDV compound. Besides, the ΔG_{bind} of RDV (−13.0 Kcal/mol) is close to the value found by Srivastava et al. [40] (−16.6 Kcal/mol), which means that our RDV model is similar to the monophosphate RDV model that was simulated previously.

For the AAK1 systems, AAK1–LKB and AAK1–NIS showed very similar ΔG_{bind} values (−41.4 and 39.2 Kcal/mol, respectively). These results suggest that NIS can bind the catalytic site of the AAK1 enzyme in a similar mode to the crystallized inhibitor (LKB). The main energetic differences were found in the ΔE_{elec} and ΔG_{polar} terms. The first term increased from −31.3 Kcal/mol to −22.0 Kcal/mol and the second decreased from 40.3 Kcal/mol to 29.1 Kcal/mol for AAK1–LKB and AAK1–NIS, respectively.

In summary, if we consider the contributions of different ΔG_{bind} components of crystal and repurposed systems, the ΔE_{vdW} is a crucial term of the difference in the binding affinity in most systems, which agrees with the previous repurposed study [35]. Interestingly, our findings for the most promising repurposed compound (OUB) suggest that it can bind the PLpro enzyme with a binding affinity driven by the electrostatic interactions, with more evidence than any other repurposed system.

A residual decomposition analysis of ΔG_{bind} (Figure 9) was carried out to improve the understanding of the features that contributed to the recognition and binding in all the simulated systems. Here, any residue that contributed to the binding free energy values below −1.20 Kcal/mol was included as an important residue in the binding process.

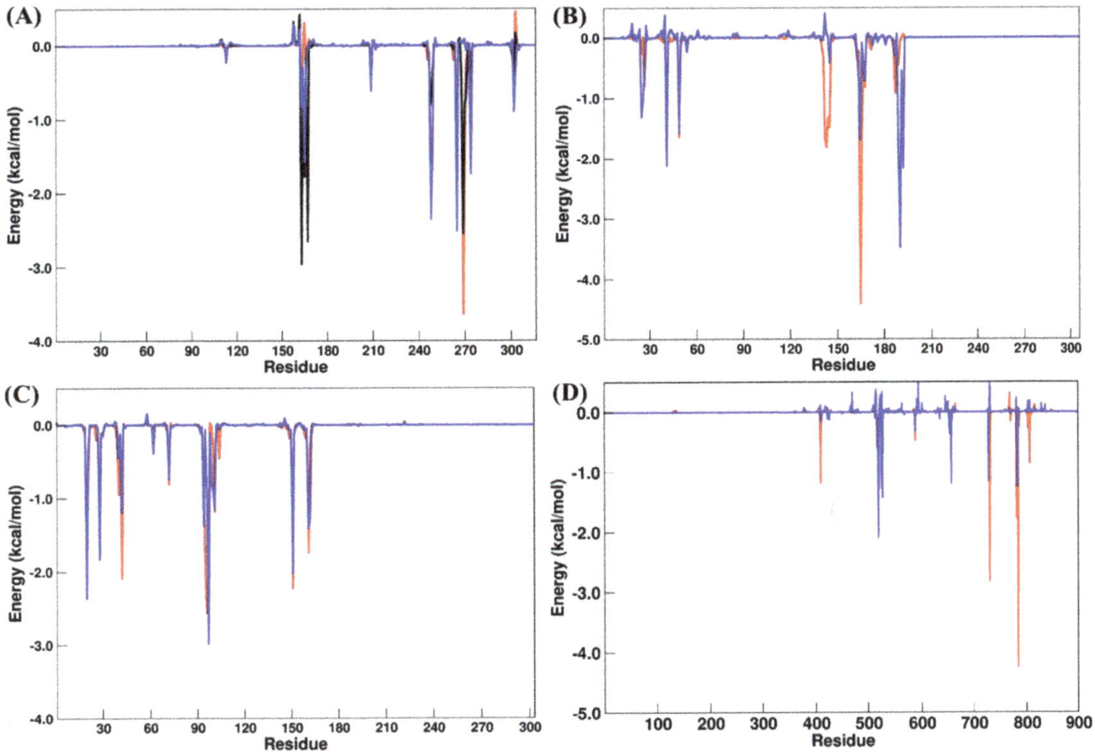

Figure 9. Residual decomposition plot for (**A**) PLpro (GRM, red; OUB, blue; SLM, black), (**B**) Mpro (X77, red; OUB, blue), (**C**) AAK1 (LKB, red; NIS, blue) and (**D**) RdRp (RDV, red; DIG, blue) systems. All values are reported in Kcal/mol.

For the PLpro systems (Table 4), the main residual differences occurred for Leu163, Asp165, and Arg167 when PLpro–GRM and PLpro–OUB were compared, while Asp165, Tyr269, and Gln270 had suitable differences in the PLpro–SLM system. These results suggest that the electrostatic shift caused by Asp165 and Arg167 explains the favorable binding energy found for the PLpro–OUB system.

Table 4. Residual decomposition analysis of binding free energies. All values are reported in Kcal/mol.

	PLpro systems		
	GRM	**OUB**	**SLM**
Leu163	−0.68	−2.96	−0.83
Asp165	0.31	−1.78	−1.58
Arg167	−0.12	−2.65	−0.02
Pro248	−1.60	−0.80	−2.35
Tyr265	−2.15	−1.75	−2.51
Tyr269	−3.64	−2.55	0.03
Gln270	−1.47	−0.81	0.09
	Mpro systems		
	X77	**OUB**	
His41	−1.55	−2.12	
Met49	−1.64	−1.56	
Asn142	−1.68	0.41	
Gly143	−1.81	0.07	
Ser144	−1.33	0.02	
Cys145	−1.48	−0.41	
Met165	−4.41	−1.69	
Glu166	−1.56	−0.09	
Gln189	−0.45	−1.98	
Thr190	−0.07	−3.47	
Gln192	0.05	−2.16	
	RdRp systems		
	RDV	**DIG**	
Ser550	−0.25	−2.08	
Val558	0.00	−1.42	
Thr688	0.03	−1.20	
Glu730	−2.82	0.30	
Cys731	−1.45	0.50	
Glu812	−1.77	0.24	
Ser815	−4.24	−1.02	
	AAK1 systems		
	LKB	**NIS**	
Leu20	−1.99	−2.36	
Val28	−1.65	−1.83	
Lys42	−2.09	−1.19	
Asp95	−2.11	−0.06	
Phe96	−2.56	−0.95	
Cys97	−1.69	−2.98	
Leu151	−2.23	−2.02	
Cys161	−1.74	−1.41	

In the Mpro systems, the increasing values for Asn142, Gly143, Ser144, Cys145, Met165, and Glu166 appear to explain the binding differences found for Mpro–X77 and Mpro–OUB, although an energetic decrease was observed for Thr190 and Gln192. On the RdRp systems, the main binding difference occurred in the electrostatic term; by considering the residual decomposition analysis, some polar residues had significant changes between the RDV and DIG systems. We can also highlight the increase in the residual energies in Glu730, Cys731, Glu812, and Ser815 from the RdRp–RDV to the RdRp–DIG system.

Finally, for the AKK1 systems, only Asp95 and Phe96 showed some significant energetic shifting, which was not enough to promote a large difference in the binding of NIS. These results suggest that both compounds (LKB and NIS) bind to the catalytic site of AKK1 with a similar binding affinity trend.

3. Computational Methods

3.1. *FDA-Approved Small Molecule Selection Rationale*

The PubChem database (https://pubchem.ncbi.nlm.nih.gov/ (accessed on 4 January 2022)) [50] (was used to retrieve the FDA-approved drugs database. Molecular Operating Environment (MOE) platform was utilized to perform energy optimization for 3D structures of each compound using the MMFF94x forcefield (with the gradient set to a root mean square (RMS) of 0.05 kcal/mo1). The structures of the drugs are shown in Scheme 1.

3.2. *Target Selection from Identified SARS-CoV-2 Protein Crystal Structures*

Although any NSP may be included as a therapeutic target, the crystal structure and reported ligand, as well as its critical role in viral infection, greatly boost the success chances. Based on this hypothesis and focusing on SARS-CoV-2 reported studies, it is worth noting the NSP structures accessible for this virus: the main protease (Mpro, NSP5) [51,52], the papain-like protease (PLpro, NSP3) [10,53], the RNA-dependent RNA polymerase (RdRp, NSP12) [54] in complex with cofactors NSP7 and NSP8, as well as a human target, adaptor-associated kinase 1 (AAK1) [14,55], which was involved in a molecular docking study.

3.3. *In Silico Docking Protocol Validation*

Before docking, native ligands in the active region of the active domains were re-docked to verify the docking process, which resulted in a docking pose with certain RMSD values. For the SARS-CoV-2 Mpro, PLpro, RdRp, and AAK1, and their respective cocrystallized ligands, the computed RMSD values were 1.483, 0.665, 0.883, and 0.321 Å, respectively. The validated docking methodologies were then utilized to investigate the ligand–protein interactions of the tested drugs at the target active site to anticipate their binding mode. The measured RMSD values of the tested drugs indicated the efficiency and validity of the docking process.

3.4. *Molecular Docking*

The molecular modeling simulations were performed using the Molecular Operating Environment (MOE, 2020.09) software. With the MMFF94x force field, all minimizations were carried out with MOE until an RMSD gradient of 0.05 Kcal/mol was reached, and the partial charges were calculated automatically. Three viral SARS-CoV-2 proteins: (i) The SARS-CoV-2 Mpro (PDB ID: 6W63, resolution: 2.16 Å), (ii) PLpro (PDB ID: 3MJ5, resolution: 2.65 Å) [56], (iii) RdRp (NSP 12, PDB ID: 7BV2, resolution: 2.50 Å) [57], in addition to a host protein (iv) AAK1 (PDB ID: 5L4Q, resolution: 1.97 Å) [58] were downloaded from the RCSB Protein Bank Database. The proteins were selected according to their therapeutic potential. First, water molecules were removed. Then, for the docking study, all protein preparation was performed in MOE with the default options.

The generated ligand conformations were docked using a validated docking methodology, and the Triangle Matcher placement method was used and scored based on London dG. The top best 50 poses were refined, and the energy was minimized in the binding site using the induced fit method and then reranked with the GBVI/WSA scoring function [59].

In all the selected systems, the cocrystallized ligands were used as a reference for the selected FDA drugs during the molecular docking procedures. Afterwards, the 3D and 2D ligand–receptor binding patterns of the selected compounds were investigated.

3.5. Molecular Dynamic (MD) Simulations and Binding Free Energy

As a powerful tool to avail in silico, the stability of the above-repurposed systems, classical molecular dynamics (MD) simulations were carried out by using the PMEMD module [60] of the Amber20 package. Preliminary, all ligands were structurally optimized at the quantum mechanics (QM) level by applying the Hartree–Fock (HF) method with 6-31G** as implemented into the Gaussian09 program [61]. Afterwards, MM charges for each molecule were computed using the restrained electrostatic potential (RESP) method [62] by using the antechamber module of AmberTools20 [63]. The GAFF [64] and ff19SB [65] MM parameter sets were chosen for ligands and enzymes, respectively. Particularly, for 3MJ5 and 7BV2 systems, the Zinc AMBER force field (ZAFF) [66] was selected to simulate their metal centers. In addition, the SWISS-MODEL web server [67] was used to add missing amino acid residues into the 7BV2 system. The protonation state of the ionizable residues of all systems, at pH = 7, was calculated on PROPKA 3.1 version [68] and protons were added by the tleap module of AmberTools20 [63]. Each system was solvated by the TIP3P water model [69], extending 8 Å away from the solute atoms. Next, the appropriate counter-ions were added in order to neutralize the charges in the solvated system.

Each solvated system was energetically minimized to avoid bad atomic contacts using the PMEMD module. The minimization was carried out in four successive stages by applying 5000 steepest descent steps followed by 5000 steps of the conjugate gradient method, where restraints were removed during the process. After successful minimization, each system was slowly heated up to 300 K over 100 ps under constant volume, where the solute was restricted with the positional restraints of 50 Kcal/mol·Å2. Next, maintaining the same solute restraints, 200 ps of MD was performed at the NPT ensemble (p = 1 atm and T = 298 K). Then, the force constant of the restraints was slowly removed over the eight stages of equilibration; each stage was carried out for 100 ps under the NPT ensemble. Finally, a 200 ns unrestrained MD simulation (named "production stage") was performed for each equilibrated system. The SHAKE algorithm [70] with the time step of 2 fs was applied for all the hydrogen bonds. The non-bonded cut-off set to 10 Å was used for the non-polar and polar interactions calculated using the Particle Mesh Ewald (PME) method [71]. The same MD protocol was used for all systems.

The structural analysis of all the simulated systems was evaluated by computing the root mean square deviation (RMSD) and root mean square fluctuation (RMSF) of the backbone atoms (Cα, N, O, C). Particularly, the RMSD calculations for each ligand were computed by considering only their respective heavy atoms. The trajectories were aligned by the main-chain atoms of the average structures from production stages by using the CPPTRAJ module [72].

The CPPTRAJ module [72] was used to extract 10 ns (a total of 1000 representative snapshots) from the production stage of the MD simulations on each system to be selected for the binding free energy calculations using the MM/GBSA approach [46–48], which was implemented into the *MMPBSA.py* module [49] of AmberTools20. The main equations of the chosen approach can be summarized as follows:

$$\Delta G_{bind} = \Delta G_{complex} - \left(\Delta G_{receptor} + \Delta G_{ligand}\right) \quad (1)$$

$$\Delta G_{bind} = \Delta E_{MM} + \Delta G_{SOLV} - T\Delta S \quad (2)$$

$$\Delta E_{MM} = \Delta E_{int} + \Delta E_{elec} + \Delta E_{vdw} \quad (3)$$

$$\Delta G_{SOLV} = \Delta G_{GB} + \Delta G_{SA} \quad (4)$$

where $\Delta G_{complex}$, $\Delta G_{receptor}$ and ΔG_{ligand} indicate the free energies of the complex, the receptor, and the ligand, respectively (Equation (1)). The ΔG_{bind} is obtained from the gas-phase MM energy (ΔE_{MM}), solvation energy (ΔG_{SOLV}) and the entropic term ($-T\Delta S$) (Equation (2)). The ΔE_{MM} includes the changes in the internal (bond, angles, and dihedral energies) (ΔE_{int}), electrostatic (ΔE_{elec}) and van der Waals (ΔE_{vdw}) contributions (Equation (3)). Here, as a single-trajectory scheme was used for the binding free energy calculations, the ΔE_{int} is equal to zero. The ΔG_{SOLV} is the sum of the polar (ΔG_{GB}) and non-polar (ΔG_{SA}) energies for ΔG_{bind} (Equation (4)). To reduce the computational cost, the entropic term ($-T\Delta S$) was not computed into the binding free energy calculations [47,48]. Furthermore, a per-residual decomposition analysis was included to avail the relative contribution of each amino acid residue [46]. This method has been frequently applied as a useful tool in SARS-CoV-2 drug design studies [35,40,41,73–76].

4. Conclusions and Prospects

A set of comprehensive docking approaches and molecular dynamics simulations were performed to find viable drugs for inhibiting SARS-CoV-2 targets. Accordingly, based on docking and the MD simulation results, Digitoxin was postulated as an effective RNA-dependent RNA polymerase inhibitor compared to Remdesivir as a known inhibitor. It was concluded that Salinomycin may act as a PLpro inhibitor, while Ouabain can share the activity with Salinomycin and Proscillaridin as a dual inhibitor of both PLpro and Mpro SARS-CoV-2, respectively. In addition, Niclosamide was proposed to be an inhibitor for adaptor-associated kinase 1. The insights provided by the present study may substantiate the valuable exploration and development of anti-SARS-CoV-2 therapeutic agents from FDA-approved drugs, to be used as leads for further drug development. Furthermore, drug repositioning can be considered an efficient tool as a source of new analogs to help defeat COVID-19.

Author Contributions: Conceptualization, W.S.Q. and J.R.A.S.; methodology, W.S.Q. and J.R.A.S.; software, W.S.Q., R.S.F. and J.R.A.S.; validation, W.S.Q., R.S.F. and J.R.A.S.; formal analysis, W.S.Q. and J.R.A.S.; investigation, W.S.Q., R.S.F. and J.R.A.S.; resources, J.R.A.S.; data curation, W.S.Q. and J.R.A.S.; writing—original draft preparation, W.S.Q. and J.R.A.S.; writing—review and editing, W.S.Q. and J.R.A.S.; visualization, W.S.Q., R.S.F. and J.R.A.S.; supervision, W.S.Q. and J.R.A.S.; funding acquisition, J.R.A.S. All authors have read and agreed to the published version of the manuscript.

Funding: This study was granted access to the computational resources of the South African Centre for High-Performance Computing (https://www.chpc.ac.za/), the Hippo at the University of KwaZulu-Natal (https://astro.ukzn.ac.za/~hippo/) and the Centro de Computação de Alto Desempenho of Universidade Federal do Pará (https://www.ccad.ufpa.br/). R.S.F. thanks the CAPES agency for the financial support of the Master's scholarship (process #88887.495149/2020-00). R.S.F. and J.R.A.S. thank PROPESP/UFPA for payment of the publication fee.

Institutional Review Board Statement: Not applicable.

Informed Consent Statement: Not applicable.

Data Availability Statement: Any data related to this work might be available from the corresponding author.

Conflicts of Interest: There are no conflict to declare.

Sample Availability: Simulated PDB files are available from the authors.

References

1. Zhu, N.; Zhang, D.; Wang, W.; Li, X.; Yang, B.; Song, J.; Zhao, X.; Huang, B.; Shi, W.; Lu, R.; et al. A novel coronavirus from patients with pneumonia in China, 2019. *N. Engl. J. Med.* **2020**, *382*, 727–733. [CrossRef] [PubMed]
2. Arshad Ali, S.; Baloch, M.; Ahmed, N.; Arshad Ali, A.; Iqbal, A. The outbreak of Coronavirus Disease 2019 (COVID-19)—An emerging global health threat. *J. Infect. Public Health* **2020**, *13*, 644–646. [CrossRef] [PubMed]
3. Guy, R.K.; DiPaola, R.S.; Romanelli, F.; Dutch, R.E. Rapid repurposing of drugs for COVID-19. *Science* **2020**, *368*, 829–830. [CrossRef]
4. WHO Coronavirus Disease (COVID-19). Available online: https://covid19.who.int/ (accessed on 4 January 2022).

5. Chakraborty, I.; Maity, P. COVID-19 outbreak: Migration, effects on society, global environment and prevention. *Sci. Total Environ.* **2020**, *728*, 138882. [CrossRef]
6. Vellingiri, B.; Jayaramayya, K.; Iyer, M.; Narayanasamy, A.; Govindasamy, V.; Giridharan, B.; Ganesan, S.; Venugopal, A.; Venkatesan, D.; Ganesan, H.; et al. COVID-19: A promising cure for the global panic. *Sci. Total Environ.* **2020**, *725*, 138277. [CrossRef]
7. Malik, J.A.; Ahmed, S.; Mir, A.; Shinde, M.; Bender, O.; Alshammari, F.; Ansari, M.; Anwar, S. The SARS-CoV-2 mutations versus vaccine effectiveness: New opportunities to new challenges. *J. Infect. Public Health* **2022**, *15*, 228–240. [CrossRef]
8. Schultze, J.L.; Aschenbrenner, A.C. COVID-19 and the human innate immune system. *Cell* **2021**, *184*, 1671–1692. [CrossRef]
9. Kiani, A.K.; Dhuli, K.; Anpilogov, K.; Bressan, S.; Dautaj, A.; Dundar, M.; Beccari, T.; Ergoren, M.C.; Bertelli, M. Natural compounds as inhibitors of SARS-CoV-2 endocytosis: A promising approach against COVID-19. *Acta Biomed.* **2020**, *91*, e2020008. [CrossRef]
10. Wu, C.; Liu, Y.; Yang, Y.; Zhang, P.; Zhong, W.; Wang, Y.; Wang, Q.; Xu, Y.; Li, M.; Li, X.; et al. Analysis of therapeutic targets for SARS-CoV-2 and discovery of potential drugs by computational methods. *Acta Pharm. Sin. B* **2020**, *10*, 766–788. [CrossRef]
11. Prajapat, M.; Sarma, P.; Shekhar, N.; Avti, P.; Sinha, S.; Kaur, H.; Kumar, S.; Bhattacharyya, A.; Kumar, H.; Bansal, S.; et al. Drug for corona virus: A systematic review. *Indian J. Pharmacol.* **2020**, *52*, 56. [CrossRef]
12. Subissi, L.; Imbert, I.; Ferron, F.; Collet, A.; Coutard, B.; Decroly, E.; Canard, B. SARS-CoV ORF1b-encoded nonstructural proteins 12–16: Replicative enzymes as antiviral targets. *Antivir. Res.* **2014**, *101*, 122–130. [CrossRef] [PubMed]
13. Pillaiyar, T.; Laufer, S. Kinases as potential therapeutic targets for anti-coronaviral therapy. *J. Med. Chem.* **2022**, *65*, 955–982. [CrossRef] [PubMed]
14. Naik, R.R.; Shakya, A.K.; Aladwan, S.M.; El-Tanani, M. Kinase inhibitors as potential therapeutic agents in the treatment of COVID-19. *Front. Pharmacol.* **2022**, *13*, 806568. [CrossRef]
15. Talevi, A. Drug repurposing. In *Comprehensive Pharmacology*; Kenakin, T.B.T.-C.P., Ed.; Elsevier: Oxford, UK, 2022; pp. 813–824. ISBN 978-0-12-820876-2.
16. Sayed, A.M.; Khalaf, A.M.; Abdelrahim, M.E.A.; Elgendy, M.O. Repurposing of some anti-infective drugs for COVID-19 treatment: A surveillance study supported by an in silico investigation. *Int. J. Clin. Pract.* **2021**, *75*, e13877. [CrossRef] [PubMed]
17. Selvaraj, C.; Panwar, U.; Dinesh, D.C.; Boura, E.; Singh, P.; Dubey, V.K.; Singh, S.K. Microsecond MD simulation and multiple-conformation virtual screening to identify potential anti-COVID-19 inhibitors against SARS-CoV-2 main protease. *Front. Chem.* **2021**, *8*, 595273. [CrossRef] [PubMed]
18. Arantes, P.R.; Saha, A.; Palermo, G. Fighting COVID-19 using molecular dynamics simulations. *ACS Cent. Sci.* **2020**, *6*, 1654–1656. [CrossRef]
19. Kandeel, M.; Abdelrahman, A.H.M.; Oh-Hashi, K.; Ibrahim, A.; Venugopala, K.N.; Morsy, M.A.; Ibrahim, M.A.A. Repurposing of FDA-approved antivirals, antibiotics, anthelmintics, antioxidants, and cell protectives against SARS-CoV-2 papain-like protease. *J. Biomol. Struct. Dyn.* **2021**, *39*, 5129–5136. [CrossRef]
20. Ibrahim, M.A.A.; Abdeljawaad, K.A.A.; Abdelrahman, A.H.M.; Hegazy, M.-E.F. Natural-like products as potential SARS-CoV-2 M pro inhibitors: In-silico drug discovery. *J. Biomol. Struct. Dyn.* **2021**, *39*, 5722–5734. [CrossRef]
21. Ibrahim, M.A.A.; Abdelrahman, A.H.M.; Mohamed, T.A.; Atia, M.A.M.; Al-Hammady, M.A.M.; Abdeljawaad, K.A.A.; Elkady, E.M.; Moustafa, M.F.; Alrumaihi, F.; Allemailem, K.S.; et al. In silico mining of terpenes from red-sea invertebrates for SARS-CoV-2 main protease (M^{pro}) inhibitors. *Molecules* **2021**, *26*, 2082. [CrossRef]
22. Ibrahim, M.A.A.; Abdelrahman, A.H.M.; Allemailem, K.S.; Almatroudi, A.; Moustafa, M.F.; Hegazy, M.-E.F. In silico evaluation of prospective anti-COVID-19 drug candidates as potential SARS-CoV-2 main protease inhibitors. *Protein J.* **2021**, *40*, 296–309. [CrossRef]
23. Ng, Y.L.; Salim, C.K.; Chu, J.J.H. Drug repurposing for COVID-19: Approaches, challenges and promising candidates. *Pharmacol. Ther.* **2021**, *228*, 107930. [CrossRef] [PubMed]
24. Kim, S.; Chen, J.; Cheng, T.; Gindulyte, A.; He, J.; He, S.; Li, Q.; Shoemaker, B.A.; Thiessen, P.A.; Yu, B.; et al. PubChem in 2021: New data content and improved web interfaces. *Nucleic Acids Res.* **2021**, *49*, D1388–D1395. [CrossRef] [PubMed]
25. Abhinand, C.S.; Nair, A.S.; Krishnamurthy, A.; Oommen, O.V.; Sudhakaran, P.R. Potential protease inhibitors and their combinations to block SARS-CoV-2. *J. Biomol. Struct. Dyn.* **2022**, *40*, 903–917. [CrossRef]
26. Daoud, S.; Alabed, S.J.; Dahabiyeh, L.A. Identification of potential COVID-19 main protease inhibitors using structure-based pharmacophore approach, molecular docking and repurposing studies. *Acta Pharm.* **2021**, *71*, 163–174. [CrossRef]
27. Ibrahim, T.M.; Ismail, M.I.; Bauer, M.R.; Bekhit, A.A.; Boeckler, F.M. Supporting SARS-CoV-2 papain-like protease drug discovery: In silico methods and benchmarking. *Front. Chem.* **2020**, *8*, 592289. [CrossRef] [PubMed]
28. Gao, Y.; Yan, L.; Huang, Y.; Liu, F.; Zhao, Y.; Cao, L.; Wang, T.; Sun, Q.; Ming, Z.; Zhang, L.; et al. Structure of the RNA-dependent RNA polymerase from COVID-19 virus. *Science* **2020**, *368*, 779–782. [CrossRef] [PubMed]
29. García-Cárceles, J.; Caballero, E.; Gil, C.; Martínez, A. Kinase inhibitors as underexplored antiviral agents. *J. Med. Chem.* **2022**, *65*, 935–954. [CrossRef]
30. Ghosh, A.K.; Takayama, J.; Rao, K.V.; Ratia, K.; Chaudhuri, R.; Mulhearn, D.C.; Lee, H.; Nichols, D.B.; Baliji, S.; Baker, S.C.; et al. Severe acute respiratory syndrome coronavirus papain-like novel protease inhibitors: Design, synthesis, protein-ligand X-ray structure and biological evaluation. *J. Med. Chem.* **2010**, *53*, 4968–4979. [CrossRef]

31. Yin, W.; Mao, C.; Luan, X.; Shen, D.-D.; Shen, Q.; Su, H.; Wang, X.; Zhou, F.; Zhao, W.; Gao, M.; et al. Structural basis for inhibition of the RNA-dependent RNA polymerase from SARS-CoV-2 by remdesivir. *Science* **2020**, *368*, 1499–1504. [CrossRef]
32. Verdonck, S.; Pu, S.-Y.; Sorrell, F.J.; Elkins, J.M.; Froeyen, M.; Gao, L.-J.; Prugar, L.I.; Dorosky, D.E.; Brannan, J.M.; Barouch-Bentov, R.; et al. Synthesis and structure-activity relationships of 3,5-disubstituted-pyrrolo[2,3-*b*]pyridines as inhibitors of adaptor-associated kinase 1 with antiviral activity. *J. Med. Chem.* **2019**, *62*, 5810–5831. [CrossRef]
33. Wojciechowski, M.; Lesyng, B. Generalized born model: Analysis, refinement, and applications to proteins. *J. Phys. Chem. B* **2004**, *108*, 18368–18376. [CrossRef]
34. Case, D.A.; Cheatham, T.E.; Darden, T.; Gohlke, H.; Luo, R.; Merz, K.M.; Onufriev, A.; Simmerling, C.; Wang, B.; Woods, R.J. The amber biomolecular simulation programs. *J. Comput. Chem.* **2005**, *26*, 1668–1688. [CrossRef] [PubMed]
35. Frisch, M.J.; Trucks, G.W.; Schlegel, H.B.; Scuseria, G.E.; Robb, M.A.; Cheeseman, J.R.; Scalmani, G.; Barone, V.; Mennucci, B.; Petersson, G.A.; et al. *Gaussian 09, Revision E. 01*; Gaussian: Wallingford, CT, USA, 2009; ISBN 9781935522027.
36. Wang, J.; Cieplak, P.; Kollman, P.A. How well does a restrained electrostatic potential (RESP) model perform in calculating conformational energies of organic and biological molecules? *J. Comput. Chem.* **2000**, *21*, 1049–1074. [CrossRef]
37. Case, D.A.; Ben-Shalom, I.Y.; Brozell, S.R.; Cerutti, D.S., III; Cruzeiro, V.W.D.; Darden, T.A.; Duke, R.E.; Ghoreishi, D.; Gilson, M.K. *AMBER 2020*; University of California: San Francisco, CA, USA, 2020.
38. Wang, J.; Wolf, R.M.; Caldwell, J.W.; Kollman, P.A.; Case, D.A. Development and testing of a general amber force field. *J. Comput. Chem.* **2004**, *25*, 1157–1174. [CrossRef]
39. Tian, C.; Kasavajhala, K.; Belfon, K.A.A.; Raguette, L.; Huang, H.; Migues, A.N.; Bickel, J.; Wang, Y.; Pincay, J.; Wu, Q.; et al. ff19SB: Amino-acid-specific protein backbone parameters trained against quantum mechanics energy surfaces in solution. *J. Chem. Theory Comput.* **2020**, *16*, 528–552. [CrossRef]
40. Peters, M.B.; Yang, Y.; Wang, B.; Füsti-Molnár, L.; Weaver, M.N.; Merz, K.M. Structural survey of zinc-containing proteins and development of the zinc AMBER force field (ZAFF). *J. Chem. Theory Comput.* **2010**, *6*, 2935–2947. [CrossRef]
41. Waterhouse, A.; Bertoni, M.; Bienert, S.; Studer, G.; Tauriello, G.; Gumienny, R.; Heer, F.T.; de Beer, T.A.P.; Rempfer, C.; Bordoli, L.; et al. SWISS-MODEL: Homology modelling of protein structures and complexes. *Nucleic Acids Res.* **2018**, *46*, W296–W303. [CrossRef]
42. Søndergaard, C.R.; Olsson, M.H.M.; Rostkowski, M.; Jensen, J.H. Improved treatment of ligands and coupling effects in empirical calculation and rationalization of pK_a values. *J. Chem. Theory Comput.* **2011**, *7*, 2284–2295. [CrossRef]
43. Jorgensen, W.L.; Chandrasekhar, J.; Madura, J.D.; Impey, R.W.; Klein, M.L. Comparison of simple potential functions for simulating liquid water. *J. Chem. Phys.* **1983**, *79*, 926–935. [CrossRef]
44. Ryckaert, J.-P.; Ciccotti, G.; Berendsen, H.J. Numerical integration of the cartesian equations of motion of a system with constraints: Molecular dynamics of n-alkanes. *J. Comput. Phys.* **1977**, *23*, 327–341. [CrossRef]
45. Darden, T.; York, D.; Pedersen, L. Particle mesh Ewald: An *N* log(*N*) method for Ewald sums in large systems. *J. Chem. Phys.* **1993**, *98*, 10089–10092. [CrossRef]
46. Roe, D.R.; Cheatham, T.E. PTRAJ and CPPTRAJ: Software for processing and analysis of molecular dynamics trajectory data. *J. Chem. Theory Comput.* **2013**, *9*, 3084–3095. [CrossRef] [PubMed]
47. Srinivasan, J.; Cheatham, T.E.; Cieplak, P.; Kollman, P.A.; Case, D.A. Continuum solvent studies of the stability of DNA, RNA, and phosphoramidate—DNA helices. *J. Am. Chem. Soc.* **1998**, *120*, 9401–9409. [CrossRef]
48. Genheden, S.; Ryde, U. The MM/PBSA and MM/GBSA methods to estimate ligand-binding affinities. *Expert Opin. Drug Discov.* **2015**, *10*, 449–461. [CrossRef] [PubMed]
49. Wang, E.; Sun, H.; Wang, J.; Wang, Z.; Liu, H.; Zhang, J.Z.H.; Hou, T. End-point binding free energy calculation with MM/PBSA and MM/GBSA: Strategies and applications in drug design. *Chem. Rev.* **2019**, *119*, 9478–9508. [CrossRef]
50. Miller, B.R.; McGee, T.D.; Swails, J.M.; Homeyer, N.; Gohlke, H.; Roitberg, A.E. MMPBSA.py: An efficient program for end-state free energy calculations. *J. Chem. Theory Comput.* **2012**, *8*, 3314–3321. [CrossRef]
51. Keretsu, S.; Bhujbal, S.P.; Cho, S.J. Rational approach toward COVID-19 main protease inhibitors via molecular docking, molecular dynamics simulation and free energy calculation. *Sci. Rep.* **2020**, *10*, 17716. [CrossRef]
52. Srivastava, M.; Mittal, L.; Kumari, A.; Asthana, S. Molecular dynamics simulations reveal the interaction fingerprint of remdesivir triphosphate pivotal in allosteric regulation of SARS-CoV-2 RdRp. *Front. Mol. Biosci.* **2021**, *8*, 639614. [CrossRef]
53. Sitthiyotha, T.; Chunsrivirot, S. Computational design of SARS-CoV-2 peptide binders with better predicted binding affinities than human ACE2 receptor. *Sci. Rep.* **2021**, *11*, 15650. [CrossRef]
54. Koulgi, S.; Jani, V.; Uppuladinne, M.V.N.; Sonavane, U.; Joshi, R. Remdesivir-bound and ligand-free simulations reveal the probable mechanism of inhibiting the RNA dependent RNA polymerase of severe acute respiratory syndrome coronavirus 2. *RSC Adv.* **2020**, *10*, 26792–26803. [CrossRef]
55. Yu, W.; Wu, X.; Zhao, Y.; Chen, C.; Yang, Z.; Zhang, X.; Ren, J.; Wang, Y.; Wu, C.; Li, C.; et al. Computational simulation of HIV protease inhibitors to the main protease (Mpro) of SARS-CoV-2: Implications for COVID-19 drugs design. *Molecules* **2021**, *26*, 7385. [CrossRef] [PubMed]
56. Silva, J.R.A.; Urban, J.; Araújo, E.; Lameira, J.; Moliner, V.; Alves, C.N. Exploring the catalytic mechanism of the RNA cap modification by nsp16-nsp10 complex of SARS-CoV-2 through a QM/MM approach. *Int. J. Mol. Sci.* **2022**, *23*, 300. [CrossRef]
57. Silva, J.R.A.; Kruger, H.G.; Molfetta, F.A. Drug repurposing and computational modeling for discovery of inhibitors of the main protease (Mpro) of SARS-CoV-2. *RSC Adv.* **2021**, *11*, 23450–23458. [CrossRef] [PubMed]

58. Jeon, S.; Ko, M.; Lee, J.; Choi, I.; Byun, S.Y.; Park, S.; Shum, D.; Kim, S. Identification of antiviral drug candidates against SARS-CoV-2 from FDA-approved drugs. *Antimicrob. Agents Chemother.* **2020**, *64*, e00819-20. [CrossRef] [PubMed]
59. Ko, M.; Jeon, S.; Ryu, W.-S.; Kim, S. Comparative analysis of antiviral efficacy of FDA-approved drugs against SARS-CoV-2 in human lung cells. *J. Med. Virol.* **2021**, *93*, 1403–1408. [CrossRef]
60. Ferreira, L.; dos Santos, R.; Oliva, G.; Andricopulo, A. Molecular docking and structure-based drug design strategies. *Molecules* **2015**, *20*, 13384–13421. [CrossRef]
61. Ibrahim, M.A.A.; Abdelrahman, A.H.M.; Hegazy, M.-E.F. In-silico drug repurposing and molecular dynamics puzzled out potential SARS-CoV-2 main protease inhibitors. *J. Biomol. Struct. Dyn.* **2021**, *39*, 5756–5767. [CrossRef]
62. Pinzi, L.; Rastelli, G. Molecular docking: Shifting paradigms in drug discovery. *Int. J. Mol. Sci.* **2019**, *20*, 4331. [CrossRef]
63. Kitchen, D.B.; Decornez, H.; Furr, J.R.; Bajorath, J. Docking and scoring in virtual screening for drug discovery: Methods and applications. *Nat. Rev. Drug Discov.* **2004**, *3*, 935–949. [CrossRef]
64. Majerová, T.; Novotný, P. Precursors of viral proteases as distinct drug targets. *Viruses* **2021**, *13*, 1981. [CrossRef]
65. Shin, D.; Mukherjee, R.; Grewe, D.; Bojkova, D.; Baek, K.; Bhattacharya, A.; Schulz, L.; Widera, M.; Mehdipour, A.R.; Tascher, G.; et al. Papain-like protease regulates SARS-CoV-2 viral spread and innate immunity. *Nature* **2020**, *587*, 657–662. [CrossRef] [PubMed]
66. Alamri, M.A.; Tahir ul Qamar, M.; Mirza, M.U.; Alqahtani, S.M.; Froeyen, M.; Chen, L.-L. Discovery of human coronaviruses pan-papain-like protease inhibitors using computational approaches. *J. Pharm. Anal.* **2020**, *10*, 546–559. [CrossRef] [PubMed]
67. Pindiprolu, S.K.S.; Kumar, C.S.P.; Golla, V.S.K.; Likitha, P.; Chandra, S.; SK, E.B.; Ramachandra, R.K. Pulmonary delivery of nanostructured lipid carriers for effective repurposing of salinomycin as an antiviral agent. *Med. Hypotheses* **2020**, *143*, 109858. [CrossRef] [PubMed]
68. Chen, H.; Wei, P.; Huang, C.; Tan, L.; Liu, Y.; Lai, L. Only one protomer is active in the dimer of SARS 3C-like proteinase. *J. Biol. Chem.* **2006**, *281*, 13894–13898. [CrossRef] [PubMed]
69. Tam, N.M.; Nam, P.C.; Quang, D.T.; Tung, N.T.; Vu, V.V.; Ngo, S.T. Binding of inhibitors to the monomeric and dimeric SARS-CoV-2 Mpro. *RSC Adv.* **2021**, *11*, 2926–2934. [CrossRef]
70. Tam, N.M.; Pham, M.Q.; Ha, N.X.; Nam, P.C.; Phung, H.T.T. Computational estimation of potential inhibitors from known drugs against the main protease of SARS-CoV-2. *RSC Adv.* **2021**, *11*, 17478–17486. [CrossRef]
71. Dömling, A.; Gao, L. Chemistry and biology of SARS-CoV-2. *Chemistry* **2020**, *6*, 1283–1295. [CrossRef]
72. Parvez, M.S.A.; Karim, M.A.; Hasan, M.; Jaman, J.; Karim, Z.; Tahsin, T.; Hasan, M.N.; Hosen, M.J. Prediction of potential inhibitors for RNA-dependent RNA polymerase of SARS-CoV-2 using comprehensive drug repurposing and molecular docking approach. *Int. J. Biol. Macromol.* **2020**, *163*, 1787–1797. [CrossRef]
73. Kadri, H.; Lambourne, O.A.; Mehellou, Y. Niclosamide, a drug with many (re)purposes. *ChemMedChem* **2018**, *13*, 1088–1091. [CrossRef]
74. Darif, D.; Hammi, I.; Kihel, A.; El Idrissi Saik, I.; Guessous, F.; Akarid, K. The pro-inflammatory cytokines in COVID-19 pathogenesis: What goes wrong? *Microb. Pathog.* **2021**, *153*, 104799. [CrossRef]
75. Karim, M.; Saul, S.; Ghita, L.; Sahoo, M.K.; Ye, C.; Bhalla, N.; Lo, C.-W.; Jin, J.; Park, J.-G.; Martinez-Gualda, B.; et al. Numb-associated kinases are required for SARS-CoV-2 infection and are cellular targets for antiviral strategies. *Antivir. Res.* **2022**, *204*, 105367. [CrossRef] [PubMed]
76. Tsai, Y.C.; Tsai, T.F. Oral disease-modifying antirheumatic drugs and immunosuppressants with antiviral potential, including SARS-CoV-2 infection: A review. *Ther. Adv. Musculoskelet. Dis.* **2020**, *12*, 1759720X2094729. [CrossRef] [PubMed]

molecules

MDPI

Article

Correlates with Vaccine Protective Capacity and COVID-19 Disease Symptoms Identified by Serum Proteomics in Vaccinated Individuals

Margarita Villar [1,2], José Miguel Urra [3,4], Sara Artigas-Jerónimo [1], Lorena Mazuecos [1], Marinela Contreras [1], Rita Vaz-Rodrigues [1], Francisco J. Rodríguez-del-Río [5], Christian Gortázar [1] and José de la Fuente [1,6,*]

1 SaBio, Instituto de Investigación en Recursos Cinegéticos IREC-CSIC-UCLM-JCCM, Ronda de Toledo 12, 13005 Ciudad Real, Spain
2 Biochemistry Section, Faculty of Science and Chemical Technologies, and Regional Centre for Biomedical Research (CRIB), University of Castilla-La Mancha, 13071 Ciudad Real, Spain
3 Immunology, Hospital General Universitario de Ciudad Real, 13005 Ciudad Real, Spain
4 Medicine School, Universidad de Castilla la Mancha (UCLM), 13005 Ciudad Real, Spain
5 Local Medical Service Horcajo de los Montes, 13110 Ciudad Real, Spain
6 Department of Veterinary Pathobiology, Center for Veterinary Health Sciences, Oklahoma State University, Stillwater, OK 74078, USA
* Correspondence: jose_delafuente@yahoo.com

Citation: Villar, M.; Urra, J.M.; Artigas-Jerónimo, S.; Mazuecos, L.; Contreras, M.; Vaz-Rodrigues, R.; Rodríguez-del-Río, F.J.; Gortázar, C.; de la Fuente, J. Correlates with Vaccine Protective Capacity and COVID-19 Disease Symptoms Identified by Serum Proteomics in Vaccinated Individuals. *Molecules* **2022**, *27*, 5933. https://doi.org/10.3390/molecules27185933

Academic Editors: Giovanni N. Roviello and Caterina Vicidomini

Received: 1 August 2022
Accepted: 7 September 2022
Published: 13 September 2022

Publisher's Note: MDPI stays neutral with regard to jurisdictional claims in published maps and institutional affiliations.

Abstract: In the last two years, the coronavirus disease 19 (COVID-19) pandemic caused by severe acute respiratory syndrome coronavirus 2 (SARS-CoV-2) has been a scientific and social challenge worldwide. Vaccines have been the most effective intervention for reducing virus transmission and disease severity. However, genetic virus variants are still circulating among vaccinated individuals with different disease symptomatology. Understanding the protective- or disease-associated mechanisms in vaccinated individuals is relevant to advances in vaccine development and implementation. To address this objective, serum-protein profiles were characterized by quantitative proteomics and data-analysis algorithms in four cohorts of uninfected and SARS-CoV-2-infected vaccinated individuals with asymptomatic, non-severe, and severe disease symptomatology. The results show that immunoglobulins were the most overrepresented proteins in infected cohorts when compared to PCR-negative individuals. The immunoglobulin profile varied between different infected cohorts and correlated with protective- or disease-associated capacity. Overrepresented immunoglobulins in PCR-positive individuals correlated with protective response against SARS-CoV-2, other viruses, and thrombosis in asymptomatic cases. In non-severe cases, correlates of protection against SARS-CoV-2 and HBV together with risk of myasthenia gravis and allergy and autoantibodies were observed. Patients with severe symptoms presented risk for allergy, chronic idiopathic thrombocytopenic purpura, and autoantibodies. The analysis of underrepresented immunoglobulins in PCR-positive compared to PCR-negative individuals identified vaccine-induced protective epitopes in various coronavirus proteins, including the spike receptor-binding domain RBD. Non-immunoglobulin proteins were associated with COVID-19 symptoms and biological processes. These results evidence host-associated differences in response to vaccination and the possibility of improving vaccine efficacy against SARS-CoV-2.

Keywords: COVID-19; proteomic; vaccine; immunology; biomarker

1. Introduction

Millions of deaths have been reported worldwide associated with coronavirus disease 19 (COVID-19), a pandemic caused by severe acute respiratory syndrome coronavirus 2 (SARS-CoV-2) (https://www.google.com/search?client=safari&rls=en&q=COVID-19+worldwide+cases&ie=UTF-8&oe=UTF-8#colocmid=/m/02j71&coasync=0).

Furthermore, the global number of deaths caused by COVID-19 may be up to four times this figure [1]. Vaccines using different platforms have been developed as the most safe and effective intervention for reducing SARS-CoV-2-virus transmission and disease severity [2,3]. However, genetic variants of the coronavirus are still circulating among vaccinated individuals with different disease symptomatology [4,5]. Understanding the protective- or disease host-associated mechanisms in vaccinated individuals is relevant to advances in vaccine development and implementation [6].

To address this challenge, understanding the protective- or disease-associated mechanisms in vaccinated individuals is relevant to advances in vaccine development and implementation. Previous proteomics studies, e.g., [7,8], have addressed response to infection and vaccination, but our study addressed for the first time the immune response to COVID-19 vaccination in uninfected and SARS-CoV-2-infected individuals with asymptomatic, non-severe, and severe disease symptomatology. In this study, serum-protein profiles were characterized by previously validated quantitative proteomics [7] in four cohorts of uninfected and SARS-CoV-2-infected vaccinated individuals with asymptomatic, non-severe, and severe disease symptomatology. The results evidence host-associated differences in response to vaccination and the possibility of advances in vaccine development and implementation against SARS-CoV-2.

2. Materials and Methods

2.1. Study Design with Serum Samples from Different Cohorts

The study design is described in Figure 1. A retrospective case-control study was conducted in patients suffering from COVID-19 and healthy controls sampled at the University General Hospital of Ciudad Real (HGUCR), Spain [8]. Individuals were confirmed as SARS-CoV-2-infected by reverse transcriptase-polymerase chain reaction (RT-PCR) and sampled between November and December 2021 (Table 1). In this study with individuals vaccinated against COVID-19, vaccine administration, clinical symptoms, and laboratory determinations associated with COVID-19 were obtained from patients' medical records to create cohorts of PCR– and PCR+ asymptomatic, non-severe, and severe individuals (Table 1). Patient symptoms can be found in Table 1. Blood samples were drawn in a vacutainer tube without anticoagulant. The tube remained at rest for 15–30 min at room temperature (RT) for clotting. Subsequently, the tube was centrifuged at 1500× g for 10 min at RT to remove the clot and obtain serum. Serum samples were heat-inactivated for 30 min at 56 °C and conserved at −20 °C until used for analysis. The use of samples and individuals' data was approved by the Ethical and Scientific Committees (University Hospital of Ciudad Real C-352 and SESCAM C-73).

Table 1. Data on serum samples included in the analysis.

Lab, Proteome ID	Age (y/o)	Sex	Sample Date (*)	Vaccine Doses	SARS-CoV-2-Neutralizing Antibodies	Dates of Vaccination
			Cohort PCR–			
16107663, C1	79	F	15.12.2021 (247 days)	Pfizer Pfizer Moderna	95.2%	13.04.2021 11.05.2021 11.11.2021
16107723, C2	54	M	15.12.2021 (323 days)	Pfizer Pfizer	32.0%	27.01.2021 22.02.2021
39385248, C3	60	F	15.12.2021 (323 days)	Pfizer Pfizer Moderna	95.3%	27.01.2021 22.02.2021 30.11.2021
39385665, C4	82	F	15.12.2021 (255 days)	Pfizer Pfizer Pfizer	92.4%	05.04.2021 26.04.2021 08.11.2021
39386122, C5	54	F	16.12.2021 (342 days)	Pfizer Pfizer	2.3% (negative)	09.01.2021 30.01.2021

Table 1. *Cont.*

Lab, Proteome ID	Age (y/o)	Sex	Sample Date (*)	Vaccine Doses	SARS-CoV-2-Neutralizing Antibodies	Dates of Vaccination
			Cohort PCR+ Asymptomatic			
16107241, A1	84	F	09.12.2021 (255 days)	Pfizer Pfizer	39.4%	30.03.2021 20.04.2021
1433003, A2	88	F	12.12.2021 (272 days)	Pfizer Pfizer Pfizer	95.6%	16.03.2021 06.04.2021 29.10.2021
161082999, A3	89	F	23.12.2021 (351 days)	Pfizer Pfizer Pfizer	51.4%	07.01.2021 28.01.2021 04.10.2021
1437141, A4	68	M	26.12.2021 (229 days)	Pfizer Pfizer Pfizer	94.1%	12.05.2021 04.06.2021 24.11.2021
88403647, A5	46	F	30.12.2021 (205 days)	Pfizer Pfizer	70.2%	09.06.2021 30.06.2021
			Cohort PCR+ Non-severe Hospital Discharge. Symptoms: fever, cough			
1429191, L1	41	F	28.11.2021 (170 days)	Pfizer	95.6%	12.06.2021
1433753, L2	19	M	14.12.2021 (140 days)	Pfizer Pfizer	75.6%	28.07.2021 18.08.2021
1433789, L3	25	F	14.12.2021 (103 days)	Pfizer	45.0%	03.09.2021
1435300, L4	26	M	20.12.2021 (168 days)	Pfizer Pfizer	62.8%	06.07.2021 27.07.2021
1435504, L5	47	M	20.12.2021 (199 days)	Pfizer Pfizer	96.1%	05.06.2021 26.06.2021
			Cohort PCR+ Severe Hospitalized. Symptoms: pneumonia, diarrhea, body weakness			
16105221, S1	79	M	10.11.2021 (209 days)	Pfizer Pfizer	7.3% (negative)	16.04.2021 07.05.2021
16106123, S2	77	F	22.11.2021 (223 days)	Pfizer Pfizer	94.6%	14.04.2021 05.05.2021
1431680, S3	91	M	08.12.2021 (251 days)	Pfizer Pfizer Pfizer	61.7%	03.03.2021 24.03.2021 03.11.2021
1432590, S4	73	M	10.12.2021 (229 days)	Pfizer Pfizer	95.5%	26.04.2021 17.05.2021
1434692, S5	75	F	17.12.2021 (225 days)	Pfizer Pfizer Moderna	92.9%	07.05.2021 28.05.2021 09.12.2021
			Cohort PCR+ ICU. Symptoms: severe bilateral pneumonia with acute respiratory distress			
1432410, U1	52	F	10.12.2021 (194 days)	Janssen Moderna	95.0%	31.05.2021 23.11.2021
1434573, U2	52	M	17.12.2021 (201 days)	Janssen Moderna	94.7%	31.05.2021 23.11.2021

(*) Days between first vaccine dose and serum sampling are shown in parentheses.

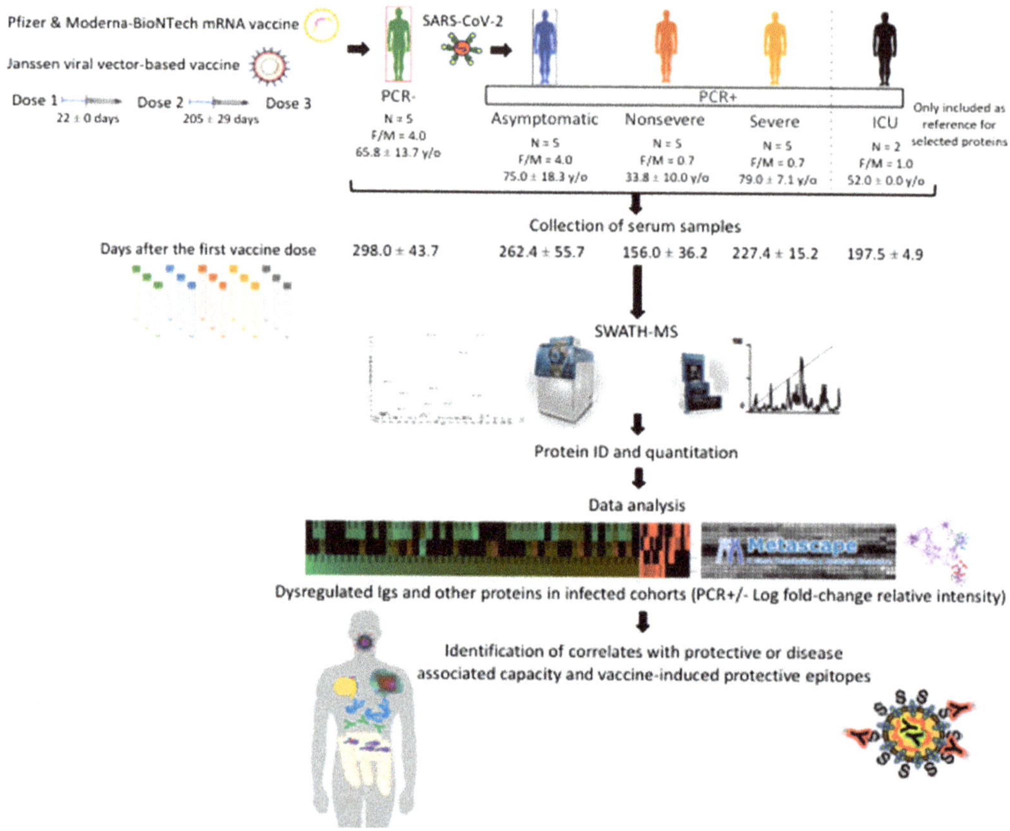

Figure 1. Experimental design and rationale. The experimental design used in our study was based on sera collected from vaccinated individuals (mostly with Pfizer and Moderna-BioNTech) and SARS-CoV-2 PCR-negative (PCR–) or infected PCR-positive (PCR+) and with asymptomatic, non-severe, and severe COVID-19 symptomatology (Table 1). Two PCR+ cases with severe symptoms and in the hospital intensive-care unit (ICU) were included only as reference for selected proteins. Sera were collected between 156 and 298 days after first vaccine-dose administration and subjected to SWATH-MS quantitative proteomics to characterize serum-protein profiles in different cohorts. The proteomics results were then translated into the identification of correlates with protective- or disease-associated capacity and vaccine-induced protective epitopes.

2.2. Serum Proteomics

The methodology and algorithms for serum proteomics were as previously described [8]. Serum samples from PCR– controls and PCR+ COVID-19 asymptomatic, non-severe, and severe individuals (n = 5 each) were individually analyzed. Two PCR+ cases with severe symptoms and in the hospital ICU were included only as reference for the selected proteins. Protein concentration in samples was determined using the BCA Protein Assay with BSA (Sigma-Aldrich) as standard. Protein-serum samples (100 µg per sample) were trypsin digested using the FASP Protein Digestion Kit (Expedeon Ltd., UK) and sequencing-grade trypsin (Promega, Madison, WI, USA) following the manufacturer's recommendations. The resulting tryptic peptides were desalted onto OMIX Pipette tips C18 (Agilent Technologies, Santa Clara, CA, USA), dried down, and stored at −20 °C until mass-spectrometry analysis. The desalted protein digests were resuspended in 2% acetonitrile, 5% acetic acid in water, and analyzed by reverse-phase liquid chromatography coupled with mass spectrometry (RP-LC-MS/MS) using an Ekspert™ nanoLC 415 system coupled online with a 6600 TripleTOF® mass spectrometer (AB SCIEX; Framingham, MA, USA) through Information-Dependent Acquisition (IDA) followed by Sequential Windowed data-independent Acquisition of the Total High-resolution Mass Spectra (SWATH-MS). The peptides were concentrated in a 0.1 × 20 mm C18 RP precolumn (Thermo Fisher Scientific, Waltham, MA, USA) with a flow rate of 2 μL/min for 10 min in solvent A. Then, peptides were separated in a 0.075 × 250 mm C18 RP column (New Objective, Woburn, MA, USA) with a flow rate of 300 nl/min. Elution was carried out in a 120 min gradient from 5% B to 30% B followed by a 15 min gradient from 30% B to 60% B (Solvent A: 0.1% formic acid in water, solvent B: 0.1% formic acid in acetonitrile) and directly injected into the mass spectrometer for analysis. For IDA experiments, the mass spectrometer was set to scanning full spectra from 350 m/z to 1400 m/z (250 ms accumulation time), followed by up to 50 MS/MS scans (100–1500 m/z). Candidate ions with a charge state between +2 and +5 and counts per second above a minimum threshold of 100 were isolated for fragmentation. One MS/MS spectrum was collected for 100 ms before adding those precursor ions to the exclusion list for 15 s (mass spectrometer operated by Analyst® TF 1.6, ABSciex®). Dynamic background subtraction was turned off. Data were acquired in high-sensitivity mode with rolling collision energy on and a collision energy spread of 5. An equal amount of the five samples for each experimental group joined together as a representative mixed sample of each of the 4 experimental groups, which were used for the generation of the reference spectral-ion library as part of SWATH-MS analysis. A total amount of 4 μg protein digests for each mixed sample was injected. For SWATH quantitative analysis, 50 independent samples (2 technical replicates from each of the 5 biological replicates for each of the 4 experimental groups and 5 technical replicates from each of the 2 biological replicates in the case of ICU samples) (6 μg each) were subjected to the cyclic data-independent acquisition (DIA) of mass spectra using the SWATH variable-window calculator (V 1.0, AB SCIEX) and the SWATH acquisition-method editor (AB SCIEX), similar to established methods [8]. A set of 50 overlapping windows was constructed (containing 1 m/z for the window overlap), covering the precursor mass range of 400–1250 m/z. For these experiments, a 50 ms survey scan (350–1400 m/z) was acquired at the beginning of each cycle, and SWATH-MS/MS spectra were collected from 100–1500 m/z for 70 ms in high-sensitivity mode, resulting in a cycle time of 3.6 s. Collision energy for each window was determined according to the calculation for a charge +2 ion centered upon the window with a collision-energy spread of 15. To create a spectral library of all the detectable peptides in the samples, the IDA MS raw files were combined and subjected to database searches in unison using ProteinPilot software v. 5.0.1 (AB SCIEX) with the Paragon algorithm. Spectra identification was performed by searching against the Uniprot human-proteome database (79,038 entries in January 2022) with the following parameters: iodoacetamide cysteine alkylation, trypsin digestion, identification focus on biological modification, and thorough ID as search effort. The detected protein threshold was set at 0.05. To assess the quality of identifications, an independent False Discovery Rate (FDR) analysis with the target-decoy approach provided by Protein Pilot™ was performed. Positive identifications were considered when identified proteins reached a 1% global FDR. The mass-spectrometry proteomics data were deposited in the Proteome Xchange Consortium via the PRIDE partner repository with the dataset identifier PXD031969 and 10.6019/PXD031969.

2.3. Quality Control of Proteomics Data

The quality of the proteomics data was controlled at multiple levels. First, a rat-ileum digest was used for the evaluation of instrument performance. Buffer A samples were run as blanks every two injections to prevent carryover. Two technical replicates were injected for each sample. For validation of serum-proteomics data, protein representation for previously identified selected biomarkers for COVID-19 and proteomics studies were used to show correlation with disease severity. An enrichment analysis was conducted using the Coronascape COVID database (https://metascape.org/COVID; [8]) to identify proteins found in our study as differentially represented in response to COVID-19 and reported in previous COVID-19 omics datasets.

2.4. Data Analysis

For SWATH processing, up to 10 peptides with 7 transitions per protein were automatically selected by the SWATH Acquisition MicroApp 2.0 in the PeakView 2.2 software with the following parameters: 15 ppm ion library tolerance, 5 min XIC extraction window, 0.01 Da XIC width, and considering only peptides with at least 99% confidence and excluding those that were shared or contained modifications. However, to ensure reliable quantitation, only proteins with 3 or more peptides available for quantitation were selected for XIC peak-area extraction and exported for analysis in the MarkerView 1.3 software (AB SCIEX). Global normalization according to the total area sums of all detected proteins in the samples was conducted (Supplementary Materials Data File S1). The Student's t-test ($p < 0.05$) was used to perform two-sample comparisons between the averaged area sums of all the transitions derived for each protein across the 10 replicate runs for each group under comparison to identify proteins that were significantly represented between groups (Supplementary Materials Data File S1). Protein representation was compared between groups by One-way ANOVA test followed by post-hoc Bonferroni and Holm multiple comparisons tests ($p < 0.05$; https://astatsa.com/OneWay_Anova_with_TukeyHSD/), and relative intensity was compared between PCR- and PCR+ cohorts by Welch's unpaired t-test ($p < 0.05$; https://www.graphpad.com/quickcalcs/ttest1/?Format=C) [8]. Data were separately analyzed for overrepresented and underrepresented proteins using the Metascape gene annotation and analysis resource (https://metascape.org/gp/index.html#/main/step1). The analytical algorithm developed using Protein BLAST sequence alignment against non-redundant protein database (nr) using compositional matrix adjustment (https://blast.ncbi.nlm.nih.gov/Blast.cgi?PROGRAM=blastp&PAGE_TYPE=BlastSearch&LINK_LOC=blasthome) and Paratome (http://www.ofranlab.org) was used for the identification of antigen-binding regions and vaccine-induced antibody-protective epitopes and correlates of identified proteins with protective- or disease-associated capacity (Supplementary Materials Data Files S2 and S4).

2.5. Antibody Neutralization Test

Antibody titers specific for the neutralization of SARS-CoV-2 virus were determined with a cPass SARS-CoV-2 neutralization antibody-detection kit (Genscript, Piscataway, NJ, USA) following the manufacturer's instructions. Briefly, 100 μL of the positive and negative controls and serum samples at 1:100 dilution and previously incubated with HRP conjugated RBD during 30 min at 37 °C were added to the 96-microwell plate coated with RBD-SARS-CoV-2 protein and incubated for 15 min at 37 °C. After washing four times with 260 μL/well of wash buffer, 100 μL/well of chromogen-substrate solution were added and incubated for 15 min at RT. Finally, the colorimetric reaction was stopped with 50 μL/well of stop solution and the absorbance was measured in a spectrophotometer (Thermo Fisher Scientific) at O.D. 450 nm. Results were evaluated by calculating the ratio between the O.D. of the sample and the O.D. of the calibrator using the following formula: Inhibition = (1 − O.D. value of sample/O.D. value of negative control) × 100.

2.6. Antibody Levels against HBV and Zika Virus

Individual sera from all cohorts included in the study were characterized for antibody levels against HBV and Zika virus using pathogen-specific ELISA tests. The ELISA kits are designed for the detection of antibodies to hepatitis B-virus surface antigen: Hepatitis B surface antigen Ab ELISA kit (AB-KA0287; Biogen Científica, Madrid, Spain) or anti-Zika Virus Non-structural Protein, Anti-Zika virus IgG ELISA kit (ab221844; Abcam, Cambridge, UK) in human serum. Antibody levels were compared between PCR− and PCR+ cohorts by Chi-squared test ($p < 0.05$). A Spearman's Rho (rs) correlation analysis was conducted between virus cross-reactive Ig levels and proteomic-protein relative intensity ($p < 0.05$; https://www.socscistatistics.com/tests/spearman/default2.aspx).

2.7. Human Autoantibody General Survey Microarray

Reactive autoantibodies were characterized with individual sera from all cohorts using the Human Autoantigens General Survey Antigen Microarray (GeneCopoeia's OmicsArray; Rockville, MD, USA), a protein microarray enabling powerful detection of autoantibodies associated with many diseases, including rheumatoid arthritis, muscular dystrophy, systemic lupus erythematosus, and type-1 diabetes. The array carries 120 superior-quality purified proteins spotted onto nitrocellulose filters, which are adhered to glass slides. Antigens known to be associated with specific autoimmune diseases are chosen based on a thorough review of peer-reviewed publications. Data interpretation: NSI, net signal intensity, averaged fluorescent-signal intensity for each antigen subtracted by local background and negative control signal; NSI-nor, NSI normalized to internal Ig controls; SBR, averaged signal-to-background ratio of each antigen; SBR-nor, SBR normalized to internal Ig controls; SNR, signal-to-noise ratio that represents the significance of the signal above the background (SNR $\geq$ 3 means the signal is significantly higher than the background). Additional information is in Supplementary Materials Data File S3.

3. Results and Discussion

3.1. Characterization of Immunoglobulin Protein Profiles and Correlation with Protective- or Disease-Associated Capacity

The experimental design used in our study was based on sera collected from SARS-CoV-2 PCR-negative (PCR−) or infected PCR-positive (PCR+) vaccinated individuals (mostly with Pfizer and Moderna-BioNTech) and with asymptomatic, non-severe, and severe COVID-19 symptomatology (Figure 1, Table 1). Two PCR+ cases with severe symptoms and in the hospital intensive-care unit (ICU) were included only as reference for selected proteins. Sera were collected in November–December 2021 before the appearance of the SARS-CoV-2 Omicron variant in Spain, and thus most PCR+ cases were probably of the Delta virus variant (Table 1). As expected, 91% of the individuals' serum samples (20 out of 22 except for samples C5 and S1; Table 1) showed neutralization antibodies against SARS-CoV-2. Nevertheless, the characterization of antibody- and non-antibody-mediated immune response to vaccination is important to understanding the response to COVID-19 vaccines. Proteomics analysis identified significantly dysregulated proteins in vaccinated and infected cohorts (asymptomatic, n = 134; non-severe, n = 117; severe, n = 230) when compared to vaccinated PCR− individuals (Figure 2A, Supplementary Materials Data File S1). More than 55% of the dysregulated proteins identified in infected cohorts when compared to PCR− individuals were immunoglobulins (Igs), mostly overrepresented in asymptomatic and severely infected cohorts (Figure 2A,B). This result is associated with response to vaccination, as supported by the finding using a similar serum-proteomics approach in healthy, unvaccinated individuals and with different COVID-19 symptomatology, in whom only 32% (60/189) of the dysregulated proteins were Igs [8].

An analytical workflow was developed to characterize selected Ig proteins identified as significantly dysregulated in vaccinated and infected cohorts when compared to vaccinated PCR− individuals (Supplementary Materials Data File S2). The use of the Paratome web server (http://www.ofranlab.org) allowed for the identification of antigen-binding regions in identified Ig light- or heavy-chain variable regions, including but not limited to complementarity-determining regions (CDRs) [9]. As has been discussed, heavy-chain complementarity-determining region 3 (HCDR3) is necessary, but insufficient for specific antibody binding [10].

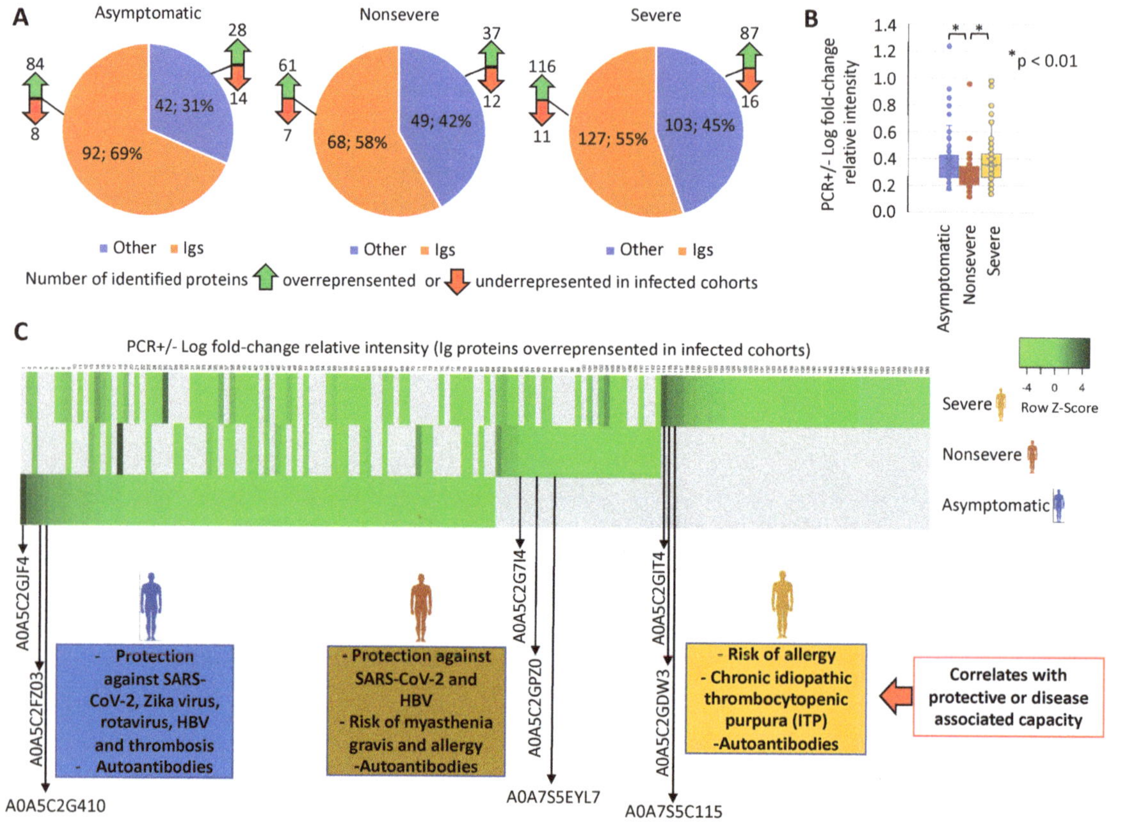

Figure 2. Serum-protein profiles and correlation of Ig overrepresented in vaccinated infected cohorts with protective- or disease-associated capacity. (**A**) Number of identified overrepresented or underrepresented Ig and non-Ig proteins in infected asymptomatic, non-severe, and severe cohorts when compared to PCR– individuals. (**B**) Change in Ig-protein levels in PCR+ cohorts. PCR+/− Log fold-change relative intensity was compared between groups by One-way ANOVA test followed by post-hoc Bonferroni and Holm multiple comparisons tests ($p < 0.05$). (**C**) Heatmap of PCR+/− Log fold-change relative intensity (Z-scored original value) for Ig proteins overrepresented in infected cohorts. Correlates with protective- or disease-associated capacity are shown for Igs highly overrepresented in PCR+ individuals.

Focusing on Igs highly overrepresented in PCR+ individuals, the results showed differences between infected cohorts (Figure 2C, Supplementary Materials Data File S2). In vaccinated and infected asymptomatic cases, predictive models associated the Igs with protection against SARS-CoV-2, Zika virus, rotavirus, Hepatitis B virus (HBV), and thrombosis. However, in cases with COVID-19 symptoms, Igs associated with protection against SARS-CoV-2 and HBV were only identified in non-severe cases, whereas Igs associated with autoantibodies and risk of allergic reactions and diseases such as myasthenia gravis and chronic idiopathic thrombocytopenic purpura (ITP) were identified only in non-severe and severe patients (Figure 2C, Supplementary Materials Data File S2). Additionally, autoantibodies were identified in all PCR+ cohorts. The analysis of mass-spectra relative intensity for selected overrepresented Ig proteins corroborated in individual samples the cohort-dependent results (Figure 3A). Then, an independent analysis of predicted biomarkers was used to validate these results for Zika virus, HBV, and a human-autoantibody general survey (Figure 3B–D).

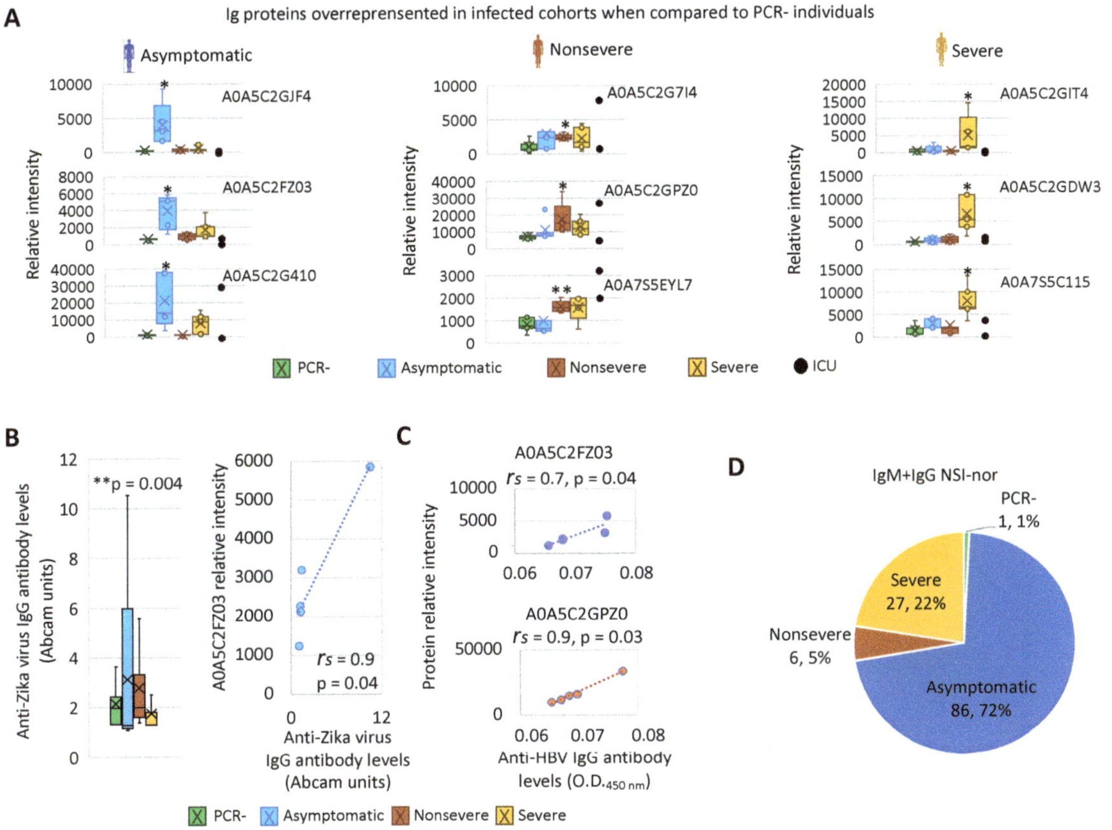

Figure 3. Validation of serum Ig proteins overrepresented in vaccinated infected individuals. (**A**) Changes in Ig-protein mass-spectra relative intensity for selected overrepresented Ig in individual samples in all PCR– and PCR+ cohorts. Relative intensity was compared between PCR– and PCR+ cohorts by Welch's unpaired t-test (* $p < 0.05$, ** $p < 0.005$; n = 5 biological replicates). Samples from individuals in the ICU were only included as reference for selected proteins. (**B**) Independent analysis of predicted biomarker A0A5C2FZ03 for anti-Zika virus Ig levels. Antibody levels were compared between PCR– and PCR+ cohorts by Chi-squared test ($p = 0.004$ for asymptomatic cohort). A Spearman's Rho correlation analysis was conducted between virus cross-reactive Ig levels and proteomics-protein relative intensity ($p < 0.05$). (**C**) Independent analysis of predicted biomarkers A0A5C2FZ03 and A0A5C2GPZ0 for anti-HBV Ig levels in asymptomatic and non-severe cohorts, respectively. A Spearman's Rho correlation analysis was conducted between HBV cross-reactive Ig levels and proteomics-protein relative intensity ($p < 0.05$). (**D**) A human-autoantibody general survey identified proteins reactive to IgM and IgG autoantibodies with significant NSI-nor values in PCR– and PCR+ cohorts. Proteins were distributed based on the highest NSI-nor average IgM+IgG value.

In agreement with serum-protein profiles and correlation of Igs overrepresented in vaccinated infected cohorts with protective-associated capacity (Figures 2C and 3A), the results showed that the anti-Zika-virus IgG levels were significantly higher in the asymptomatic cohort only (Figure 3B). Furthermore, a significant positive correlation between Zika-virus cross-reactive Ig levels and relative intensity of associated A0A5C2FZ03 protein was obtained (Figure 3B, Supplementary Materials Data File S2). For HBV, the sensitivity of the surface-antigen ELISA (positive values > 0.034) did not allow for identification of positive individuals in PCR+ cohorts and allowed identification of only one individual in the PCR–

cohort (C2 in Table 1, value = 0.079). Nevertheless, as predicted for serum-protein profiles (Figures 2C and 3A), a significant positive correlation was obtained between mean HBV ELISA O.D. 450 nm values and relative intensity of associated proteins A0A5C2FZ03 and A0A5C2GPZ0 only in asymptomatic and non-severe individuals, respectively (Figure 3C, Supplementary Materials Data File S2).

Autoantibodies in COVID-19 patients have been recently correlated with increased antiviral humoral and inflammatory immune responses [11]. In our study regarding autoantibodies, the survey identified 120 target proteins with significant signal-to-noise ratio (SNR $\geq$ 3; Figure 3D, Supplementary Materials Data File S3). The net signal intensity for each antigen subtracted by the local background and negative control signal and normalized to internal Ig controls (NSI-nor; Supplementary Materials Data File S3) showed a distribution by different cohorts with the highest NSI-nor average IgM+IgG value (IgM+IgG NSI-nor; Figure 3D). In accordance with serum-protein profiles, autoantibodies were identified mostly in PCR+ cohorts with only one protein (Histone H1), with the highest IgM+IgG NSI-nor value in PCR– individuals (Figures 3D and 4, Supplementary Materials Data File S3). As predicted by our analysis (Supplementary Materials Data File S2), most of the identified proteins reactive to autoantibodies were involved in the regulation of the immune system and/or associated with different diseases (Figure 4, Supplementary Materials Data File S3). For example, nuclear pore-membrane glycoprotein GP210 identified with the highest IgM+IgG NSI-nor value in the severe cohort is a prognostic marker in patients with primary biliary cirrhosis [12]. Another protein with significant signal-to-noise ratio in all PCR– and PCR+ cohorts (SNR $\geq$ 4.5) and the highest IgM+IgG NSI-nor in severe patients (SNR > 12) was the Type-1 angiotensin II receptor (AGTR; Supplementary Materials Data File S3), which during SARS-CoV-2 infection can recognize and internalize the soluble angiotensin-converting enzyme 2 (ACE2)–coronavirus spike protein complex through dynamin 2-dependent endocytosis [13]. In accordance with our results, these autoantibodies are associated with an unfavorable COVID-19 disease course [14].

The Igs identified as underrepresented in infected cohorts and thus with higher relative levels in PCR– individuals (Figure 5A, Supplementary Materials Data File S2) were used for the identification of vaccine-induced protective epitopes using a designed analytical workflow (Figure 5B, Supplementary Materials Data File S2). The results showed that the protective epitopes were not only identified in the SARS-CoV-2 spike (S) receptor-binding domain (RBD) associated with vaccine-protective capacity [15], but also in other virus proteins such as envelope small-membrane glycoprotein M (ORF3a), membrane-protein E, and nucleocapsid phosphoprotein N (ORF1ab) (Figure 5C, Supplementary Materials Data File S2). A correlation analysis was conducted between SARS-CoV-2-neutralizing antibodies (Table 1) and mass-spectra relative intensity of identified Igs with vaccine-induced protective epitopes (Supplementary Materials Data Files S1 and S2). The results showed no significant correlation for all cohorts together (R^2 = 0.182) but revealed a positive correlation in PCR– individuals in whom these proteins were overrepresented (R^2 = 0.826), thus providing support to the predicted protective epitopes in response to vaccination (Figure 5D).

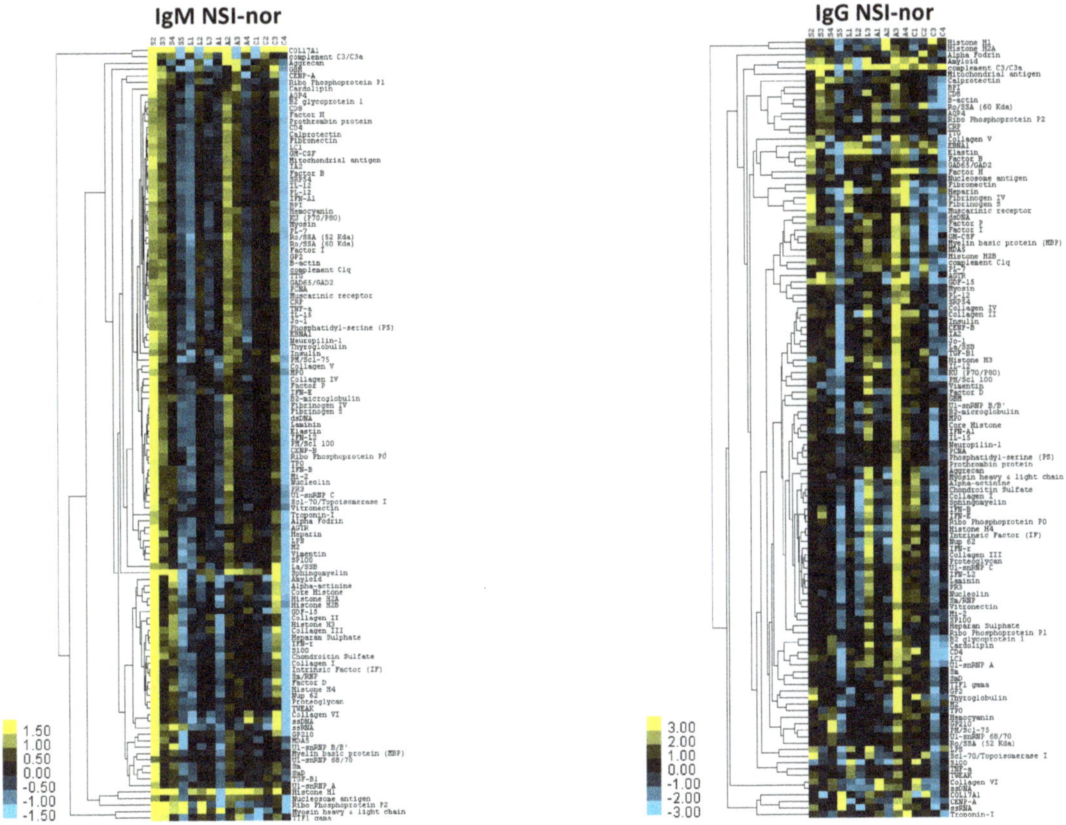

Figure 4. Heatmap of IgM and IgG autoantibodies overrepresented in vaccinated infected cohorts. According to the NSI-nor value, log2 (NSI+1) is calculated, the data are normalized, and a heat map is generated. The antigens are clustered according to Euclidean distance. Additional information is in Supplementary Materials Data File S3. Abbreviations: S, severe; L, non-severe: A, asymptomatic; C, PCR– (Table 1).

These results further advance our knowledge on the antibody response in vaccinated uninfected (fully protected) and vaccinated SARS-CoV-2-infected (partially protected) individuals associated with host factors such as age, comorbidities, and coronavirus infection [16–19]. Whereas vaccinated PCR– individuals developed a protective response mediated by Igs against multiple SARS-CoV-2 proteins to prevent infection, PCR+ individuals showed overrepresented Ig profiles associated with COVID-19 symptomatology with protective Igs to control virus infection and thrombosis in asymptomatic cases and limited or no protective response against SARS-CoV-2 with Ig-associated risk of allergy and other diseases in non-severe and severe patients.

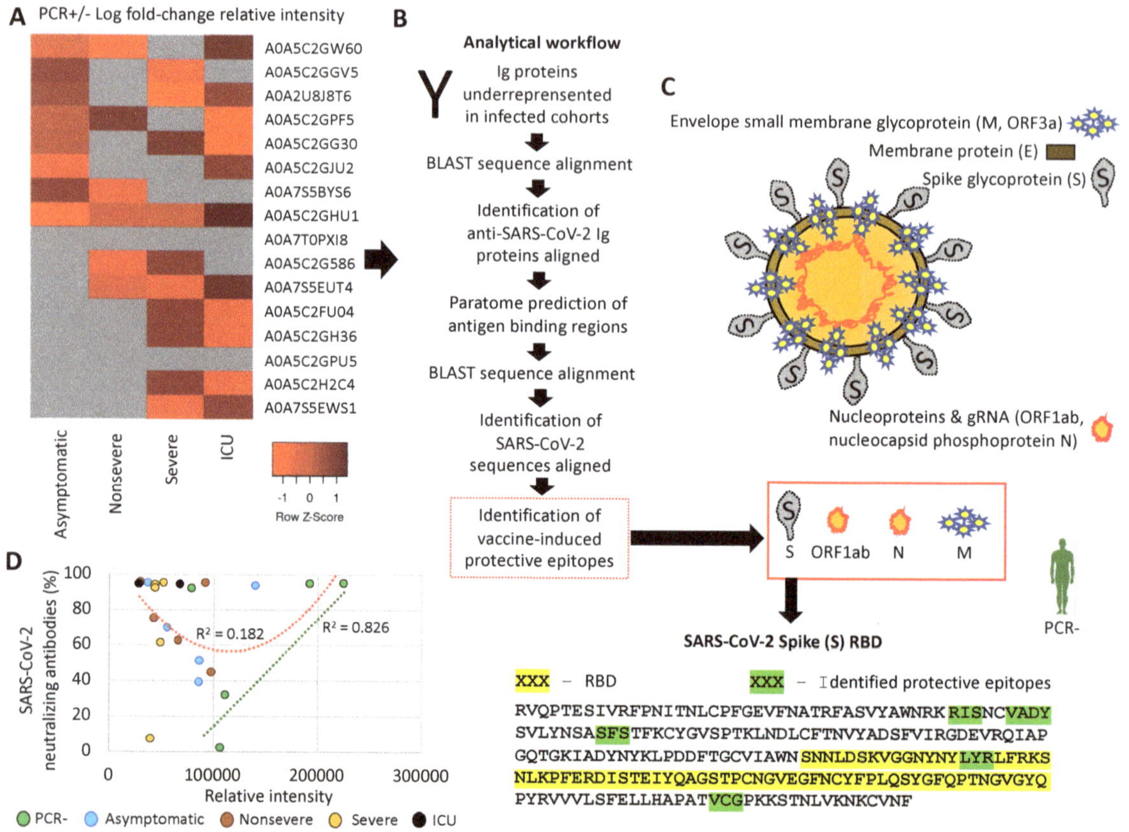

Figure 5. Serum-protein profiles of Ig underrepresented in vaccinated infected cohorts and identification of vaccine-induced protective epitopes. (**A**) Heatmap of PCR+/− Log fold-change relative intensity (Z-scored original value) for Ig proteins underrepresented in infected cohorts. (**B**) Analytical workflow developed for the dentification of vaccine-induced protective epitopes. (**C**) Identification of SARS-CoV-2 proteins with predicted reactive epitopes to Ig underrepresented in infected cohorts and thus overrepresented in PCR− individuals. All methods and results are disclosed in Supplementary Materials Data File S2. (**D**) Correlation analysis between SARS-CoV-2-neutralizing antibodies (Table 1) and mass-spectra relative intensity of identified Igs with vaccine-induced protective epitopes (Supplementary Materials Data Files S1 and S2).

3.2. Characterization of Non-Ig Protein Profiles and Correlation with COVID-19

As reported in previous proteomics studies [7,8], identified dysregulated non-Ig proteins and biological processes in vaccinated infected PCR+ cohorts when compared to vaccinated PCR− individuals were associated with SARS-CoV-2 infection and COVID-19 (Figures 6A,B and 7, Supplementary Materials Data Files S1 and S4). As expected, PCR+/− Log fold-change relative intensity was higher in individuals with severe symptoms (Figure 6A,B). Accordingly, protein–protein-interaction networks and components for non-Ig proteins over and underrepresented in infected cohorts showed a higher representation in the severe cohort when compared to PCR− cases (Figure 7). Gene-ontology (GO) categories with overrepresented proteins involved in the regulation of complement and coagulation cascades and antibody-mediated complement activation were the most represented in protein–protein interactions (Figure 7; identified with yellow stars). Hyperactivation of the complement and coagulation systems are associated with the clinical syndrome of COVID-19 [20].

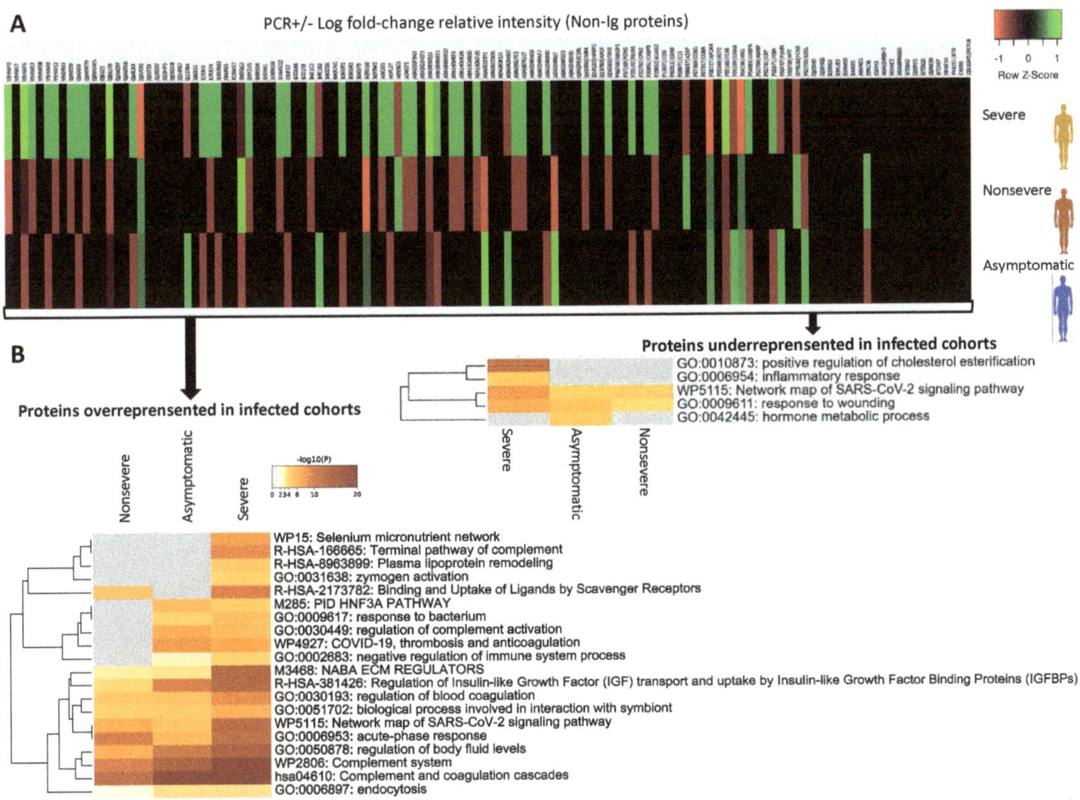

Figure 6. Multiple differential representation and enrichment ontology clusters for non-Ig proteins. (**A**) Heatmap of PCR+/− Log fold-change relative intensity (Z-scored original value). (**B**) Enriched GO/KEGG ontology clusters for proteins over- and underrepresented in infected cohorts when compared to PCR− cases. Accumulative hypergeometric *p*-values and enrichment factors were calculated and used for filtering. Remaining significant terms were then hierarchically clustered into a tree based on Kappa-statistical similarities among their gene memberships. Then, the 0.3 kappa score was applied as the threshold to cast the tree into term clusters. The term with the best *p*-value within each cluster was selected as its representative term and displayed in a dendrogram.

The other GO identified in the protein–protein-interaction network of overrepresented proteins was the insulin-like growth factor (IGF) pathway as seen in Figure 7. Although an association has been proposed between low IGF1 levels and poor outcome in patients with COVID-19 [21], an epidemiological study provided evidence that higher IGF-1 concentrations are associated with a lower risk of COVID-19 mortality [22]. The results of our study suggested that activation of the IGF pathway may occur in response to vaccination by regulating immune-cell homeostasis to reduce risk for COVID-19 mortality [23].

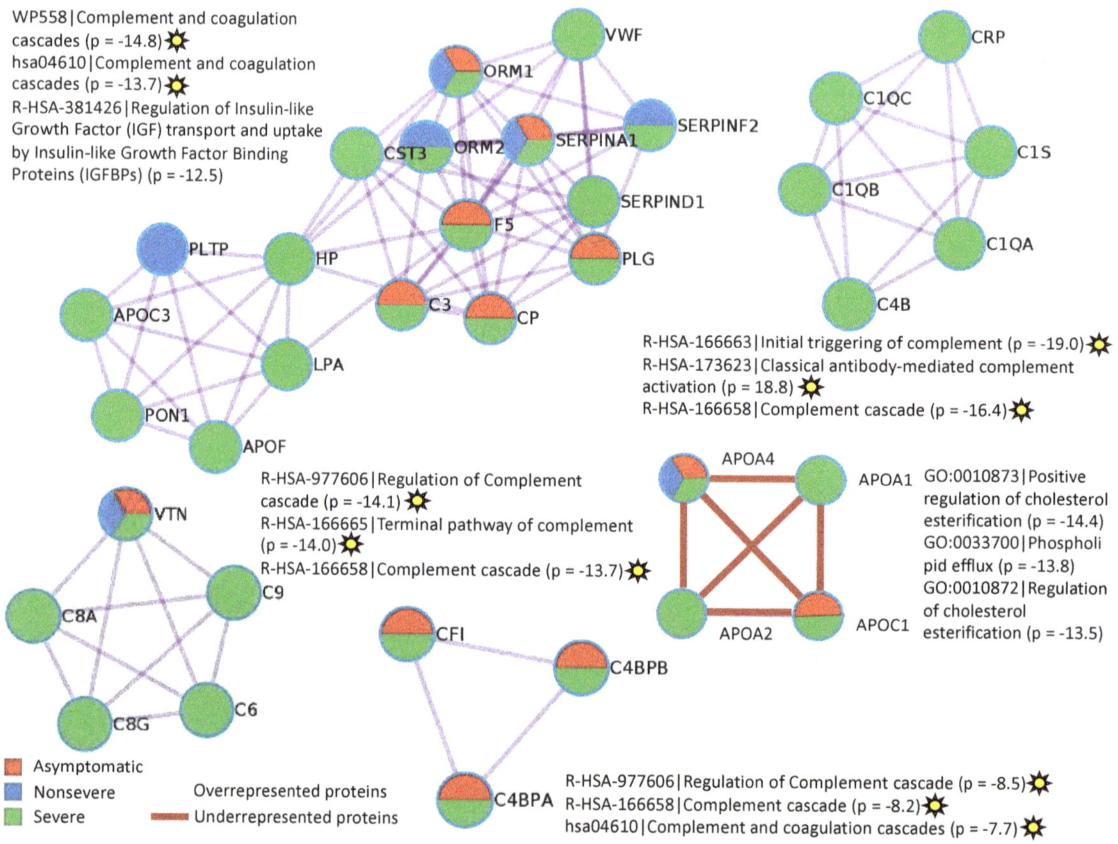

Figure 7. Protein–protein-interaction networks and components for non-Ig proteins over- and underrepresented in infected cohorts when compared to PCR– cases. The MCODE algorithm was applied to networks to identify neighborhoods where proteins are densely connected. GO enrichment analysis was applied to the original protein–protein-interaction network and its MCODE network components to extract their "biological meanings," where the top three best *p*-value terms were retained. GOs with proteins involved in the regulation of complement and coagulation cascades and antibody-mediated complement activation are identified with yellow stars.

Another finding of our study was related to severe-cohort overrepresented proteins in the biological process involved in interaction with symbiont (GO:0051702) (Figure 6B, Supplementary Materials Data File S4). One of the proteins identified in this biological process, Apolipoprotein E isoform 1 (APOE1; A0A0S2Z3D5), was overrepresented in severe (Log fold-change = 0.157) and UCI (Log fold-change = 0.081) patients (Supplementary Materials Data File S1). The expression of ApoE proteins, including APOE1, is critical for the assembly of infectious Hepatitis C virus (HCV) in a strain-specific and cell-type dependent manner [24]. Related to COVID-19, higher disease risk has been associated with *apoE4* genetic variants [25], but this is the first possible implication of ApoE1 in this process. Therefore, APOE1-protein levels and genetic variants may be a biomarker associated with disease severity in vaccinated and SARS-CoV-2-infected individuals.

As in recent studies [26], the interacting underrepresented proteins in vaccinated and infected cohorts were apolipoproteins APOA1, APOA2, APOA4, and APOC1 involved in the regulation of cholesterol esterification and phospholipid efflux (Figure 7). Higher levels of APOA1 have been correlated with protection from COVID-19 severity [27]. Furthermore, cholesterol esterification may counteract the normally exacerbating effect of cholesterol on coronavirus cytopathology [28]. Consequently, our results suggest that higher levels of some apolipoproteins in PCR– individuals may be associated with a vaccine-protective effect.

4. Conclusions

In summary, novel findings of the study include (a) characterization of Ig and non-Ig protein profiles in vaccinated uninfected (fully protected) and vaccinated SARS-CoV-2-infected (partially protected) individuals with identification of disease and protection-associated biomarkers; (b) identification of candidate-protective epitopes not only in SARS-CoV-2 RBD but also in glycoprotein M (ORF3a), membrane protein E, and nucleocapsid phosphoprotein N (ORF1ab); (c) analysis of autoantibody profiles that are associated with an unfavorable COVID-19 disease course even after vaccination; and (d) prediction on non-Ig serum biomarkers associated with vaccine-protective capacity or disease severity in vaccinated and SARS-CoV-2-infected individuals.

The main limitation of this study is that serum-proteomics analysis was conducted with five samples from each cohort, which may have reduced the effect of case-by-case differences in serum-protein representation. Nevertheless, the results of this study using a serum-proteomics approach to characterize host-associated factors to COVID-19-vaccine response suggest protective- and disease-associated mechanisms in vaccinated individuals. Despite differences in individual age, sex, vaccine provider, and doses, the results were consistent between different cohorts. These results may lead to studies with a higher number of individuals and including different vaccine formulations to improve vaccine efficacy and implementation against SARS-CoV-2.

Supplementary Materials: The following supporting information can be downloaded at: https://www.mdpi.com/article/10.3390/molecules27185933/s1, Data File S1. Serum proteomics analysis; Data File S2. Analysis of immunoglobulin proteins underrepresented and overrepresented in infected cohorts when compared to PCR– individuals; Data File S3. Human-autoantibody general survey in PCR– and PCR+ cohorts; Data File S4. Analysis of selected non-immunoglobulin proteins underrepresented and overrepresented in infected cohorts when compared to PCR– individuals.

Author Contributions: Conceptualization, J.d.l.F., J.M.U., M.V., and C.G.; methodology, M.V., and J.d.l.F.; formal analysis, M.V., J.d.l.F., J.M.U., F.J.R.-d.-R., and J.d.l.F.; investigation, M.V., S.A.-J., M.C., R.V.-R., and L.M.; writing—original draft preparation, J.d.l.F.; writing—review and editing, J.d.l.F., J.M.U., M.V., and C.G.; visualization, J.d.l.F.; supervision, J.d.l.F., M.V., and C.G.; funding acquisition, J.d.l.F., and C.G. All authors have read and agreed to the published version of the manuscript.

Funding: This research was funded by the Ministerio de Ciencia e Innovación/Agencia Estatal de Investigación MCIN/AEI/10.13039/501100011033, Spain, and EU-FEDER (Grant BIOGAL PID2020-116761GB-I00); and the Junta de Comunidades de Castilla-La Mancha (JCCM), Spain, and EU-FEDER (grant MYCOTRAINING SBPLY/19/180501/000174).

Institutional Review Board Statement: The study was conducted in accordance with the Declaration of Helsinki, and approved by the Ethical and Scientific Committees (University Hospital of Ciudad Real C-352 and SESCAM C-73).

Informed Consent Statement: Informed consent was obtained from all subjects involved in the study. Informed consent has been obtained from the patient(s) to publish this paper.

Data Availability Statement: All data are available in the main text or the Supplementary Materials. The mass-spectrometry proteomics data have been deposited in the ProteomeXchange Consortium via the PRIDE partner repository with the dataset identifier PXD031969 and 10.6019/PXD031969.

Acknowledgments: We acknowledge the contribution of other members of our SaBio group on discussing some of the results of this study.

Conflicts of Interest: The authors declare no conflict of interest. The funders had no role in the design of the study; in the collection, analyses, or interpretation of data; in the writing of the manuscript; or in the decision to publish the results.

Sample Availability: Samples of the compounds are available from the authors.

References

1. Adam, D. The pandemic's true death toll: Millions more than official counts. *Nature* **2022**, *601*, 312–315. [CrossRef] [PubMed]
2. Chaudhary, J.K.; Yadav, R.; Chaudhary, P.K.; Maurya, A.; Kant, N.; Rugaie, O.A.; Haokip, H.R.; Yadav, D.; Roshan, R.; Prasad, R.; et al. Insights into COVID-19 Vaccine development based on immunogenic structural proteins of SARS-CoV-2, host immune responses, and herd immunity. *Cells* **2021**, *10*, 2949. [CrossRef] [PubMed]
3. Ceacareanu, A.C.; Wintrob, Z.A.P. Summary of COVID-19 vaccine-related reports in the vaccine adverse event reporting system. *J. Res. Pharm. Pract.* **2021**, *10*, 107–113. [CrossRef] [PubMed]
4. Adesokan, A.; Obeid, M.A.; Lawal, A.F. SARS-CoV-2: Vaccinology and emerging therapeutics; challenges and future developments. *Ther. Deliv.* **2022**, *13*, 117–183. [CrossRef]
5. Mohapatra, R.K.; Kuppili, S.; Kumar Suvvari, T.; Kandi, V.; Behera, A.; Verma, S.; Kudrat-E-Zahan Biswal, S.K.; Al-Noor, T.H.; El-Ajaily, M.M.; Sarangi, A.K.; et al. SARS-CoV-2 and its variants of concern including Omicron: Looks like a never ending pandemic. *Chem. Biol. Drug Des.* **2022**, *99*, 769–788. [CrossRef]
6. Li, J.; Wang, Y.; Liu, Y.; Zhang, Z.; Zhai, Y.; Dai, Y.; Wu, Z.; Nie, X.; Du, L. Polymorphisms and mutations of ACE2 and TMPRSS2 genes are associated with COVID-19: A systematic review. *Eur. J. Med. Res.* **2022**, *27*, 26. [CrossRef]
7. Amiri-Dashatan, N.; Koushki, M.; Rezaei-Tavirani, M. Mass spectrometry-based proteomics research to fight COVID-19: An expert review on hopes and challenges. *OMICS* **2022**, *26*, 19–34. [CrossRef]
8. Villar, M.; Urra, J.M.; Rodríguez-Del-Río, F.J.; Artigas-Jerónimo, S.; Jiménez-Collados, N.; Ferreras-Colino, E.; Contreras, M.; Fernández de Mera, I.G.; Estrada-Peña, A.; Gortázar, C.; et al. Characterization by quantitative serum proteomics of immune-related prognostic biomarkers for COVID-19 symptomatology. *Front. Immunol.* **2021**, *12*, 730710. [CrossRef]
9. Kunik, V.; Ashkenazi, S.; Ofran, Y. Paratome: An online tool for systematic identification of antigen-binding regions in antibodies based on sequence or structure. *Nucleic Acids Res.* **2012**, *40*, W521–W524. [CrossRef]
10. D'Angelo, S.; Ferrara, F.; Naranjo, L.; Erasmus, M.F.; Hraber, P.; Bradbury, A.R.M. Many routes to an Antibody Heavy-Chain CDR3: Necessary, yet insufficient, for specific binding. *Front. Immunol.* **2018**, *9*, 395. [CrossRef]
11. Taeschler, P.; Cervia, C.; Zurbuchen, Y.; Hasler, S.; Pou, C.; Tan, Z.; Adamo, S.; Raeber, M.E.; Bächli, E.; Rudiger, A.; et al. Autoantibodies in COVID-19 correlate with antiviral humoral responses and distinct immune signatures. *Allergy* **2022**. [CrossRef] [PubMed]
12. Uchida, M.; Tashiro-Itoh, T.; Matsuda, Y.; Ishihara, K.; Asakura, H. Autoantibodies against a 210 kDa glycoprotein of the nuclear pore complex as a prognostic marker in patients with primary biliary cirrhosis. *J. Gastroenterol. Hepatol.* **1998**, *13*, 257–265.
13. Yeung, M.L.; Teng, J.L.L.; Jia, L.; Zhang, C.; Huang, C.; Cai, J.P.; Zhou, R.; Chan, K.H.; Zhao, H.; Zhu, L.; et al. Soluble ACE2-mediated cell entry of SARS-CoV-2 via interaction with proteins related to the renin-angiotensin system. *Cell* **2021**, *184*, 2212–2228.e12. [CrossRef]
14. Miedema, J.; Schreurs, M.; van der Sar-van der Brugge, S.; Paats, M.; Baart, S.; Bakker, M.; Hoek, R.; Dik, W.A.; Endeman, H.; Van Der Velden, V.; et al. Antibodies against Angiotensin II Receptor Type 1 and Endothelin A Receptor are associated with an unfavorable COVID19 disease course. *Front. Immunol.* **2021**, *12*, 684142. [CrossRef]
15. Niu, L.; Wittrock, K.N.; Clabaugh, G.C.; Srivastava, V.; Cho, M.W. A structural landscape of neutralizing antibodies against SARS-CoV-2 Receptor Binding Domain. *Front. Immunol.* **2021**, *12*, 647934. [CrossRef] [PubMed]
16. Bastard, P. Why do people die from COVID-19? *Science* **2022**, *375*, 829–830. [CrossRef]
17. Schiaffini, R.; Campana, A.; Deodati, A.; Peschiaroli, E.; Lanzillotta, M.F.; Fierabracci, A. SARS-CoV-2 infection as possible downstream disease precipitator in autoantibody-positive insulin-dependent diabetes mellitus: A case report. *Ital. J. Pediatr.* **2022**, *48*, 33. [CrossRef] [PubMed]
18. Dotan, A.; David, P.; Arnheim, D.; Shoenfeld, Y. The autonomic aspects of the post-COVID19 syndrome. *Autoimmun. Rev.* **2022**, *21*, 103071. [CrossRef] [PubMed]
19. Juanes-Velasco, P.; Landeira-Viñuela, A.; García-Vaquero, M.L.; Lecrevisse, Q.; Herrero, R.; Ferruelo, A.; Góngora, R.; Corrales, F.; Rivas, J.L.; Lorente, J.A.; et al. SARS-CoV-2 Infection Triggers Auto-Immune Response in ARDS. *Front. Immunol.* **2022**, *13*, 732197. [CrossRef]
20. Afzali, B.; Noris, M.; Lambrecht, B.N.; Kemper, C. The state of complement in COVID-19. *Nat. Rev. Immunol.* **2022**, *22*, 77–84. [CrossRef]
21. Ilias, I.; Diamantopoulos, A.; Botoula, E.; Athanasiou, N.; Zacharis, A.; Tsipilis, S.; Jahaj, E.; Vassiliou, A.G.; Vassiliadi, D.A.; Kotanidou, A.; et al. COVID-19 and Growth Hormone/Insulin-Like Growth Factor 1: Study in critically and non-critically ill patients. *Front. Endocrinol.* **2021**, *12*, 644055. [CrossRef] [PubMed]
22. Fan, X.; Yin, C.; Wang, J.; Yang, M.; Ma, H.; Jin, G.; Song, M.; Hu, Z.; Shen, H.; Hang, D. Pre-diagnostic circulating concentrations of insulin-like growth factor-1 and risk of COVID-19 mortality: Results from UK Biobank. *Eur. J. Epidemiol.* **2021**, *36*, 311–318. [CrossRef] [PubMed]
23. Yoon, I.S.; Park, H.; Kwak, H.W.; Woo Jung, Y.; Nam, J.H. Macrophage-derived insulin-like growth factor-1 affects influenza vaccine efficacy through the regulation of immune cell homeostasis. *Vaccine* **2017**, *35*, 4687–4694. [CrossRef]

24. Weller, R.; Hueging, K.; Brown, R.J.P.; Todt, D.; Joecks, S.; Vondran, F.W.R.; Pietschmann, T. Hepatitis C virus strain-dependent usage of Apolipoprotein E modulates assembly efficiency and specific infectivity of secreted virions. *J. Virol.* **2017**, *91*, e00422-17. [CrossRef] [PubMed]
25. Hubacek, J.A. Effects of selected inherited factors on susceptibility to SARS-CoV-2 infection and COVID-19 progression. *Physiol. Res.* **2021**, *70*, S125–S134. [CrossRef]
26. Nuñez, E.; Orera, I.; Carmona-Rodríguez, L.; Paño, J.R.; Vázquez, J.; Corrales, F.J. Mapping the Serum Proteome of COVID-19 Patients; Guidance for Severity Assessment. *Biomedicines* **2022**, *10*, 1690. [CrossRef]
27. Ulloque-Badaracco, J.R.; Hernandez-Bustamante, E.A.; Herrera-Añazco, P.; Benites-Zapata, V.A. Prognostic value of apolipoproteins in COVID-19 patients: A systematic review and meta-analysis. *Travel Med. Infect. Dis.* **2021**, *44*, 102200. [CrossRef]
28. Cervin, M.; Anderson, R. Modulation of coronavirus-mediated cell fusion by homeostatic control of cholesterol and fatty acid metabolism. *J. Med. Virol.* **1991**, *35*, 142–149. [CrossRef]

molecules

MDPI

Article

Transcriptomics, Cheminformatics, and Systems Pharmacology Strategies Unveil the Potential Bioactives to Combat COVID-19

Sivakumar Adarshan [1], Sakthivel Akassh [2,†], Krishnakumar Avinash [2,†], Mathivanan Bharathkumar [2,†], Pandiyan Muthuramalingam [2,3,4,*], Hyunsuk Shin [3,4,*], Venkidasamy Baskar [5], Jen-Tsung Chen [6,*], Veluswamy Bhuvaneshwari [7] and Manikandan Ramesh [1]

1 Department of Biotechnology, Science Campus, Alagappa University, Karaikudi 630003, India
2 Department of Biotechnology, Sri Shakthi Institute of Engineering and Technology, Coimbatore 641062, India
3 Division of Horticultural Science, College of Agriculture and Life Sciences, Gyeongsang National University, Jinju 52725, Korea
4 Department of GreenBio Science, Gyeongsang National University, Jinju 52725, Korea
5 Department of Oral and Maxillofacial Surgery, Saveetha Dental College and Hospitals, Saveetha Institute of Medical and Technical Sciences, Chennai 602105, India
6 Department of Life Sciences, National University of Kaohsiung, Kaohsiung 811, Taiwan
7 Department of Biotechnology, PSGR Krishnammal College for Women, Coimbatore 641004, India
* Correspondence: pandianmuthuramalingam@gmail.com (P.M.); shinpomo@gnu.ac.kr (H.S.); jentsung@nuk.edu.tw (J.-T.C.)
† These authors contributed equally to this work.

Citation: Adarshan, S.; Akassh, S.; Avinash, K.; Bharathkumar, M.; Muthuramalingam, P.; Shin, H.; Baskar, V.; Chen, J.-T.; Bhuvaneshwari, V.; Ramesh, M. Transcriptomics, Cheminformatics, and Systems Pharmacology Strategies Unveil the Potential Bioactives to Combat COVID-19. *Molecules* **2022**, *27*, 5955. https://doi.org/10.3390/molecules27185955

Academic Editors: Giovanni N. Roviello and Caterina Vicidomini

Received: 16 August 2022
Accepted: 10 September 2022
Published: 13 September 2022

Abstract: Coronavirus disease (COVID-19) is a viral disease caused by the SARS-CoV-2 virus and is becoming a global threat again because of the higher transmission rate and lack of proper therapeutics as well as the rapid mutations in the genetic pattern of SARS-CoV-2. Despite vaccinations, the prevalence and recurrence of this infection are still on the rise, which urges the identification of potential global therapeutics for a complete cure. Plant-based alternative medicine is becoming popular worldwide because of its higher efficiency and minimal side effects. Yet, identifying the potential medicinal plants and formulating a plant-based medicine is still a bottleneck. Hence, in this study, the systems pharmacology, transcriptomics, and cheminformatics approaches were employed to uncover the multi-targeted mechanisms and to screen the potential phytocompounds from significant medicinal plants to treat COVID-19. These approaches have identified 30 unique COVID-19 human immune genes targeted by the 25 phytocompounds present in four selected ethnobotanical plants. Differential and co-expression profiling and pathway enrichment analyses delineate the molecular signaling and immune functional regulations of the COVID-19 unique genes. In addition, the credibility of these compounds was analyzed by the pharmacological features. The current holistic finding is the first to explore whether the identified potential bioactives could reform into a drug candidate to treat COVID-19. Furthermore, the molecular docking analysis was employed to identify the important bioactive compounds; thus, an ultimately significant medicinal plant was also determined. However, further laboratory evaluation and clinical validation are required to determine the efficiency of a therapeutic formulation against COVID-19.

Keywords: COVID-19; cheminformatics; molecular docking; phytocompounds; systems pharmacology

1. Introduction

Coronavirus disease (COVID-19), a contagious viral disease that emerged as a pandemic in 2020, is still a global threat because of the higher transmission rate. The causative agent for this disease is the SARS-CoV-2 virus, which belongs to the family of Coronaviridae [1]. Dry cough, pneumonia, fever, and fatigue are the common symptoms of this disease [2]. COVID disease is frequently associated with a lung infection and respiratory illness, which are the hallmarks of this disease [3]. COVID-19 has become a deadly world health crisis. Presently, the global prevalence of this disease has crossed 603 million cases with more than 6.4 million deaths (WHO—https://covid19.who.int/; https://www.cgtn.com/special/The-latest-on-the-COVID-19-pandemic.html, accessed on 7 September 2022) [4]. Vaccines play a major role in preventing this disease, but re-infection of this disease is still prevalent [5,6]. Despite the advanced medicinal technologies and facilities, a complete cure for this disease is still not identified. Furthermore, targeting the viral proteins is a major bottleneck due to the rapid mutations of the virus. Hence, identification of the human targets and novel therapeutic drugs are needed immediately.

Medicinal plants comprise plenty of biologically active pharmaceutical agents, most significantly in the era of disease treatments [7]. Medicinal plants have long been considered a massive storehouse of potent resources for humankind [1,8–10]. Exploring plant bioactive molecules could be a potential strategy to identify new therapeutic molecules, including anti-viral drugs, due to their structural and chemical diversity. Furthermore, plant bioactives have been utilized for pharmacotherapeutic purposes around the globe for a long time [1,9,11]. Some traditional Indian medicinal plants such as *Ammi visnaga, Amaranthus viridis, Terminalia bellirica,* and *Piper betle* were used to combat respiratory tract diseases [12–15]. Furthermore, these medicinal plants have diverse pharmacological activities such as immune-modulator, anti-viral, anti-microbial, anti-inflammatory, anti-tumorigenic, anti-cancer (breast, ovarian, cervical cancer), anti-mutagenic, antioxidant, analgesic, and anti-depressant [12–15].

One unique way to learn in-depth about how active biomolecules perform their therapeutic function is to impute the gene networks regulated by medicinal plant bioactive molecules [16,17]. In drug discovery, the one drug–one target–one disease strategy has become highly inefficient. Though single target methods might be a useful approach for single gene disorders, for multiple gene associated diseases, such one-target strategies are not fruitful [16]. Therefore, the concept of developing a multi-target treatment strategy against complex diseases, including COVID-19, COPD, and Alzheimer's diseases, is emerging in drug discovery. Regarding this, systems pharmacology unravels the disease mechanisms as potential networks targeted by synergistic, multiple drugs. Recently, the use of network/systems pharmacology for an in-depth understanding of the mode of action of bioactive compounds has become the most popular [1,9,10].

Considering the importance of Indian traditional medicine and network pharmacology, how the bioactive molecules will act and what all of their essential human targets are still remain major roadblocks. The main aim of this holistic study is to explore the prominent molecular mechanism and pharmaceutical activities of bioactive molecules and important medicinal plants against COVID-19. In addition, emerging advancements in systems pharmacology, transcriptomics, and important analytical tools such as cheminformatics and molecular docking analyses have paved the way to understanding the molecular mechanism of Indian traditional medicine to combat this deadly disease. Hence, the present study focuses on the transcriptomics, cheminformatics, molecular docking, and systems pharmacology approaches to identify the differentially expressed COVID-19-associated genes and the potential bioactive compounds from crucial medicinal plants such as *A. visnaga, A. viridis, T. bellirica,* and *P. betle,* which can be employed in the formulation of a cure from COVID-19, after clinical validation. This study would serve as an essential tool for dissecting novel biomolecules and their importance in traditionally used medicinal plants.

2. Materials and Methods

2.1. Selection of Medicinal Plants

Medicinal plants that are easily available in abundance, ethnobotanically important, and commonly utilized in Indian households have been selected, which include *A. visnaga, A. viridis, T. bellirica,* and *P. betle* [12–15].

2.2. Retrieval of Phytocompounds

Phytocompounds are the bioactive compounds present in plants with the characteristics of therapeutic potential. Phytocompounds present in the selected medicinal plants were identified using a literature survey and web sources [12–15]. The canonical SMILES for those phytocompounds were collected from the PubChem database (https://pubchem.ncbi.nlm.nih.gov/, accessed on 23 February 2022) [18].

2.3. Mining of Human Targets

Canonical SMILES obtained from the PubChem database were used as an input to obtain the compound-specific human targets through the SwissTargetPrediction tool (www.swisstargetprediction.ch/, accessed on 23 February 2022) [19]. In addition, reported 130 SARS-CoV-2-associated genes (Supplementary Table S1) were collected [1,20–22] for manual comparative analysis to find the unique genes.

2.4. Over-Representation Analysis (ORA)

ORA analysis is one of the most common statistical methods for determining over-represented genes in the subset of obtained data [23]. Collated receptors were provided as the input onto Network Analyst (www.networkanalyst.ca/, accessed on 12 March 2022) to identify the involvement of these genes in various molecular activities [24].

2.5. Compound-Target-Network (C-T-N)

C-T-N was constructed using the Cytoscape v3.8.2 plugin [25], which helps to investigate the deeper mechanisms of phytocompounds' activity. This network helps to understand the multiple interacting partners of the phytocompounds, thus providing the collective potentials of those compounds [26]. Furthermore, the understanding of the interactions between the potential target genes was then visualized by GeneMANIA (https://genemania.org/, accessed on 12 March 2022) [27].

2.6. Analyses of Whole Blood Human Transcriptome of Healthy Control and SARS-CoV-2 Infection

The identified unique genes were imported to COVID19db (http://hpcc.siat.ac.cn/covid19db/home, accessed on 2 September 2022) [28] and a complete method was used to obtain the differential expression heatmap plot against the human COVID-19 transcriptomic datasets of GEO (Gene Expression Omnibus) accession of healthy controls (GSM4622633, GSM4622634, GSM4622635, GSM4622636, GSM4622637) and COVID-19-infected (GSM4622702, GSM4622703, GSM4622704, GSM4622705, GSM4622706) whole blood tissue. Further, GO (Gene Ontology) enrichment analysis of these unique COVID-19-associated genes was also performed against these datasets in the COVID19db differential expression module [28] with significant parameter threshold values, such as log2FC cutoff ≥ 1, p-value cutoff ≤ 0.05, dysregulation type—all, GO-p-value cutoff ≤ 0.1, and GO-q-value cutoff ≤ 0.3 to attain the GO biological process and molecular functions according to the enriched GO terms.

2.7. Pathway Enrichment Analysis of Unique COVID-19 Genes

Biological pathways enrichment (Kyoto Encyclopedia of Genes and Genomes-KEGG) analysis of COVID-19-associated unique genes was carried out through g:Profiler (https://biit.cs.ut.ee/gprofiler/gost, accessed on 4 September 2022) [29] against "Homo sapiens" and a KEGG term ID with an adjusted p-value < 0.05 was considered as significant. The KEGG pathway terms with adjusted p-values were supposed to have potent effects on combating deadly COVID-19, and the pathway terms were illustrated by an adjusted p-value from low to high.

2.8. Corrplot Analysis

The unique COVID-19-responsible genes were uploaded to the corrplot co-expression module of COVID19db (http://hpcc.siat.ac.cn/covid19db/home, accessed on 2 September 2022) [28] against the above-mentioned COVID-19 transcriptomic datasets of GEO accession of healthy controls and COVID-19-infected whole blood tissue was employed with a circle visualization method to understand the correlationships among each two players. Further, a p-value cutoff was to set as a threshold for the significant correlation in accordance with the inbuilt database statistical program.

2.9. Box and Volcano Plot Analyses

The maximum phytocompound-linked and crucial COVID-19-pathway-associated unique genes were selected manually and subjected to the box plot differential expression module of COVID19db (http://hpcc.siat.ac.cn/covid19db/home, accessed on 2 September 2022) [28] against the above-mentioned GEO accession of healthy controls and COVID-19-infected whole blood tissue, executed with an inbuilt ANOVA statistical method to understand the expression of the players involved. Similarly, volcano plot analysis was also employed in the COVID19db [28] expression module with two significant parameters, such as log2FC and p-value cutoffs of ≥ 1 and ≤ 0.05, respectively, to unveil the players' expression level.

2.10. Active Compound Property

The Molinspiration tool (https://www.molinspiration.com/, accessed on 5 April 2022) [30] was employed to identify the features of these compounds, including a number of violations (nVio), enzyme inhibitory activity (Ei), Kinase inhibitory activity (Ki), Protease inhibitory activity (Pi), GPCR ligand activity (GPCR), and enzymes and nuclease receptors (Ncr). Among these features, nVio is the significant property in determining the credibility of using these compounds as a potent drug [1,9].

2.11. Molecular Docking

COVID-19-associated genes identified through differential expression profiling and their respective proteins' 3D structures were retrieved from PDB (protein data bank), and further energy was minimized by chimera [31]. Finally, molecular docking was performed for the compounds with their respective ligands using Autodock Vina 1.1 [32] to identify the potential compound that can be employed in the treatment of COVID-19 disease.

2.12. Toxicity and Drug Likeliness Properties

The potential phytocompounds identified using molecular docking were further subjected to the toxicity prediction analysis. AdmetSAR (http://lmmd.ecust.edu.cn/admetsar2/, accessed on 4 September 2022) [33] and SwissADME (http://www.swissadme.ch/, accessed on 4 September 2022) [34] online tools were employed to screen the toxic properties and health effects on human organs [35]. In addition, the drug likeliness properties of these compounds were analyzed using the SwissADME tool.

3. Results

3.1. Collation of Phytocompounds

A total of 34 phytocompounds were scrutinized from the selected medicinal plants, such as *A. visnaga*, *A. viridis*, *T. bellirica*, and *P. betle* (Table 1). The canonical SMILES and the structure for the selected phytocompounds are provided in Supplementary Table S2.

3.2. Human Targets and Unique COVID-Associated Genes Mining

Phytocompounds targeting human receptors were identified through SwissTargetPrediction. Study results revealed that out of 34 selected phytocompounds, 25 compounds targeting 375 human targets (9 compounds, namely AMM, KLOL, GALLO, CHAVIA, CHAVIME, CAPE, FPIN, LIME, and SPRO were eliminated from the further studies since these compounds do not have potential binding targets). Based on the gene probability score, genes are selected and shown in Table 2, which lists the gene features for each of the compounds along with the Uniprot ID, chromosome number, and orthologous information of the target. For identifying the unique COVID-19-associated genes obtained by the human targets and already reported 130 COVID-19-associated genes were matched and identified the 30 common/unique genes and the list of all the compounds with their unique active targets is provided in the Supplementary Table S3. It was observed that the 25 compounds were notably targeting these 30 unique COVID-19-associated genes.

Table 1. List of phytocompounds with their abbreviations.

S.No.	Common Name	Scientific Name	Compounds	Abbreviations
1	Toothpick-plant	*A. visnaga*	γ-pyrones (furanochromone) Khellin Visnagin Khellinol Ammiol Khellol Khellinin Coumarin	PYR KHE VGN KHNOL AMM KLOL KHEL CM
2	Slender amaranth	*A. viridis*	Quercetin Lutein Rutin Beta carotene Amasterol Squalene Spinasterol Polyprenol Phytol	QUE LUE RU BCAR AMST SQU SPIN POPRE PHY
3	Bibhitaki	*T. bellirica*	Corilagin Chebulagic acid Galloylpunicalagin Cardenolides	COR CHA GALLO CARD
4	Betel leafs	*P. betle*	Chavibetal Caryophyllene Chavibetol acetate Allylpyrocatechol Diacetate Chavibetal methyl ether Campene β-Pinene Eugenol u-limonene α-Pinene 1,8-Cineol Saprobe Allylpyrocatechol Monoacetate	CHAVI CAPH CHAVIA ALYD CHAVIME CAPE βPIN EU LIME αPIN CIN SPRO ALYM

Table 2. Features of compound-targeted potential receptors.

S.No.	Compound	Target	UniProt ID	Chr. No.	Orthologs
1	Pyrones	*NFKB1*	P19838	4	*Nfkb1*
2	Khellin	*HSP90AA1*	P07900	14	*Hsp90aa1*
3	Visnagin	*PTGS2*	Q6ZYK7	1	*Ptgs2*
4	Khellinol	*PTGS2*	Q6ZYK7	1	*Ptgs2*
5	Khellinin	*MMP1*	P03956	11	*Mmp1a*

Table 2. *Cont.*

S.No.	Compound	Target	UniProt ID	Chr. No.	Orthologs
6	Coumarin	*NFKB1*	P19838	4	*Nfkb1*
7	Quercetin	*EGFR*	Q504U8	7	*Egfr*
8	Lutein	*VDR*	P11473	12	*Vdr*
9	Rutin	*PTGS2*	Q6ZYK7	1	*Ptgs2*
10	Beta carotene	*MAPK14*	Q16539	6	*Mapk14*
11	Amasterol	*VDR*	P11473	12	*Vdr*
12	Squalene	*CCR5*	P51681	3	*Ccr5*
13	Spinasterol	*VDR*	P11473	12	*Vdr*
14	Polyprenol	*VDR*	P11473	12	*Vdr*
15	Phytol	*CCR1*	P32246	3	*Ccr1l1*
16	Corilagin	*PTGS2*	Q6ZYK7	1	*Ptgs2*
17	Chebulagic acid	*PTGS2*	Q6ZYK7	1	*Ptgs2*
18	Cardenolides	*MAPK14*	Q16539	6	*Mapk14*
19	Chavibetal	*MAPK14*	Q16539	6	*Mapk14*
20	Caryophyllene	*CCR5*	P51681	3	*Ccr5*
21	Allylpyrocatechol Diacetate	*TYK2*	P29597	19	*Tyk2*
22	Eugenol	*TLR7*	Q9NYK1	X	*Tlr7*
23	α-Pinene	*MAPK14*	Q16539	6	*Mapk14*
24	1,8-Cineol	*HSP90AA1*	P07900	14	*Hsp90aa1*
25	Allylpyrocatechol Monoacetate	*TLR7*	Q9NYK1	X	*Tlr7*

3.3. ORA Enrichment

ORA analysis using the Network Analyst tool predicted the involvement of these genes in numerous important pathways, such as the regulation of cytokine production, interleukin 8 production, and the viral reproductive process, which are important for the regulation of viruses (Figure 1). The activity of the selected phytocompounds on these targets may lessen the risk and severity of COVID-19 infection.

3.4. Compound-Target-Network Analysis

The C-T-N analysis was performed using Cytoscape v3.8.2 to visualize the interactions of 25 compounds targeting their 30 unique COVID-19-linked human targets (Figure 2). This interaction analysis helps to understand the multiple-target potentiality of these compounds and justifies the activity of these compounds as a potent drug to treat COVID-19.

3.5. Cross-Talks of Genes

Molecular interactions between 30 potential target genes revealed the complex interactions between the target genes and were visualized by GeneMANIA (Figure 3). These signaling cross-talks represent various types of interacting evidence, such as co-expression, and physical and genetic interactions.

3.6. Transcriptomic and Enrichment Profiling of Unique COVID-19-Associated Genes

The unique genes and their expression profiling were imputed from COVID19db. The heatmap plot exhibits that 30 unique genes were differentially regulated and distinguishes the COVID-19-infected patients from the healthy controls (Figure 4). Further, the heatmap plot indicates the diverse physiological and pathological conditions of these unique genes. Molecular gene enrichment of these 30 COVID-19-associated unique genes were involved in various biological processes and molecular functions. These players were predicted to be involved in the significant biological regulation of nucleosome organization, humoral immune response, defense response, anti-microbial response, and neutrophil degranulation, etc. (Figure 5). They were also predicted to be involved in different molecular functions such as RAGE receptor, glycosaminoglycan, heparin, icosanoid, long-chain fatty acid binding activities, etc. (Figure 6).

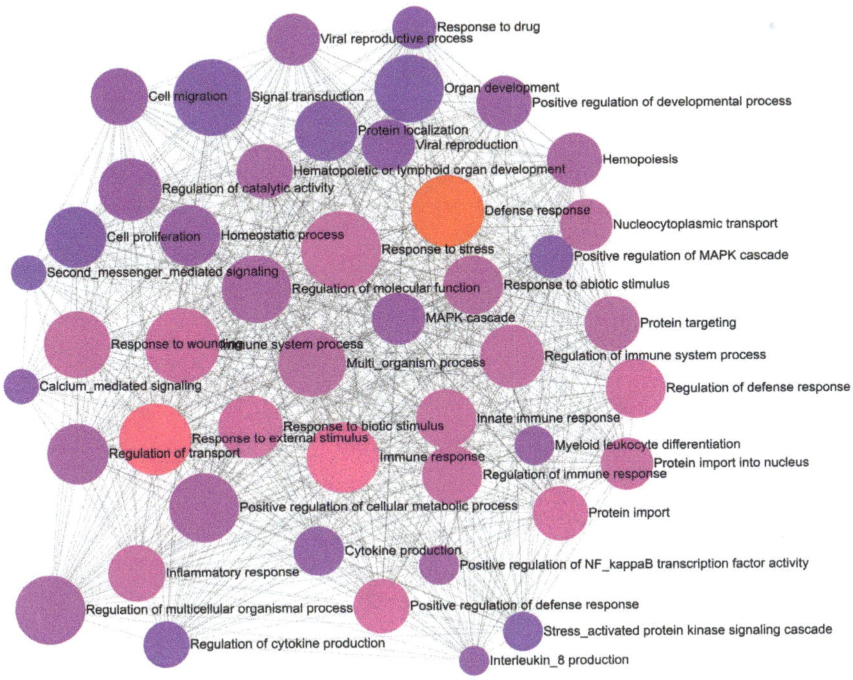

Figure 1. ORA predicts the involvement of genes in various activities.

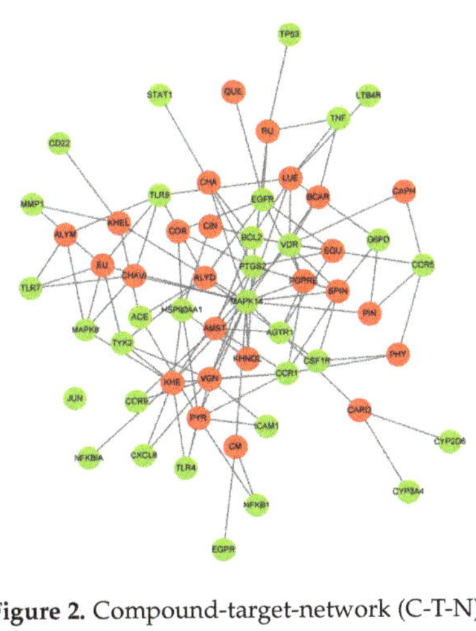

Figure 2. Compound-target-network (C-T-N) analysis depicts the interaction between the phytocompounds and human COVID-19 immune targets. Green color represents human COVID-19 immune targets and the red color indicates compounds.

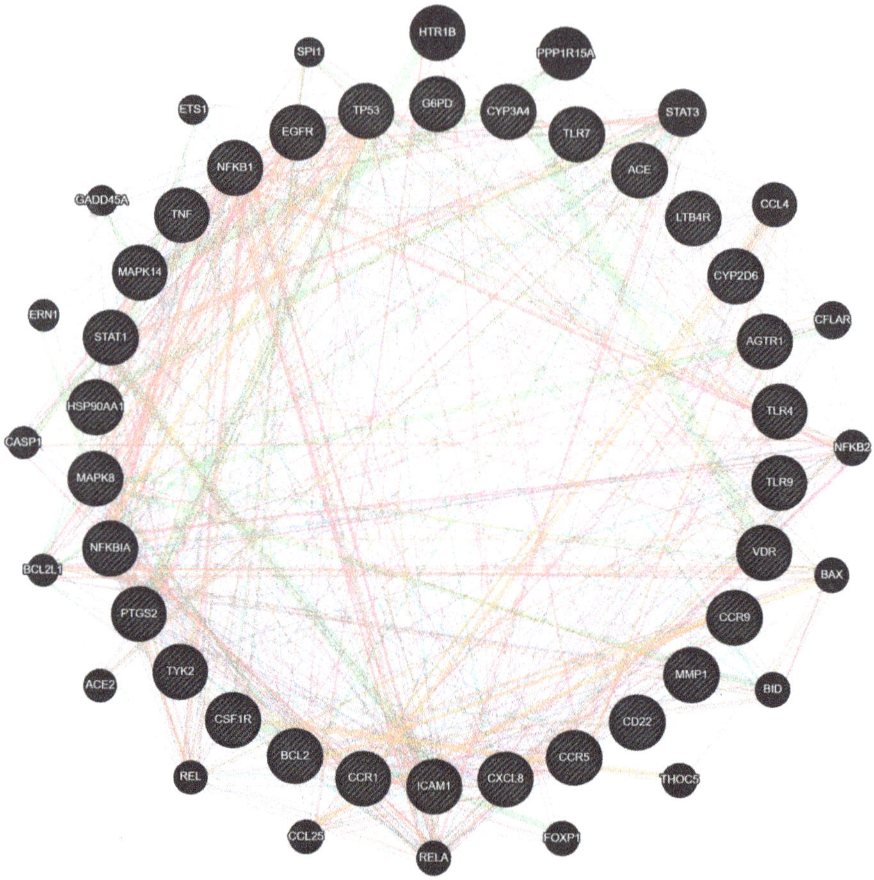

Figure 3. Visualization of COVID-19-associated unique gene interaction. Colored lines between the genes indicate various types of interacting evidence: periwinkle—co-expression; orange—physical interaction; green—genetic interaction; blue—localization; sky blue—involvement in pathways.

3.7. Analysis of KEGG Pathways

A total of 68 KEGG pathway terms were recognized. These pathways were ranked by adjusted *p*-value—0.05 and the enrichment conditions of COVID-19-linked 30 unique genes are represented in Figure 7. These genes were predominantly involved in COVID-19, toll-like receptor signaling pathway, NOD-like receptor signaling pathway, IL-17 signaling pathway, AGE–RAGE signaling pathway in diabetic complications, NF-kappa B signaling pathway, TNF signaling pathway, MAPK signaling pathway, pathways in cancer, and so on (Figure 7). Further, the results revealed that enriched pathways of the bioactives against COVID-19 were associated with cancers, various immune responses, and other cell cycle processes. The pathway results suggested that identified bioactives may strongly perform an anti-viral effect via regulating these pathways related to COVID-19-associated targets as exposed.

3.8. Correlationship Expression Analysis

The COVID-19-associated unique genes and their co-expression correlationship profiling were imputed by COVID19db. Corrplot showed that 30 unique genes are the closely associated genes that have a significant correlation with each other (Figure 8). Further, the association of this immune profile might be the significant immune deregulator mechanism induced by deadly COVID-19 infection, but further validation through in vitro and in vivo is a pre-requisite.

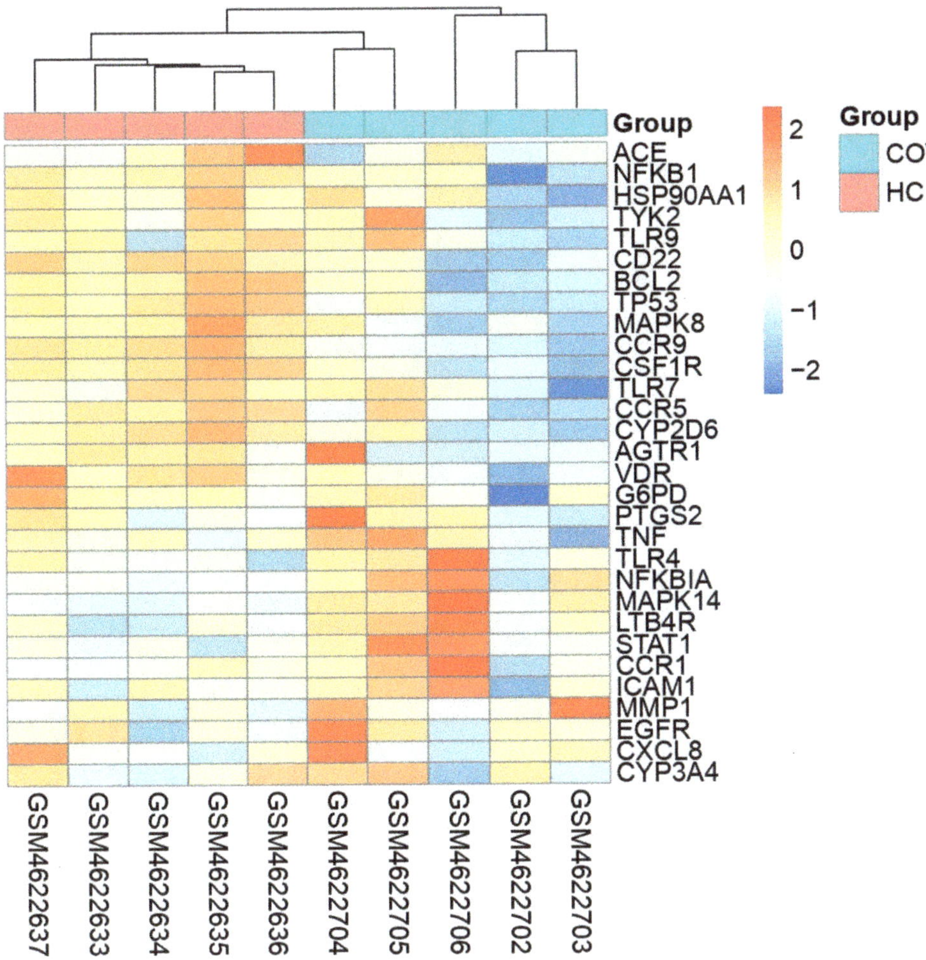

Figure 4. The heatmap plot denoting the 30 COVID-19-associated unique genes that are differentially regulated between the COVID-19-infected patients and healthy controls. The colored scale bar at the right indicates relative expression, where −2 and 2 represent down and up-regulation, respectively. Red color—up-regulation; blue color—down-regulation; yellow color—stable/non-significant expression. COV—COVID-19; HC—Healthy Control.

3.9. Box and Volcano Plot Analysis of Unique SARS-CoV-2-Associated Genes

Mitogen-activated protein kinase 14 (*MAPK14*), angiotensin-converting enzyme (*ACE*), toll-like receptor (*TLR4*), and *MAPK8* are the main players that are directly linked to the majority of the phytocompounds and played a major role in COVID-19-associated pathways. Box and volcano plots were performed for these players and suggested differential expression between different physiological and pathological conditions. Both box (Figure 9) and volcano (Figure 10) plots represent that the *MAPK14, ACE, TLR4,* and *MAPK8* genes were dysregulated after the COVID-19 infection.

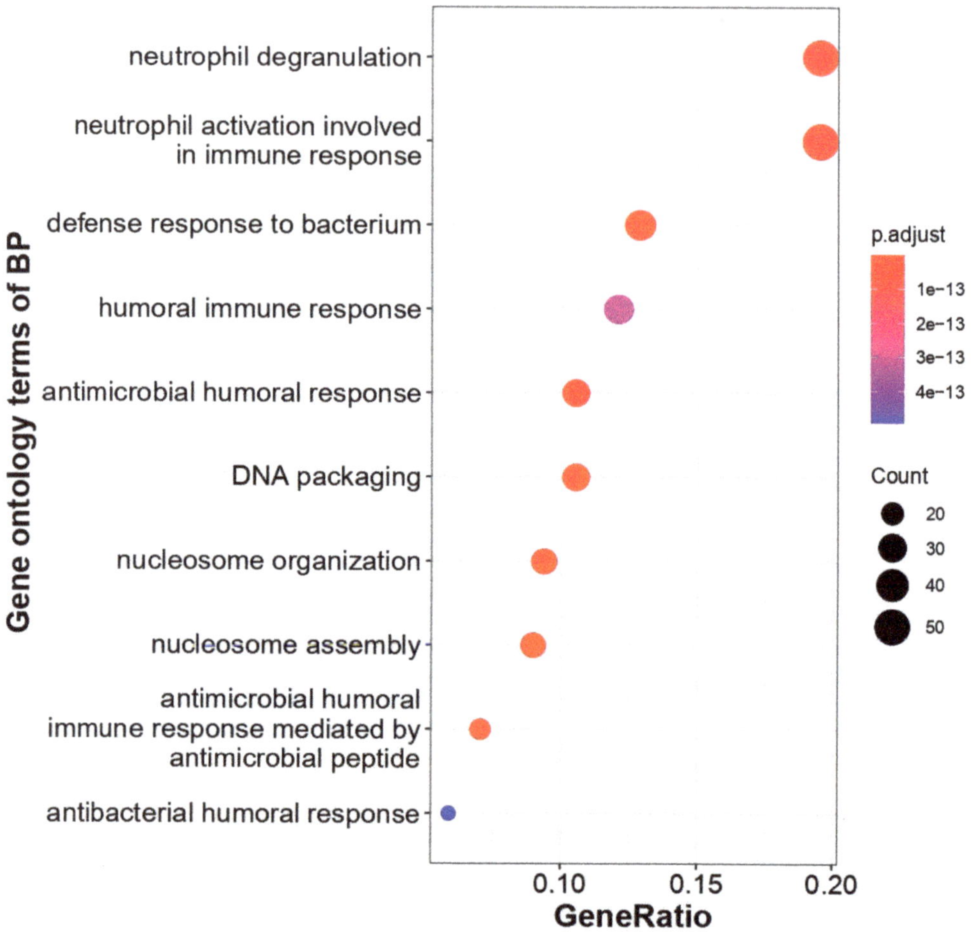

Figure 5. GO enrichments of COVID-19-associated unique genes with their biological processes. The number of unique COVID-19-associated genes falling in each GO biological process term is directly proportional to the count ball size. The balls are color-shaded according to the significant enrichment level (log2FC cutoff $\geq$ 1, *p*-value cutoff $\leq$ 0.05, dysregulation type—all, GO-*p*-value cutoff $\leq$ 0.1, GO-*q*-value cutoff $\leq$ 0.3).

3.10. Active Compound Property

Properties such as nVio, GPCR, Ncr, Ki, Pi, and Ei of these 25 compounds were identified using the molinspiration tool (Table 3). Phytocompounds that had zero nVio scores were considered highly significant compounds. This analysis identifies 14 compounds as significant with zero nVio score and only these compounds were employed for further analysis.

3.11. Molecular Docking Analysis

MAPK14, ACE, TLR4, and MAPK8 genes are present in all the compounds, and these genes are associated with COVID-19 pathways. Molecular docking was performed for these common genes with their potentially targeting compounds (Figure 11). The results are given in Table 4.

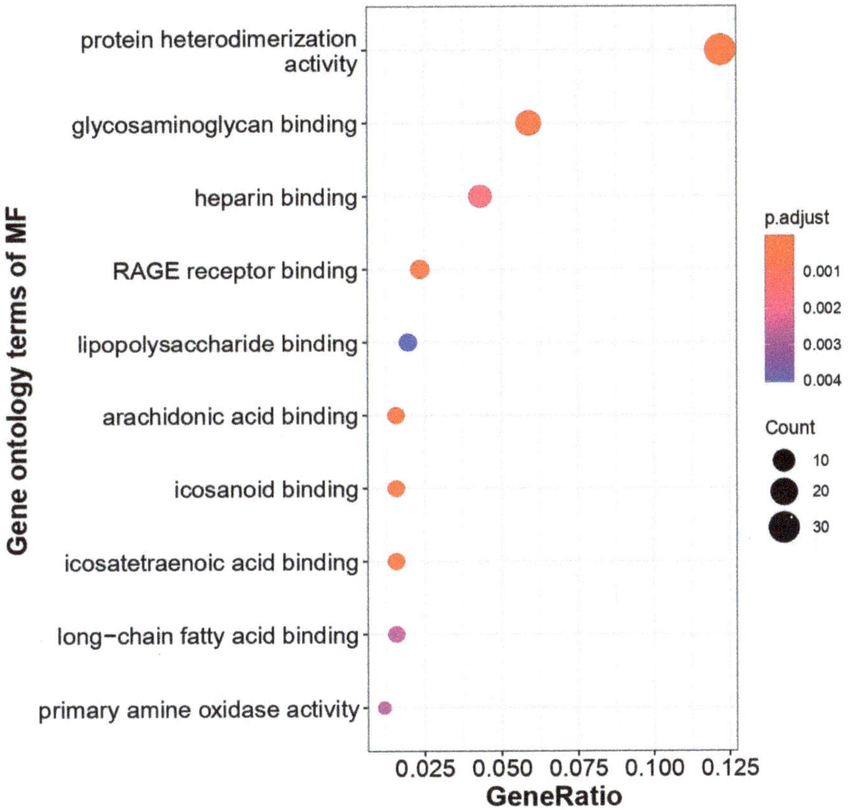

Figure 6. COVID-19-associated unique genes and their enriched molecular function. These unique genes' fall in each GO molecular function term is directly proportional to the count ball size. The balls are color-shaded according to the significant enrichment level (log2FC cutoff ≥ 1, *p*-value cutoff ≤ 0.05, dysregulation type—all, GO-*p*-value cutoff ≤ 0.1, GO-*q*-value cutoff ≤ 0.3).

Table 3. Bio-properties of phytocompounds.

S. No.	Compounds	GPCR Ligand	Kinase Inhibitor	Nuclear Receptor Ligand	Protease Inhibitor	Enzyme Inhibitor	No. of Violations
1	Pyrones	−0.63	−0.95	−1.12	−1.28	−0.23	0
2	Khellin	−0.36	−0.51	−0.51	−0.64	−0.07	0
3	Visnagin	−0.55	−0.79	−0.79	−0.92	−0.28	0
4	Khellinol	−0.39	−0.64	−0.49	−0.77	0.01	0
5	Khellinin	0.07	−0.23	−0.37	−0.06	0.29	0
6	Coumarin	−1.44	−1.57	−1.42	−1.43	−0.58	0
7	Quercetin	−0.06	0.28	0.36	−0.25	0.28	0
8	Lutein	0.03	−0.25	0.47	−0.03	0.28	2
9	Rutin	−0.05	−0.14	−0.23	−0.07	0.12	3
10	Beta carotene	−0.04	−0.15	0.4	−0.06	0.17	2
11	Amasterol	0.09	−0.31	0.88	−0.14	0.61	1
12	Squalene	0.04	−0.1	0.19	−0.03	0.16	1
13	Spinasterol	0.18	−0.3	0.68	0.06	0.53	1
14	Polyprenol	0.14	−0.14	0.37	0.02	0.33	1
15	Phytol	0.11	−0.32	0.35	0	0.31	1

Table 3. *Cont.*

S. No.	Compounds	GPCR Ligand	Kinase Inhibitor	Nuclear Receptor Ligand	Protease Inhibitor	Enzyme Inhibitor	No. of Violations
16	Corilagin	−0.11	−0.45	−0.44	−0.03	−0.15	3
17	Chebulagic acid	−3.5	−3.72	−3.68	−3.23	−3.56	3
18	Cardenolides	0.11	−0.48	0.39	−0.17	0.63	0
19	Chavibetal	−0.86	−1.14	−0.78	−1.29	−0.41	0
20	Caryophyllene	−0.34	−0.78	0.13	−0.6	0.19	1
21	Allylpyrocatechol Diacetate	−0.46	−0.75	−0.25	−0.67	−0.18	0
22	Eugenol	−0.86	−1.14	−0.78	−1.29	−0.41	0
23	α-Pinene	−0.48	−1.5	−0.62	−0.85	−0.34	0
24	1,8-Cineol	−0.93	−1.6	−1.07	−0.9	−0.15	0
25	Allylpyrocatechol Monoacetate	−0.7	−1.04	−0.39	−0.95	−0.23	0

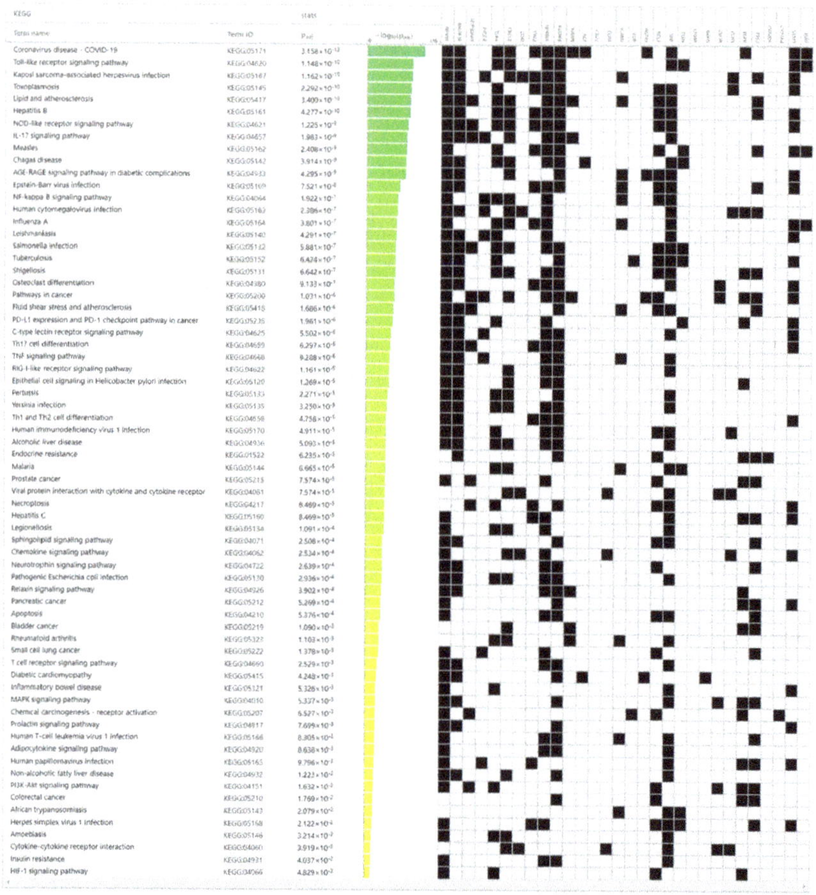

Figure 7. Biological pathway enrichment analysis of identified bioactives against COVID-19. KEGG enrichment pathway analysis identification result according to an adjusted *p*-value of 0.05 and the ordinate indicates the −log10 (adjusted *p*-value) of the KEGG terms.

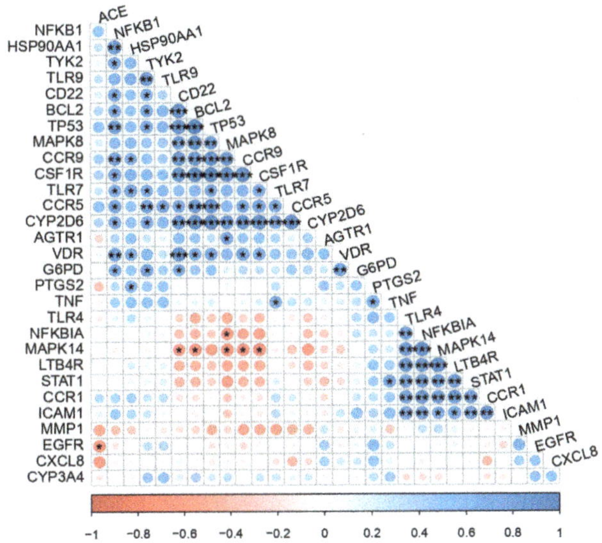

Figure 8. Corrplot indicates the correlationships between the unique COVID-19-associated genes. Blue color—positive correlations, red color—negative correlation, blank/white boxes—insignificant. The size of the circle and the color intensity are directly proportional to the correlation coefficients. In addition, p-value cutoff (*: $p \leq 0.05$; **: $p \leq 0.01$; ***: $p \leq 0.001$) is to set a threshold for the significant correlation.

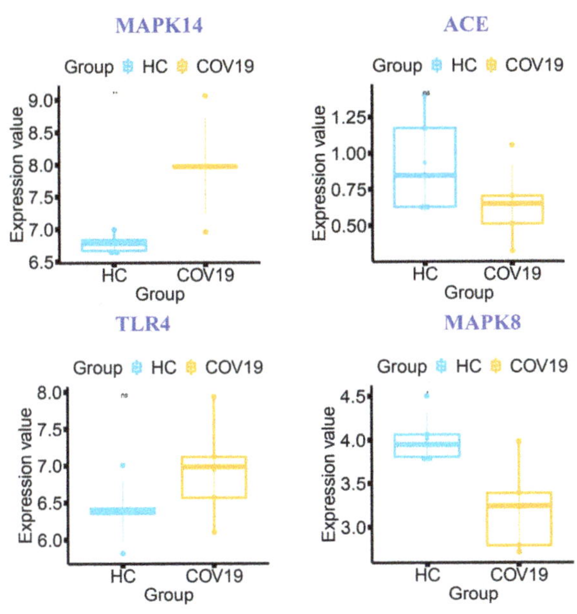

Figure 9. Box plots encode the 4 COVID-19-associated gene expression levels in COVID-19-infected compared with healthy controls (HCs). The plots indicate that the players were dysregulated after the SARS-CoV-2 infection. HCs are marked in blue, and COVID-19-infected tissues are marked in yellow. Significant difference of the expression value was calculated by inbuilt ANOVA and significant **: $p \leq 0.01$; ns: $p \geq 0.05$; *: $p \leq 0.05$ values.

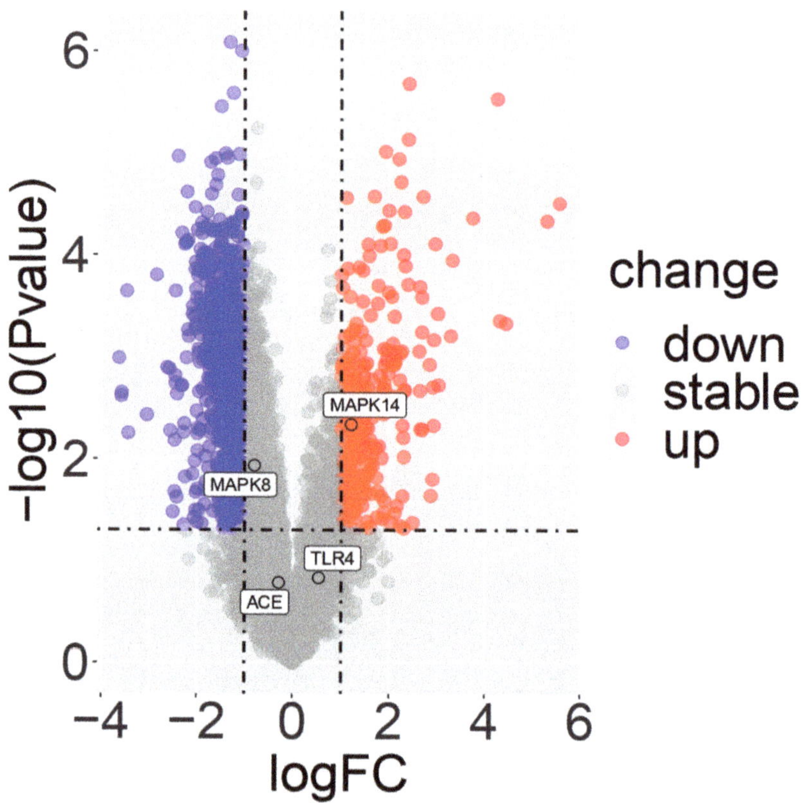

Figure 10. Volcano plots encode the 4 COVID-19-associated gene expression levels in COVID-19-infected whole blood tissue transcriptome datasets. Scattered balls encode genes. The x-axis indicates the log2 fold change levels between HC and COVID-19-infected, whereas the y-axis is adjusted *p*-value based on –log10. The plots indicate that the players were dysregulated after the SARS-CoV-2 infection. Red color—up-regulation, blue color—down-regulation, gray color—stable/non-significant. Genes revealed the differentially expressed players in accordance with significant thresholds.

3.12. Identification of Significant Medicinal Plant

Molecular docking analysis revealed that the compound khellin interacts with all four COVID-19-responsible genes with significant binding energy and compounds, namely, coumarin, khellinin, khellinol, and visnagin possess significant binding energy with the COVID-19-responsible gene *MAPK14*. These compounds were present in the plant *A. visnaga* (toothpick-plant). Hence, this plant was predicted as a significant medicinal plant and its bioactives can be utilized in the treatment of COVID-19.

3.13. Analysis of Toxicity and Drug Likeliness Properties

The toxicity properties for the significant phytocompounds selected through molecular docking analysis were carried out using AdmetSAR and SwissADME software. Analyzing the toxicity characteristics of phytocompounds is highly essential for it to qualify as a potential drug candidate in the human body. The hERG I and II inhibition action, LD50 (rat), skin toxicity, and carcinogenicity properties of these compounds were determined using AdmetSAR software. Solubility and drug likeliness properties (Lipinski's rule of 5) were determined using SwissADME software. The results of these analyses are tabulated in Supplementary Table S4.

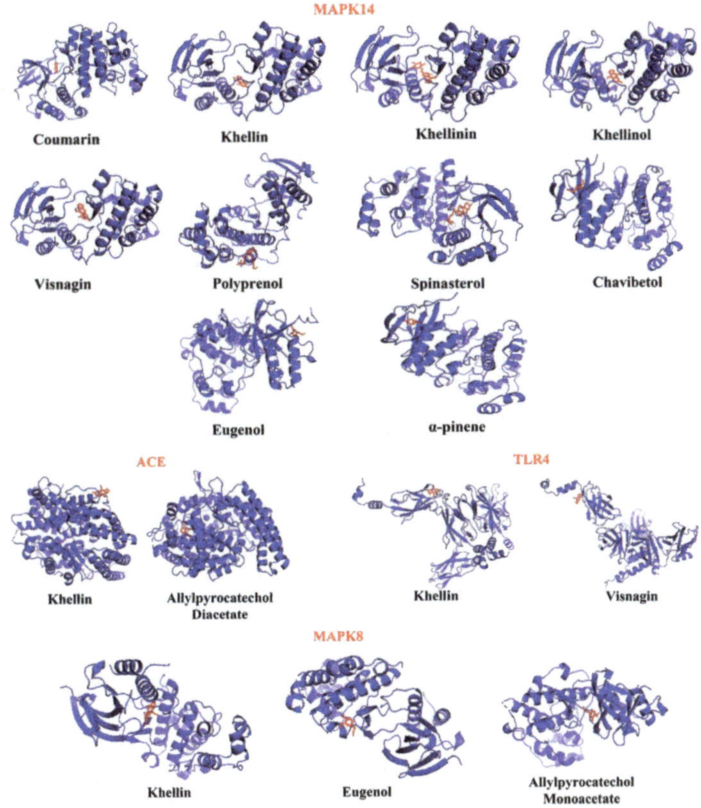

Figure 11. Molecular docking analysis: three-dimensional (3D) representation of interaction patterns of potentially bioactive compounds and COVID-19-associated MAPK14, ACE, TLR4, MAPK8 genes.

Table 4. The binding affinity of compounds docked against human target receptors.

Receptor	Ligand	Binding Affinity (kcal/mol)
MAPK14	Coumarin	−6.1
	Khellin	−6.1
	Khellinin	−7.9
	Khellinol	−7
	Visnagin	−6.8
	Polyprenol	−2.8
	Spinasterol	−6
	Chavibetol	−5.9
	Eugenol	−5.8
	α-pinene	−6
ACE	Allylpyrocatechol Diacetate	−5.5
	Khellin	−6.4
TLR4	Khellin	−5.7
	Visnagin	−6.3
MAPK8	Khellin	−6.4
	Allylpyrocatechol Monoacetate	−5.5
	Eugenol	−5.3

4. Discussion

COVID-19 is a viral pandemic that predominantly affects the lungs [36] and requires comprehensive research to identify a global treatment method. Since modern medicines cannot control this disease, the search for an efficient alternative medicinal source becomes highly important. Alternative medicines are becoming a trend all over the world and, at present, nearly more than half of the world population has started using alternative medicines [37]. Plants form the basis of various successful formulations to combat various illnesses, but the mode of action and molecular mechanisms remain not fully classified [38]. Thus, this study utilizes the new arena in the name of transcriptomics, cheminformatics, and systems pharmacology to identify potential bioactives from medicinal plants and uncovers the compound associated with human COVID-19 immune targets for treating COVID-19 disease. In addition, it helps to predict the activity of these compounds in the molecular aspects. Moreover, the plant-derived bioactives are less toxic compared to synthetic drug molecules. Therefore, the current findings prescribe the bioactive molecules used to combat SARS-CoV-2 by stimulating the immune regulations.

In the present study, four medicinal plants predominantly used in India were selected and 34 bioactives were identified from the selected plants. Among 34 compounds, 25 compounds were found to possess strong interactions with 30 unique and COVID-19-associated human receptors. Subsequently, the network was constructed using Cytoscape 3.8.2 to predict the therapeutical potentials of these compounds.

In our previous study, we highlighted that multiple genes and pathways are involved in disease progression, so targeting an individual gene does not provide relevant results [10]. Hence, targeting multiple receptors at the same time might provide an efficient therapeutical effect. In this respect, the C-T-N analysis revealed multiple targets of the selected compounds. Since numerous molecular and biological factors are involved in COVID-19 disease, identifying compounds with multiple targets can provide a better solution.

Transcriptomic analyses were performed to understand the differential expression pattern of unique genes between COVID-19-infected patients and healthy controls by using publicly available transcriptomic datasets in COVID19db. A heatmap plot was generated based on the transcriptomic datasets' (GEO accessions) expression intensities of the 30 COVID-19-associated unique genes that were differentially expressed and clearly distinguished the COVID-19-infected patients from their controls. Notably, differential expression and co-expression of these players were imputed, and they are involved in various molecular activities against COVID-19. Functional regulations and attributes of these genes were demonstrated by gene and pathway enrichments, corrplot analysis, and box and volcano plots. The results of differential expression and co-expression analyses delineate that these unique genes are more specific to COVID-19. In addition, the study showed differential expression regulation of these unique genes between different physiological and pathological conditions.

Interestingly, transcriptomic analyses confirm that these unique COVID-19 players are linked with many other viral infections such as hepatitis B and C, pertussis, influenza A, measles, tuberculosis, human papilloma virus, rheumatoid arthritis, herpes simplex virus infection, and so on. The correlation between the host immune responses to these viral infections and COVID-19 is far from clear. Suppositionally, a spectrum of drugs is already available for treating these viral infections and may have therapeutic properties against COVID-19 infection [1,39,40].

As described in KEGG enrichment analysis, these 30 unique genes were putatively involved in the regulation of various immune-response-associated pathways such as NOD-like receptor, NF-kappa B, toll-like receptor, HIF-1 signaling, IL-17, TNF, PI3K-Akt, T-cell receptor, and MAPK signaling pathways. In addition, identified bioactive compounds that target the COVID-19-associated unique genes and also inhibit their functions and their results, altering the above-mentioned pathways, and inducing the production of anti-COVID-19 antibodies through antigen-presenting cells.

Further, the pharmacological features, such as GPCR, Pi, Ki, Ncr, Ei, and nVio influenced the oral bioavailability, solubility, and permeability of the drug and were predicted through experimentally validated computational approaches in accordance with rule-of-5 (Ro5) drug discovery. Compounds that satisfied the decided criteria were selected for further studies [41,42]. Targets of those compounds were compared with the COVID-19-specific targets to identify the potential genes and were subjected to molecular docking analysis.

Finally, molecular docking analysis identified *A. visnaga* as a highly significant medicinal plant to treat COVID-19, which houses compounds such as khellin, khellinin, khellinol, and visnagin. Among the compounds, khellin interacts with all of the selected genes (*MAPK14, ACE, TLR4*, and *MAPK8*), and Khellinin interacts with *MAPK14* with the highest binding affinity of -7.9 kcal/mol. These compounds may exert potent pharmacological activity to combat SARS-CoV2 through associated immune-modulation, anti-inflammatory, and anti-viral effects. Employing the molecular docking and systems network pharmacological strategies to uncover the molecular mechanisms, anti-SARS-CoV2/COVID-19 effects of these potential bioactive molecules could be shown to be altered by key bioactives and some corresponding genes such as *MAPK14, ACE, TLR4*, and *MAPK8*. These are significant genes that are directly correlated with COVID-19 infection [1,20–22]. In general, *MAPK14* is one of the important components of the MAP kinase signaling pathway, which activates the transcription factors directly when evoked by physical stress or cytokines. *MAPK8* is also involved in several activities similar to *MAPK14* and also in various other cellular processes such as apoptosis, proliferation, and differentiation. *TLR4* is important in innate immune response mechanisms against antigens and stimulates pro-inflammatory responses. Consequently, an interaction between TLR4 and SARS-CoV2 spike proteins could be the reason behind the COVID-19 immunopathological expression [1]. The activity of ACE leads to an increase in vasoconstriction by inactivating the vasodilator agent, bradykinesia. In addition, the main cellular receptor for SARS-CoV2 is ACE, the expression of ACE genes (ACE1 and ACE2), which induce exposure to infection. Further, this gene has been associated with respiratory tract disease and plays a prominent role in the severity of SARS-CoV2 [43]. Toxicity and drug likeliness analysis results made evident that none of the selected phytocompounds possess toxicity properties. These compounds do not violate the Lipinski rule of 5, which clears the primary evaluation for their drug likeliness activity. Further, solubility is one of the significant parameters to determine the pharmacological response of the drug [44] and all the selected compounds were found to be soluble. Thus, these compounds can be considered as a potential drug candidate, but the imputed results are still required to be experimentally validated rigorously. These results claim that bioactive compounds in *A. visnaga* plants have dynamic interactions with COVID-19-responsible genes, thus possessing significant therapeutical actions.

5. Conclusions

The present study reveals the therapeutic potentials of commonly used Indian medicinal plants against the deadly COVID-19 pandemic. In short, amidst the selected four medicinal plants, *A. visnaga*, commonly known as toothpick-plant, was predicted as a potent medicinal plant against COVID-19. Bioactive compounds in this plant target the significant COVID-19 responsible receptors such as MAPK14, ACE, TLR4, and MAPK8, which makes this plant an ideal candidate for treating this deadly infection. Traditionally, this plant has multiple medicinal properties and is generally prescribed in Ayurvedic formulations to dilate blood vessels without affecting blood pressure. In some parts of the world, this plant was used against kidney and heart ailments. In a line, this study explores the possibility of the usage of *A. visnaga* in COVID-19 treatment and also unravels the various biological processes and their signaling interactions that will aid the way to open the COVID-19 advanced research sluicegates in a combination of Indian Ayurveda and modern medicine. Further, the study also serves as a notable pioneer for researchers and budding scientists working in the era of medicinal plant biology/omics by employing systems pharmacology approaches.

Further, this pilot study hypothesizes that the formulation of phytocompounds present in *A. visnaga* can be used in the treatment of COVID-19 disease. However, these predictions need to be evaluated using in vitro and in vivo experimentations to design antagonists for COVID-19-responsible human targets, which ultimately may result in the cure of COVID-19 disease. In addition, this study undoubtedly provides a conceptual shift in the development of drug discovery and creates an impact on the globalization and modernization of plant bioactive molecules. Our holistic study lays the groundwork for enabling further research on the molecular perspective of Indian traditional medicinal plants in disease treatment, immunobiology of various diseases including SARS-CoV2, and the applications of systems pharmacology in novel drug discovery.

Supplementary Materials: The following supporting information can be downloaded at: https://www.mdpi.com/article/10.3390/molecules27185955/s1, Table S1: List of COVID-19-associated genes collected from the literature; Table S2: Bioactive compounds and their canonical SMILES and structures; Table S3: Identified 30 unique/common genes associated with COVID-19; Table S4: Toxicity and drug likeliness properties of potential compounds

Author Contributions: P.M. and H.S. conceived and designed the experiments. S.A. (Sivakumar Adarshan), S.A. (Sakthivel Akassh), K.A., M.B. and P.M. performed the experiments. S.A. (Sivakumar Adarshan), P.M., H.S., V.B. (Venkidasamy Baskar), J.-T.C., V.B. (Veluswamy Bhuvaneshwari) and M.R. analyzed the results. S.A. (Sivakumar Adarshan) and P.M. wrote the manuscript. P.M., V.B. (Venkidasamy Baskar), J.-T.C. and M.R. reviewed and edited the manuscript. P.M., H.S. and J.-T.C. approved the final version of the manuscript. All authors have read and agreed to the published version of the manuscript.

Funding: This research received no external funding.

Data Availability Statement: All data generated or analyzed during this study are included in this article and its Supplementary Information files.

Conflicts of Interest: The authors declare that they have no conflicts of interest with the contents of this article.

Sample Availability: Samples of the compounds are not available from the authors.

Abbreviations

COVID-19	Coronavirus disease-19
C-T-N	Compound-target-network
ORA	Over-representation analysis
WHO	World Health Organization
SARS-CoV2	Severe acute respiratory syndrome coronavirus 2

References

1. Muthuramalingam, P.; Jeyasri, R.; Valliammai, A.; Selvaraj, A.; Karthika, C.; Gowrishankar, S.; Pandian, S.K.; Ramesh, M.; Chen, J.-T. Global Multi-Omics and Systems Pharmacological Strategy Unravel the Multi-Targeted Therapeutic Potential of Natural Bioactive Molecules against COVID-19: An In Silico Approach. *Genomics* **2020**, *112*, 4486–4504. [CrossRef] [PubMed]
2. Wu, Y.-C.; Chen, C.-S.; Chan, Y.-J. The Outbreak of COVID-19: An Overview. *J. Chin. Med. Assoc.* **2020**, *83*, 217–220. [CrossRef] [PubMed]
3. Tang, D.; Comish, P.; Kang, R. The Hallmarks of COVID-19 Disease. *PLoS Pathog.* **2020**, *16*, e1008536. [CrossRef] [PubMed]
4. WHO. WHO Coronavirus (COVID-19) Dashboard. Available online: https://covid19.who.int/ (accessed on 7 September 2022).
5. Forni, G.; Mantovani, A. COVID-19 vaccines: Where we stand and challenges ahead. *Cell Death Differ.* **2021**, *28*, 626–639. [CrossRef]
6. Peltan, I.D.; Beesley, S.J.; Webb, B.J.; Lopansri, B.K.; Sinclair, W.; Jacobs, J.R.; Brown, S.M. Evaluation of potential COVID-19 recurrence in patients with late repeat positive SARS-CoV-2 testing. *PLoS ONE* **2021**, *16*, e0251214. [CrossRef]
7. Pal, S.K.; Shukla, Y. Herbal medicine: Current status and the future. *Asian Pac. J. Cancer Prev.* **2003**, *4*, 281–288.

8. Rehman, A.; Ashfaq, U.A.; Shahid, F.; Noor, F.; Aslam, S. The Screening of phytochemicals against NS5 Polymerase to treat Zika Virus infection: Integrated computational based approach. *Comb. Chem. High Through. Screen.* **2021**, *25*, 738–751. [CrossRef]
9. Jeyasri, R.; Muthuramalingam, P.; Suba, V.; Ramesh, M.; Chen, J.-T. Bacopa Monnieri and Their Bioactive Compounds Inferred Multi-Target Treatment Strategy for Neurological Diseases: A Cheminformatics and System Pharmacology Approach. *Biomolecules* **2020**, *10*, 536. [CrossRef]
10. Adarshan, S.; Muthuramalingam, P.; Jeyasri, R.; Lakshmi, M.A.; Sathishkumar, R.; Pandian, S.K.; Shin, H.; Chen, J.T.; Ramesh, M. *Vitex negundo* L. derived specialized molecules unveil the multi-targeted therapeutic avenues against COPD: A systems pharmacology approach. *Front. Biosci.-Landmark* **2022**, *27*, 87. [CrossRef]
11. Tan, N.; Gwee, K.A.; Tack, J.; Zhang, M.; Li, Y.; Chen, M.; Xiao, Y. Herbal medicine in the treatment of functional gastrointestinal disorders: A systematic review with meta-analysis. *J. Gastroenterol. Hepatol.* **2020**, *35*, 544–556. [CrossRef]
12. Al-Snafi, A.E. Chemical constituents and pharmacological activities of *Ammi majus* and *Ammi visnaga*. A Review. *Int. J. Pharm.* **2013**, *3*, 10.
13. Ragasa, C.Y.; Austria, J.P.M.; Subosa, A.F.; Torres, O.B.; Shen, C.-C. Chemical Constituents of Amaranthus Viridis. *Chem. Nat. Compd.* **2015**, *51*, 146–147. [CrossRef]
14. Sobeh, M.; Mahmoud, M.F.; Hasan, R.A.; Abdelfattah, M.A.O.; Osman, S.; Rashid, H.-O.; El-Shazly, A.M.; Wink, M. Chemical Composition, Antioxidant and Hepatoprotective Activities of Methanol Extracts from Leaves of *Terminalia Bellirica* and *Terminalia Sericea* (Combretaceae). *PeerJ* **2019**, *7*, e6322. [CrossRef]
15. Shah, S.K.; Garg, G.; Jhade, D.; Patel, N. Piper Betle: Phytochemical, Pharmacological and Nutritional Value in Health Management. *Int. J. Pharm. Sci. Rev. Res.* **2016**, *38*, 181–189.
16. Casas, A.I.; Hassan, A.A.; Larsen, S.J.; Gomez-Rangel, V.; Elbatreek, M.; Kleikers, P.W.; Guney, E.; Egea, J.; López, M.G.; Baumbach, J.J. From single drug targets to synergistic network pharmacology in ischemic stroke. *Proc. Natl. Acad. Sci. USA* **2019**, *116*, 7129–7136. [CrossRef]
17. Noor, F.; Tahir ul Qamar, M.; Ashfaq, U.A.; Albutti, A.; Alwashmi, A.S.; Aljasir, M.A. Network Pharmacology Approach for Medicinal Plants: Review and Assessment. *Pharmaceuticals* **2022**, *15*, 572. [CrossRef]
18. Kim, S.; Thiessen, P.A.; Bolton, E.E.; Chen, J.; Fu, G.; Gindulyte, A.; Han, L.; He, J.; He, S.; Shoemaker, B.A.; et al. PubChem Substance and Compound Databases. *Nucleic Acids Res.* **2016**, *44*, D1202–D1213. [CrossRef]
19. Daina, A.; Michielin, O.; Zoete, V. SwissTargetPrediction: Updated Data and New Features for Efficient Prediction of Protein Targets of Small Molecules. *Nucleic Acids Res.* **2019**, *47*, W357–W364. [CrossRef]
20. Kumar, S. COVID-19: A Drug Repurposing and Biomarker Identification by Using Comprehensive Gene-Disease Associations through Protein-Protein Interaction Network Analysis. *Preprints* **2020**, 2020030440. [CrossRef]
21. Li, R.; Wu, K.; Li, Y.; Liang, X.; Tse, W.K.F.; Yang, L.; Lai, K.P. Revealing the Targets and Mechanisms of Vitamin A in the Treatment of COVID-19. *Aging* **2020**, *12*, 15784–15796. [CrossRef]
22. Zheng, H.; Cao, J.J. Angiotensin-Converting Enzyme Gene Polymorphism and Severe Lung Injury in Patients with Coronavirus Disease 2019. *Am. J. Pathol.* **2020**, *190*, 2013–2017. [CrossRef]
23. Karp, P.D.; Midford, P.E.; Caspi, R.; Khodursky, A. Pathway size matters: The influence of pathway granularity on over-representation (enrichment analysis) statistics. *BMC Genom.* **2021**, *22*, 191. [CrossRef]
24. Hevey, D. Network Analysis: A Brief Overview and Tutorial. *Health Psychol. Behav. Med.* **2018**, *6*, 301–328. [CrossRef]
25. Shannon, P.; Markiel, A.; Ozier, O.; Baliga, N.S.; Wang, J.T.; Ramage, D.; Amin, N.; Schwikowski, B.; Ideker, T. Cytoscape: A software environment for integrated models of biomolecular interaction networks. *Genome Res.* **2003**, *13*, 2498–2504. [CrossRef]
26. Syahrir, N.H.A.; Sumarheni; Hasbi, M. Compound-Target Prediction and Network-Target Analysis on Jamu Formula. *J. Phys. Conf. Ser.* **2021**, *1752*, 012028. [CrossRef]
27. Franz, M.; Rodriguez, H.; Lopes, C.; Zuberi, K.; Montojo, J.; Bader, G.D.; Morris, Q. GeneMANIA update 2018. *Nucleic Acids Res.* **2018**, *46*, W60–W64. [CrossRef]
28. Zhang, W.; Zhang, Y.; Min, Z.; Mo, J.; Ju, Z.; Guan, W.; Zeng, B.; Liu, Y.; Chen, J.; Zhang, Q.; et al. COVID19db: A comprehensive database platform to discover potential drugs and targets of COVID-19 at whole transcriptomic scale. *Nucleic Acids Res.* **2022**, *50*, D747–D757. [CrossRef]
29. Raudvere, U.; Kolberg, L.; Kuzmin, I.; Arak, T.; Adler, P.; Peterson, H.; Vilo, J. g: Profiler: A web server for functional enrichment analysis and conversions of gene lists (2019 update). *Nucleic Acids Res.* **2019**, *47*, W191–W198. [CrossRef]
30. Rajkumar, T.; Shirisha, P.; Faisal, S.S.; Hussain, S.A.; Nagaiah, T.; Sulthana, A. Prediction of molecular properties, bioactivity & docking of novel hydrazines. *Int. J. Pharm. Life Sci.* **2016**, *2*, 757–768.
31. Pettersen, E.F.; Goddard, T.D.; Huang, C.C.; Couch, G.S.; Greenblatt, D.M.; Meng, E.C.; Ferrin, T.E. UCSF Chimera—A visualization system for exploratory research and analysis. *J. Comput. Chem.* **2004**, *25*, 1605–1612. [CrossRef]
32. Trott, O.; Olson, A.J. AutoDock Vina: Improving the speed and accuracy of docking with a new scoring function, efficient optimization, and multithreading. *J. Comput. Chem.* **2010**, *31*, 455–461. [CrossRef] [PubMed]
33. Yang, H.; Lou, C.; Sun, L.; Li, J.; Cai, Y.; Wang, Z.; Li, W.; Liu, G.; Tang, Y. admetSAR 2.0: Web-service for prediction and optimization of chemical ADMET properties. *Bioinformatics* **2019**, *35*, 1067–1069. [CrossRef] [PubMed]
34. Daina, A.; Michielin, O.; Zoete, V. SwissADME: A free web tool to evaluate pharmacokinetics, drug-likeness and medicinal chemistry friendliness of small molecules. *Sci. Rep.* **2017**, *7*, 42717. [CrossRef] [PubMed]
35. Parasuraman, S. Toxicological screening. *J. Pharmacol. Pharmacother.* **2011**, *2*, 74.

36. Sadhukhan, P.; Ugurlu, M.T.; Hoque, M.O. Effect of COVID-19 on Lungs: Focusing on Prospective Malignant Phenotypes. *Cancers* **2020**, *12*, 3822. [CrossRef]
37. Akter, S.; Hasan, M.N.; Rokeya, B.; Akhter, H.; Gazi, M.S.; Sabrin, F.; Kim, S.S. Alternative Medicine: A Recent Overview. In *Alternative Medicine*; IntechOpen: London, UK, 2021.
38. Ahmad, S.; Zahiruddin, S.; Parveen, B.; Basist, P.; Parveen, A.; Parveen, R.; Ahmad, M. Indian medicinal plants and formulations and their potential against COVID-19–preclinical and clinical research. *Front. Pharmacol.* **2021**, *11*, 2470. [CrossRef]
39. Favalli, E.G.; Ingegnoli, F.; De Lucia, O.; Cincinelli, G.; Cimaz, R.; Caporali, R. COVID-19 infection and rheumatoid arthritis: Faraway, so close! *Autoimmun. Rev.* **2020**, *19*, 102523. [CrossRef]
40. Gao, J.; Tian, Z.; Yang, X. Breakthrough: Chloroquine phosphate has shown apparent efficacy in treatment of COVID-19 associated pneumonia in clinical studies. *Biosci. Trends* **2020**, *14*, 72–73. [CrossRef]
41. Lipinski, C.A.; Lombardo, F.; Dominy, B.W.; Feeney, P.J. Experimental and computational approaches to estimate solubility and permeability in drug discovery and development settings. *Adv. Drug Deliv. Rev.* **1997**, *23*, 3–25. [CrossRef]
42. Veber, D.F.; Johnson, S.R.; Cheng, H.Y.; Smith, B.R.; Ward, K.W.; Kopple, K.D. Molecular properties that influence the oral bioavailability of drug candidates. *J. Med. Chem.* **2002**, *45*, 2615–2623. [CrossRef]
43. Gómez, J.; Albaiceta, G.M.; García-Clemente, M.; López-Larrea, C.; Amado-Rodríguez, L.; Lopez-Alonso, I.; Hermida, T.; Enriquez, A.I.; Herrero, P.; Melón, S.; et al. Angiotensin-Converting Enzymes (ACE, ACE2) Gene Variants and COVID-19 Outcome. *Gene* **2020**, *762*, 145102. [CrossRef]
44. Coltescu, A.R.; Butnariu, M.; Sarac, I. The importance of solubility for new drug molecules. *Biomed. Pharmacol. J.* **2020**, *13*, 577–583. [CrossRef]

molecules

MDPI

Article

In Silico Identification and Validation of Organic Triazole Based Ligands as Potential Inhibitory Drug Compounds of SARS-CoV-2 Main Protease

Vishma Pratap Sur [1], Madhab Kumar Sen [2] and Katerina Komrskova [1,3,*]

1 Laboratory of Reproductive Biology, Institute of Biotechnology of the Czech Academy of Sciences, BIOCEV—Biotechnology and Biomedicine Centre of the Academy of Sciences and Charles University, Prumyslova 595, 252 50 Vestec, Czech Republic; VishmaPratap.Sur@ibt.cas.cz
2 Department of Agroecology and Crop Production, Faculty of Agrobiology, Food and Natural Resources, Czech University of Life Sciences Prague, Kamycka 1176, 165 00 Prague, Czech Republic; senm@af.czu.cz
3 Department of Zoology, Faculty of Science, Charles University, Vinicna 7, 128 44 Prague, Czech Republic
* Correspondence: katerina.komrskova@ibt.cas.cz

Abstract: The SARS-CoV-2 virus is highly contagious to humans and has caused a pandemic of global proportions. Despite worldwide research efforts, efficient targeted therapies against the virus are still lacking. With the ready availability of the macromolecular structures of coronavirus and its known variants, the search for anti-SARS-CoV-2 therapeutics through in silico analysis has become a highly promising field of research. In this study, we investigate the inhibiting potentialities of triazole-based compounds against the SARS-CoV-2 main protease (M^{pro}). The SARS-CoV-2 main protease (M^{pro}) is known to play a prominent role in the processing of polyproteins that are translated from the viral RNA. Compounds were pre-screened from 171 candidates (collected from the DrugBank database). The results showed that four candidates (Bemcentinib, Bisoctrizole, PYIITM, and NIPFC) had high binding affinity values and had the potential to interrupt the main protease (M^{pro}) activities of the SARS-CoV-2 virus. The pharmacokinetic parameters of these candidates were assessed and through molecular dynamic (MD) simulation their stability, interaction, and conformation were analyzed. In summary, this study identified the most suitable compounds for targeting Mpro, and we recommend using these compounds as potential drug molecules against SARS-CoV-2 after follow up studies.

Keywords: SARS-CoV-2; main protease; triazole; docking; MD simulation; drug

Citation: Sur, V.P.; Sen, M.K.; Komrskova, K. In Silico Identification and Validation of Organic Triazole Based Ligands as Potential Inhibitory Drug Compounds of SARS-CoV-2 Main Protease. *Molecules* **2021**, *26*, 6199. https://doi.org/10.3390/molecules26206199

Academic Editors: Giovanni N. Roviello and Caterina Vicidomini

Received: 10 September 2021
Accepted: 12 October 2021
Published: 14 October 2021

1. Introduction

Reports suggest that the SARS-CoV-2 virus penetrates target tissues by manipulating two important proteins present on the surface of cells. The two key proteins are transmembrane serine protease 2 (TMPRSS2) and angiotensin-converting enzyme 2 (ACE2). The SARS-CoV-2 virus belongs to the β category of human coronaviruses [1–3], and its genomic organization is similar to that of other coronaviruses [4]. The viral genomic RNA (27–32 Kb) codes both structural and non-structural proteins. The structural proteins include membrane (M), envelope (E), nucleocapsid (N), hemagglutinin-esterase (HE), and spike (S) proteins. These proteins are known to facilitate the transmission and replication of viruses in host cells [5]. The replicase gene (ORF1a) and protease gene (ORF1b) encode polyprotein1a (pp1a) and polyprotein1ab (pp1ab). These polyproteins are further processed by Papain-like protease (PLpro) and Chymotrypsin-like protease (3CLpro) to generate nonstructural proteins (nsp) [3,6]. The main protease (M^{pro}) is an essential enzyme, which plays a vital role in the lifecycle of the virus and can therefore be used in research efforts to identify potential target drugs. Additionally, since no proteases with M^{pro}-like cleaving characteristics are found in humans, any potential protease inhibitors are likely to be nontoxic to humans.

The testing of broad-spectrum antiviral drugs is currently in process. However, despite unprecedented research efforts, efficient targeted therapies (which could provide a long-term solution to COVID-19) have still not been identified. Computer-aided drug discovery (CADD) methodologies have been widely used during the past decade and are a powerful tool to study protein-drug and protein-protein interactions. In recent developments, CADD methodologies are being used as a key resource for drug discovery to mitigate the COVID-19 pandemic [7–9]. Cava et al. have identified potential drug candidates that could impact the spread of COVID-19, such as: nimesulide, fluticasone propionate, and thiabendazole. Cava et al. used in silico gene-expression profiling to study the mechanisms of the ACE2 and its co-expressed genes [10]. Wang et al. conducted virtual screening of authorized drugs along with those that are in clinical trials to identify drug candidates against 3CLpro [11]. Liang et al., used molecular dynamics simulation to reveal the binding stability of an α-ketoamide inhibitor within the SARS-CoV-2 main protease (M^{pro}) [12]. Gaudêncio and Florbela used CADD methodologies to screen natural marine products to identify effective ligands with SARS-CoV-2 main protease (M^{pro}) with inhibiting potential [13]. Another potential approach is drug repurposing, which includes the screening of pre-existing drug compounds with anti-SARS-CoV-2 properties, which is followed by target identification and functional and structural characterization of any targeted enzymes. Finally, after successful screening and characterization, clinical trials can commence. In addition to the drug molecules, there are reports on applications of nanomaterials, such as metal-based, two-dimensional, and colloidal nanoparticles and nanomicelles, for antiviral and virus sensing applications [14–17]. Despite their small size and selective nature, nanoparticles have proved to be effective against wide range of pathogens, including bacteria and viruses. However, some metal-based nanoparticles have also been reported to have non-specific bacterial toxicity mechanisms, thereby reducing the chances of developing resistance as well as expanding the spectrum of antimicrobial activity [18]. Although the interest in designing nanomaterial-based, non-traditional drugs is growing, more advanced research is required to uncover their full potentials for being considered as promising agents against SARS-CoV-2.

To date, no specialized drugs are available on the market to cure COVID-19. Over recent years, the triazole group-based ligands have attracted the interest of the scientific community due to their comprehensive and multipurpose medicinal applications. Reports have been published stating that this group of ligands have potential antiviral, antibacterial, antifungal, antiparasitic and anti-inflammatory applications. Moreover, owing to the nature of their chemical properties, this group of ligands can be easily synthesized [19–21]. The triazole group-based ligands could be a potential drug-candidate for use against the SARS-CoV-2 virus [22,23]. Efforts to develop efficient therapeutic strategies against COVID-19 are still in progress.

In this study, we had evaluated the potential of the triazole ligands as effective antiviral agents. We identified the most suitable anti-SARS-CoV-2 candidate chemicals (based on their molecular docking scores), which were then further analyzed for positive ADMET properties. Scientists across the world are researching different antiviral compounds, to identify those with the highest potential effectivity against SARS-CoV-2 as well as having low or no toxicity for humans. Our results suggest that the recommended drugs in this study may be candidates for use in the treatment of COVID-19. Even though triazole ligands are already clinically approved drugs, they would still require clinical trials prior to repurposing as anti-COVID-19 medicines (Figure 1).

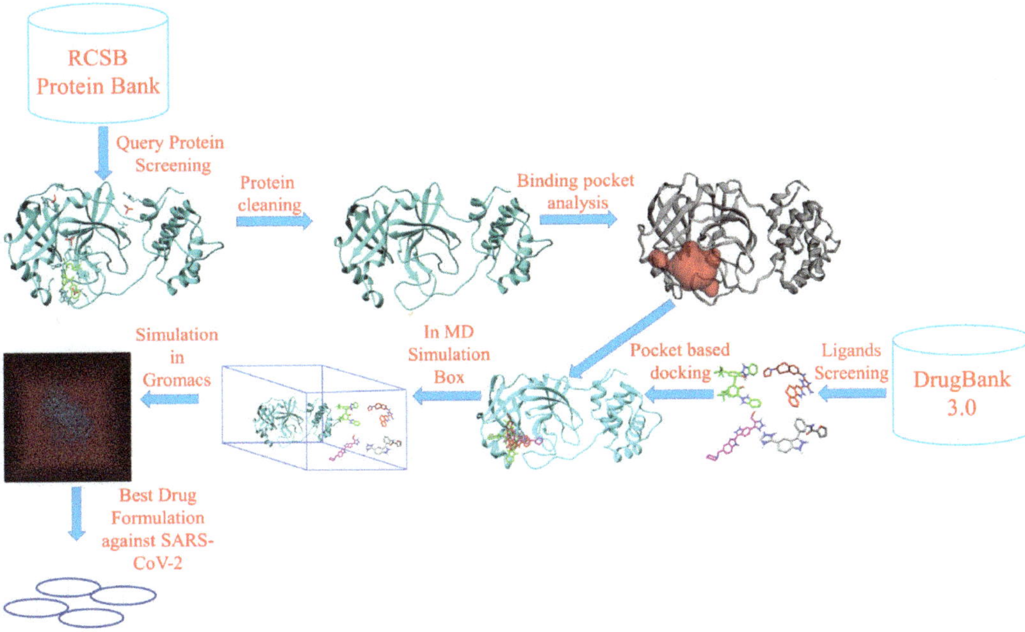

Figure 1. Schematic diagram of the workflow.

2. Results

2.1. Structural Analysis

The protein structure used for the molecular docking and simulation studies is shown in Figure 2. The binding pocket volume and surface area were determined through the CASTp webserver, utilizing previous findings [24]. A binding pocket was predicted at the surface as well as in the interior of proteins. The binding pocket volume of M^{pro} was 402.7 (SA) (Figure 3), which signifies an optimum space for ligand binding. All the participating residues are listed in Supplementary Table S2.

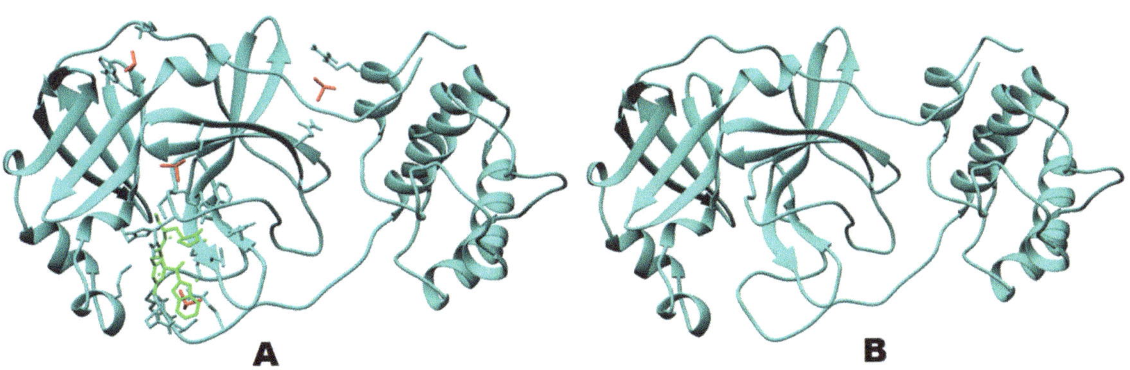

Figure 2. Protein structures: (**A**). before docking studies and (**B**). after cleaning of ligand and additional molecules, used for further docking and MD simulation.

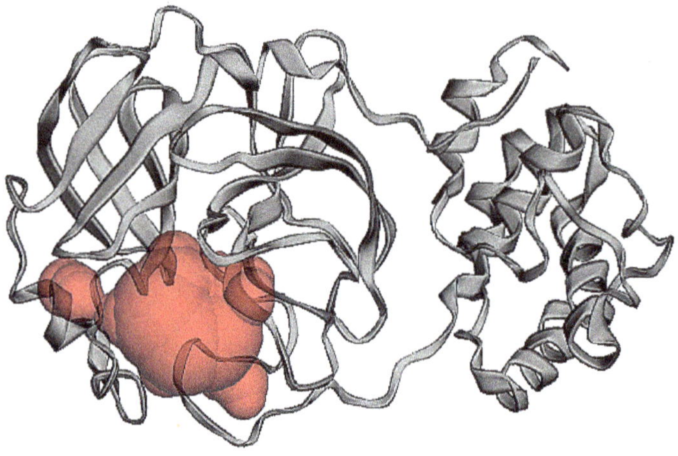

Figure 3. Binding pocket analysis (predicted by CASTp software).

2.2. Molecular Docking

To identify a potential SARS-CoV-2 protease inhibitor, the structure-based molecular docking approach was performed on 171 triazole based compounds. These selected compounds have therapeutic potential against cancer, infectious diseases, and some other diseases. All 171 compounds were docked with the SARS-CoV-2 (M^{pro}) chain A using target specific docking (pre-identified pocket with CastP). Out of 171 compounds, 27 compounds gave a docking score of −10.2 to −8 kcal/mol (Figures S1 and S2 and Table S3). The list of compounds, based on their binding energies (PyRx based Vina scores) of the highest ranked position of the docked ligand with SARS-CoV-2 main protease, are shown in Table 1 and Supplementary Table S3.

Table 1. Organic triazole compounds used for further analysis for molecular interactions in the SARS-CoV-2 main protease.

Triazole Based Compounds	Binding Affinity Values (kcal/mol)	No. of H-bonds	H-bonds and Interacting Residues	No. of Other Interactions	Other Interaction and Interacting Residues
Bemcentinib (DB12411)	−10.2	2	Ser46, Thr26	1	Met49
Bisoctrizole (DB11262)	−9.0	2	Cys44, Gln189	1	Leu50
PYIITM (DB07213)	−8.8	4	His41 (3), Thr45 (1)	2	Met49, Cys44
NIPFC (DB07020)	−8.8	2	Cys44, Asn142	1	Met49

Four best ligand molecules were selected based on the top hit criteria and were further analyzed for molecular interactions with SARS-CoV-2 (M^{pro}) (Table 1, Figure S13). The ligands are 1-{3,4-diazatricyclo[9.4.0.0^{2,7}]pentadeca-1(15),2(7),3,5,11,13-hexaen-5-yl}-*N*3-[(7S)-7-(pyrrolidin-1-yl)-6,7,8,9-tetrahydro-5Hbenzo[7]annulen-2-yl]-1*H*-1,2,4-triazole-3,5-diamine (Bemcentinib;DB12411), 2-(2*H*-1,2,3-benzotriazol-2-yl)-6-{[3-(2*H*-1,2,3-benzotriazol-2-yl)-2-hydroxy-5-(2,4,4-trimethylpentan-2-yl)phenyl]methyl}-4-(2,4,4-trimethylpentan-2-yl)phenol (Bisoctrizole;DB11262), (5-{3-[5-(Piperidin-1-Ylmethyl)-1h-Indol-2-Yl]-1h-Indazol-6-Yl}-2h-1,2,3-Triazol-4-Yl)methanol (PYIITM;DB07213), *N*-{3-[5-(1*H*-1,2,4-triazol-3-yl)-1*H*-indazol-3-yl]phenyl}furan-2-carboxamide (NIPFC;DB07020). Bemcentinib (DB12411 an investigational drug for the treatment of non-small cell lung cancer) (Figure S1A,E) showed the highest binding energy, −10.2 kcal/mol, with the SARS-CoV-2 M^{pro} (Table 1). The results showed two hydrogen bonds with two main protease residues, Ser46, Thr26. Bemcentinib also showed one hydrophobic interaction (Pi-Alkyl) with Met49, residues of the SARS-CoV-2 M^{pro} enzyme (Figure 4, and Table 1).

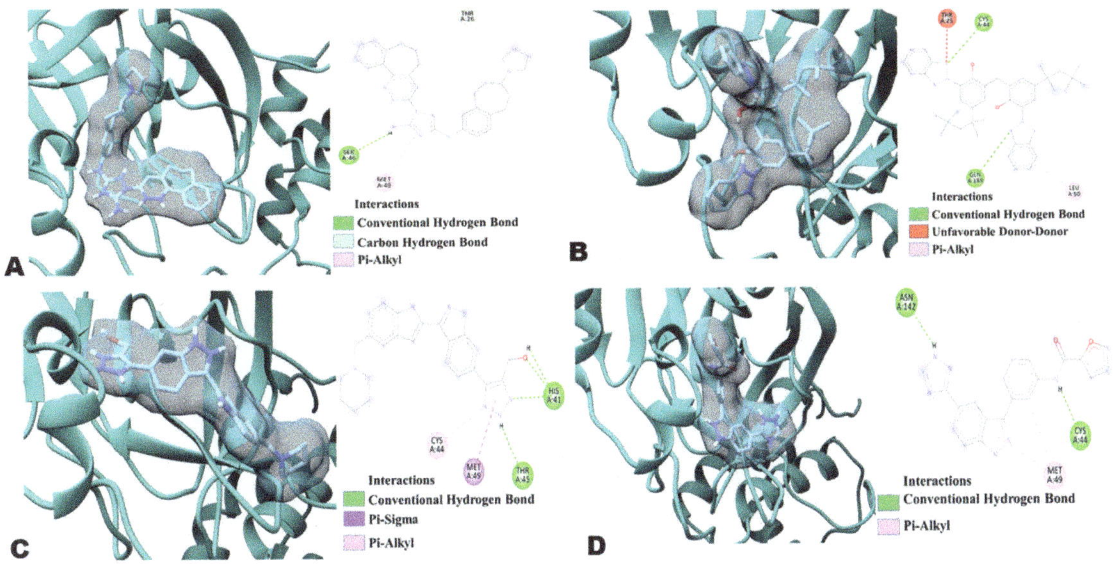

Figure 4. Molecular docking analysis of M^{pro} system in complex with (**A**). Bemcentinib (**B**). Bisoctriazole (**C**). PYIITM and (**D**). NIPFC.

In terms of highest binding energy, the other three potent organic triazole based compounds were Bisoctrizole (DB11262), PYIITM (DB07213), and NIPFC (DB07020) (Table 1, Table S3, Supplementary Figure S1). Bisoctrizole (DB11262 is a benzotriazole-based organic molecule that absorbs, reflects, and scatters both UV-A and UV-B rays) showed −9 kcal/mol binding energy against SARS-CoV-2 M^{pro} (Table 1). The interaction study showed two hydrogen bonds with M^{pro} residues, Cys44 and Gln189. Bisoctrizole also showed one unfavorable donor-donor interaction with residue Thr25 and one hydrophobic interaction (Pi-Alkyl) with Leu50 (Figure 4 and Table 1).

PYIITM (DB07213) showed −8.8 kcal/mol binding energy against SARS-CoV-2 M^{pro} (Table 1). The interaction study showed four hydrogen bonds with M^{pro} residues, three with His41, and one with Thr45, while PYIITM showed one electrostatic interaction (Pi Sigma) with residue Met49 and one hydrophobic interaction (Pi-Alkyl) with Cys44 (Figure 4 and Table 1).

NIPFC (DB07020) also showed −8.8 kcal/mol binding energy against SARS-CoV-2 M^{pro} (Table 1). The interaction study showed two hydrogen bonds with M^{pro} residues, Cys44 and Asn142, also on NIPFC, showed one hydrophobic interaction (Pi-Alkyl) with Met49 (Figure 4 and Table 1).

In our study, the ligands 11a and 11b (crystalized ligand structure used as inhibitor of M^{pro} in previous study) [25] were also docked against M^{pro} for assessment purposes. The 11a and 11b inhibitory ligands docking scores is low (−7.2 kcal/mol and −7.5 kcal/mol, Table S5), whereas our best triazole ligands showed binding affinities of −10.2 kcal/mol (Bemcentinib (DB12411)), −9 kcal/mol (Bisoctrizole:DB11262), −8.8 kcal/mol (PYIITM:DB07213), and −8.8 kcal/mol (NIPFC:DB07020). A previous study suggests that 17 (Thr25, Thr26, His41, Cys44, Met49, Phe140, Asn142, Gly143, Cys145, His163, His164, Met165, Glu166, Pro168, Asp187, Arg188, Gln189) amino acids were participating or present in the M^{pro}

and inhibitory ligands interaction [25]. Our protein–ligand interaction study suggested that seven amino acids (Thr25, Thr26, His41, Cys44, Met49, Asn142, Gln189) were involved in M^{pro} inhibition. Interestingly, these amino acids are also involved in M^{pro}–Bemcentinib, M^{pro}–Bisoctrizole, M^{pro}–PYIITM, and M^{pro}–NIPFC interaction, which indicates that all four triazole based ligands have binding affinity with amino acids, which play crucial roles in M^{pro} inhibition. In these terms, it can be concluded that Bemcentinib, Bisoctrizole, PYIITM, and NIPFC can be used as potential M^{pro} inhibitors.

2.3. Absorption, Distribution, Metabolism, Excretion, and Toxicity (ADMET) Analysis

Based on highest docking score, four ligands were selected for pharmacokinetics, including: the Lipinski rule of five, drug likeness, and ADMET analysis. Results obtained from the Lipinski rule of five are listed in Supplementary Table S4. PYIITM (DB07213) and NIPFC (DB07020) satisfied all the Lipinski rule parameters. Whereas the other two compounds, Bemcentinib (DB12411) and Bisoctrizole (DB11262), violated two Lipinski rules, previous studies suggested that, with two violations, compounds could be used as orally active antiviral agents [26]. However, all four compounds show favorable drug-likeness properties (Supplementary Table S4 and Supplementary Figure S3). ADMET properties of the four selected compounds were analyzed by a free pkCSM (http://biosig.unimelb.edu.au/pkcsm/prediction, accessed on 28 February 2021) web tool.

2.3.1. Absorption

Drug absorption is mainly analyzed through the water solubility of compounds, cell permeability using colon carcinoma (Caco-2) cell line, human intestinal absorption, skin permeability, and whether the molecule is a P-glycoprotein substrate or inhibitor [27]. The compound water solubility reflects the compound solubility in water at 25 °C. All the selected compounds are moderately soluble in water (Table 2). Caco-2 cell permeability and human intestinal absorption determine the ultimate bioavailability; a drug having a value of more than 0.90 is considered readily permeable [26]. Bemcentinib (DB12411) showed particularly good permeability, whereas Bisoctrizole (DB11262) and PYIITM (DB07213) showed moderate permeability (Table 2), but NIPFC (DB07020) showed negligible permeability.

Table 2. ADMET pharmacokinetics; absorbance and distribution parameters.

Compounds/ Ligands	Water Solubility log mol/L	Caco-2 Permeability log 10^{-6} cm/s	Human Intestinal Absorption (%)	P-glycoprotein Substrate	P-glycoprotein I Inhibitor	P-glycoprotein II Inhibitor	VDss (log L/kg)	Fraction Unbound (Human)
Bemcentinib	−3.166	1.336	100	Yes	Yes	Yes	0.755	0.179
Bisoctrizole	−2.929	1.489	100	No	No	Yes	−1.227	0.437
PYIITM	−2.889	0.877	80.603	Yes	No	Yes	−0.083	0.161
NIPFC	−2.871	0.355	84.718	Yes	No	Yes	−0.557	−0.557

The human intestine is the primary site for drug absorption. A previous study suggested that a molecule with >30% absorbency is considered readily absorbed [27]. In silico absorbance analysis showed that Bemcentinib (DB12411) and Bisoctrizole (DB11262) have a 100% absorbance rate in the human intestine (Table 2), whereas the other compounds, PYIITM (DB07213) and NIPFC (DB07020), achieve a >80% absorbance rate. This clearly indicates that all the organic triazole based ligands have a high absorbance rate in the human intestine. All compounds were substrates for P-glycoprotein, except Bisoctrizole (DB11262). All four compounds were P-glycoprotein II inhibitors. Only Bemcentinib (DB12411) showed inhibition against P-glycoprotein I (Table 2).

2.3.2. Distribution

The distribution was calculated using the following parameters: human volume of distribution, human fraction unbound in plasma, blood-brain barrier, and central nervous system permeability. In the bloodstream, drugs are generally transported in a free or unbound state or in a partly reversibly bound state. However, irrespective of the transportation state, the steady-state volume of distribution (VDss) remains one of the key pharmacokinetic parameters that must be considered when designing a drug dose range. VDss can be defined as the theoretical volume of a particular drug dose, which vary and give a similar blood plasma concentration. Generally, the greater the VDss value, the more a drug is distributed in tissue rather than plasma. However, for antibiotics and antivirals, more wide-ranging tissue distribution is desirable [27]. VDss is considered low if the log of the VDss value is lower than −0.15, while a value >0.45 is considered high [27]. Of the four compounds in question, Bemcentinib (DB12411) showed the highest distribution value, followed by PYIITM (DB07213) (Table 2). Bisoctrizole (DB11262) showed the lowest distribution value of the four compounds. The effectiveness of a drug may vary depending on the limit to which it can bind to blood proteins. The more effective the binding of the drug with blood proteins, the more efficiently the drug compounds can transverse the cellular membrane [27]. Fraction unbound to human plasma ranges between 0.02 to 1.0 [28]. All compounds showed a high fraction unbound value to human plasma, except NIPFC (DB07020) (Table 2).

2.3.3. Metabolism

The metabolism of a drug depends upon the molecule being a Cytochrome P450 substrate or inhibitor. Bemcentinib (DB12411) showed moderate inhibition (CYP2C19, CYP3A4) of the cytochrome enzymes, whereas Bisoctrizole (DB11262) showed non-inhibitory properties against all enzymes (Table 3). PYIITM (DB07213) showed inhibition activity against only CYP1A2, whereas NIPFC (DB07020) showed inhibition against all cytochrome enzymes (Table 3). The results indicate that the Bisoctrizole (DB11262), PYIITM (DB07213), and Bemcentinib (DB12411) will be metabolized by the action of the cytochrome enzymes. On the other hand, NIPFC (DB07020) will not be metabolized by the cytochrome enzymes due to its inhibitory nature against all cytochrome enzymes.

Table 3. ADMET pharmacokinetics; metabolism and excretion parameters.

Compounds/ Ligands	CYP2D6 Substrate	CYP3A4 Substrate	CYP1A2 Inhibitor	CYP2C19 Inhibitor	CYP2C9 Inhibitor	CYP2D6 Inhibitor	CYP3A4 Inhibitor
Bemcentinib (DB12411)	No	Yes	No	Yes	No	No	Yes
Bisoctrizole (DB11262)	No	Yes	No	No	No	No	No
PYIITM (DB07213)	Yes	Yes	Yes	No	No	No	No
NIPFC (DB07020)	Yes	Yes	Yes	Yes	Yes	Yes	Yes

2.3.4. Excretion

Organic cation transporter 2 (OCT2) belongs to the category of renal uptake transporters, which are known to play important roles during deposition and clearing of drugs from the kidneys [28]. Excretion depends on factors such as total clearance and whether the molecule is a renal OCT2 substrate. None of the triazole compounds act as a substrate for Renal OCT2 and can be removed from the body through the renal system. Except PYIITM (DB07213), all the selected compounds show total clearance of less than log (CLtot) 1 mL/min/kg (Table 4).

Table 4. ADMET pharmacokinetics; toxicity parameters.

Compounds/ Ligands	AMES Toxicity	Total Clearance log ml/ min/kg	Renal OCT2 Substrate	Max. Tolerated Dose (Human)	Oral Rat Acute Toxicity (LD50)	Skin Sensitization	Minnow Toxicity
Bemcentinib (DB12411)	Yes	0.920	No	0.181	2.995	No	1.920
Bisoctrizole (DB11262)	No	−1.185	No	0.429	3.115	No	−5.882
PYIITM (DB07213)	No	1.088	No	0.529	2.517	No	1.985
NIPFC (DB07020)	No	0.305	No	0.602	2.890	No	3.334

2.3.5. Toxicity

A negative AMES result indicates that the molecule is non-mutagenic and non-carcinogenic. None of the selected triazole compounds showed AMES toxicity except Bemcentinib (DB12411) (Table 4). Bemcentinib (DB12411) is under investigation as an anti-cancer drug against small lung tumors. The maximum recommended tolerance dose (MRTD) provides an estimate of the toxic dose in humans. MRTD values less than or equal to log 0.477 (mg/kg/day) is considered low [28]. Bemcentinib (DB12411) and Bisoctrizole (DB11262) had low toxicity to humans whereas PYIITM (DB07213) and NIPFC (DB07020) showed toxicity (Table 4). All four triazole compounds were not skin sensitive (Table 4). A molecule with a high oral rat acute toxicity (LD50) value is less lethal than the lower LD50 value [27,29]. For a given molecule, the LD50 is the amount that causes the death of 50% of the test animals [27,29]. All the selected ligands showed high oral rat acute toxicity (LD50) value (Table 4). The lethal concentration values (LC50) represent the concentration of a molecule necessary to cause 50% of fathead minnow death. For a given molecule, if the log LC50 < 0.5 mM (log LC50 < −0.3), then it is regarded as having high acute toxicity [29,30]. All three triazole compounds showed a satisfactory score that indicated that they are less toxic, except for Bisoctrizole (DB11262) (Table 4).

2.4. *In Silico Antiviral Prediction*

Bemcentinib showed more than 50.34% antiviral activity against all tested viruses, with 60.71% antiviral activity against HIV (Supplementary Table S5); Bisoctriazole showed more than 61.38% antiviral activity against all tested viruses, with more than 60.32% activity against HIV; and PYIITM showed more than 62.49% antiviral activity against all tested viruses, with 48.11% antiviral activity against HIV. NIPFC showed more than 36% antiviral activity against all tested viruses, with 60.61% antiviral activity against HIV (Supplementary Table S6). Based on antiviral prediction, it can be concluded that Bemcentinib, Bisoctriazole, and PYIITM can be used as potent antiviral drugs against the SARS-CoV-2 virus (Supplementary Table S5), because previous case and clinical studies suggested that some antiviral drugs mostly used for HIV showed effects against SARS-CoV-2 virus [31,32].

2.4.1. MD Simulation and Analysis

Based on the best docking score four top hit molecules, Bemcentinib (−10.2 kcal/mol), Bisoctriazole (−9 kcal/mol), PYIITM (DB07213) (−8.8 kcal/mol), and NIPFC (DB07020) (−8.8 kcal/mol) were selected for MD simulation studies (with all-atoms). The dynamic features of the protease-inhibitor interactions were analyzed based on various parameters, such as RMSD, RMSF, Rg, H-bonds, SASA, and interaction energy.

2.4.2. RMSD Analysis

To determine M^{pro} docked complex conformation stability with drug compounds, Bemcentinib (−10.2 kcal/mol), Bisoctriazole (−9 kcal/mol), PYIITM (−8.8 kcal/mol), and NIPFC (DB07020), the backbone root mean square deviation (Cα-RMSD) were computed, as shown in Figure 5. The result shows that the RMSD trajectory of M^{pro}–Bemcentinib was equilibrated during 0–5 ns and remained steady with a RMSD value ~2.0 ± 0.2 Å at the end of simulation at 40 ns (Figure 5A), which indicates very stable structural complexity of the M^{pro}–Bemcentinib complex. Likewise, the RMSD plot of the M^{pro}–Bisoctriazole complex showed a reasonably stable structure during the 40 ns stimulation process. Mpro–Bisoctriazole complex exhibited RMSD ~1.7 Å (Figure 5A). Similarly, M^{pro}–PYIITM and M^{pro}–NIPFC RMSD plots showed RMSD values ~1.6 Å and ~1.75 Å, respectively, which clearly indicates the structural stability of M^{pro}–PYIITM and M^{pro}–NIPFC complexes. (Figure 5A). All the RMSD values indicate a very stable structural conformation of the M^{pro} protein with all four ligand compounds.

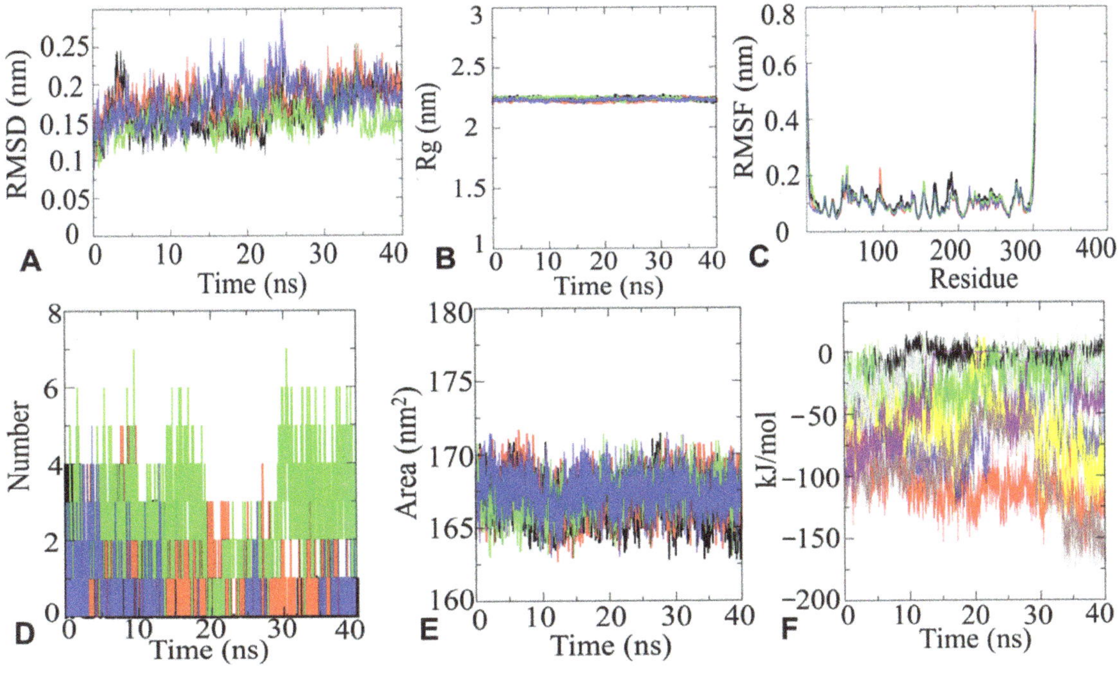

Figure 5. (**A**). RMSD plot of the M^{pro} system in complex with Bemcentinib, Bisoctriazole, PYIITM, and NIPFC. Here, black line defines Bemcentinib, red line defines Bisoctriazole, green line defines PYIITM, and blue line defines NIPFC. (**B**). Rg plot of the M^{pro} system in complex with Bemcentinib, Bisoctriazole, PYIITM, and NIPFC, which clearly indicates the compactness of the protein in the complex with ligand compounds. Here, black line defines Bemcentinib, red line defines Bisoctriazole, green line defines PYIITM, and blue line defines NIPFC. (**C**). RMSF analysis plot for SARS-CoV-2 main protease system in complex with Bemcentinib, Bisoctriazole, PYIITM, and NIPFC. Here, black line defines Bemcentinib, red line defines Bisoctriazole, green line defines PYIITM, and blue line defines NIPFC. (**D**). Hydrogen bond dynamics between SARS-CoV-2 Mpro in complex with Bemcentinib, Bisoctriazole, PYIITM, and NIPFC. Here, black line defines Bemcentinib, red line defines Bisoctriazole, green line defines PYIITM, and blue line defines NIPFC. (**E**). SASA plot for SARS-CoV-2 main protease system in complex with Bemcentinib, Bisoctriazole, PYIITM, and NIPFC. Here, black line defines Bemcentinib, red line defines Bisoctriazole, green line defines PYIITM, and blue line defines NIPFC. (**F**). Interaction energy plot for SARS-CoV-2 main protease system in complex with Bemcentinib, Bisoctriazole, PYIITM, and NIPFC. Here, black line defines Bemcentinib, red line defines Bisoctriazole, green line defines PYIITM, and blue line defines NIPFC.

2.4.3. Rg Analysis

Additionally, the conformation stability of the M^{pro}–ligand was evaluated by the radius of gyration (Rg). The Rg parameter is used by computational biologists to describe the structural compactness of proteins. To examine the structural compactness and integrity of M^{pro}–ligand bound complexes, the radius of gyration (Rg) is calculated for each system [33,34]. From Figure 5, it can be observed that the structure of M^{pro}–Bemcentinib, M^{pro}–Bisoctriazole, M^{pro}–PYIITM, and M^{pro}–NIPFC stabilized around an Rg value 22.5 Å ± 0.1 Å, and it can be seen that there was no structural drift (Figure 5B). The structural compactness of M^{pro}–drug complexes calculated by Rg analyses suggested stable molecular interaction with all four compounds, which are stabilized in 22.5 Å ± 0.1 Å (Figure 5B).

2.4.4. RMSF Analysis

The RMSF plots of M^{pro}–Bemcentinib, M^{pro}–Bisoctriazole, M^{pro}–PYIITM, and M^{pro}–NIPFC represent that the amino acid residues belonging to termini (N-and C-terminal) and loops have an average atomic fluctuation >1.5 Å (Figure 5C). In divergence, the conformational dynamics of stable secondary structure, α-helices, and β-sheets (interacting protein residues with the ligand compounds) remain stable during the whole simulation process, providing an indication of the stability of molecular interactions of M^{pro} with triazole based ligand compounds. The average atomic fluctuations were measured using RMSF plots, which suggested that all four M^{pro}–drug complexes showed similar 3D binding patterns, which clearly indicates that all four triazole based compounds were well accommodated at the binding pocket of M^{pro} with favorable molecular interactions.

2.4.5. H-Bonds Analysis

Furthermore, the time evolution plot of hydrogen bond occupancy (H-bonds) between target SARS-CoV-2 main protease and inhibitors was computed. H-bonds are also designated as the "master key of molecular recognition" due their crucial role in ligand binding and enzyme catalysis. Although H-bonds are weaker bonds compared to covalent bonds, their flexibility makes them the most important physical interaction in systems of bio-compounds in aqueous solution. They are critical for maintaining the shape and stability of protein structure. In the case of M^{pro}–Bemcentinib interactions, initially, four H-bonds were detected; however, over time, the number of H-bonds reduced. No H-bonds were obtained from approximately 24–32 ns. After this time, some spikes for H-bonds were identified. Finally, at 40 ns, one H-bond was detected, which came close to supporting our docking interaction data. In the case of M^{pro}–Bisoctriazole, initially, four H-bonds were detected; thereafter, the number of H-bonds varied from two to three, which strongly supports our docking calculations. In the case of PYIITM and M^{pro}, we detected four to five H-bonds, and NIPFC maintained two hydrogen bonds throughout the simulation time, which strongly agreed with our docking interaction calculations (Figure 5D).

2.4.6. SASA Analysis

Hydrophobic interactions can be considered determinants of protein conformational dynamics. Protein conformational dynamics are known to guarantee the structural stability of molecular interactions [34,35]. Computation of the solvent-accessible surface area (SASA) is an important parameter when studying changes in structural features of M^{pro}–Bemcentinib, M^{pro}–Bisoctriazole, M^{pro}–PYIITM, and M^{pro}–NIPFC complexes. The proper functioning of protein–ligand complexes depend on how well the protein maintains its fold during the interactions. Figure 5E (black line) shows that the complex structure SARS-CoV-2 M^{pro} occupied with the Bemcentinib had an average SASA value of 166.25 nm^2 ± 2 nm^2. The complex structures SARS-CoV-2 M^{pro} occupied with Bisoctriazole, PYIITM, and NIPFC had an average SASA value of 168.50 nm^2 ± 2 nm^2 (Figure 5E red, gree, blue line). Almost no change in orientation in the protein surface was detected for the molecular interaction of SARS-CoV-2 M^{pro} with Bisoctriazole, PYIITM, and NIPFC. However, in the case of interaction with Bemcentinib, a negligible decrease in the protein accessible area was detected, which is an indication of insignificant orientational change in the protein surface. Thus, the SASA investigation for all four complexes suggested no significant changes in the conformational dynamics of M^{pro}–Bemcentinib, M^{pro}–Bisoctriazole, M^{pro}–PYIITM and M^{pro}–NIPFC complexes.

2.4.7. Interaction Energy Analysis

The short-range electrostatic (Coul-SR) and van der Waals/hydrophobic (LJ-SR) interaction energies between M^{pro}–Bemcentinib, M^{pro}–Bisoctriazole, M^{pro}–PYIITM, and M^{pro}–NIPFC complexes explained promising electrostatic as well as hydrophobic interactions. For M^{pro}–Bemcentinib, average values of Coul-SR, -7.19 ± 3.2 kJ/mol, and LJ-SR, -109.162 ± 4.9 kJ/mol, were observed. For M^{pro}–Bisoctriazole, a Coul-SR of -25.37 ± 4 kJ/mol and an LJ-SR of -67.22 ± 6.1 kJ/mol were observed. M^{pro}–PYIITM complex exerts a Coul-SR of -61.02 ± 6.3 kJ/mol and an LJ-SR of -94.07 ± 1.3 kJ/mol. M^{pro}–NIPFC complexes showed a Coul-SR of -11.21 ± 5.4 kJ/mol and an LJ-SR of -30.76 ± 1.2 kJ/mol (Figure 5F). This suggested that the role of hydrophobic interaction was more important than the electrostatic interactions [36] in stabilizing the complex, a conclusion that is also supported by previous experimental data.

3. Materials and Methods

3.1. Target and Ligand Preparation

The crystal structure of SARS-CoV-2 main protease in complex with an inhibitor 11b (PDB-ID: 6M0K at resolution 1.80 Å, R-Value Free: 0.193, R-Value Work: 0.179 and R-Value Observed: 0.180) was retrieved from RCSB PDB database (http://www.rcsb.org/pdb, accessed on 27 February 2021) and used in the present study. The inhibitor 11b was removed from the structure with Chimera 1.15 for docking studies. The 3D SDF structure library of 171 triazole based compounds was downloaded from the DrugBank 3.0 database (https://go.drugbank.com/; accessed on 27 January 2021). All compounds were then imported into Open Babel software (Open Babel development team, Cambridge, UK) using the PyRx Tool and were exposed to energy minimization. The energy minimization was accomplished with the universal force field (UFF) using the conjugate gradient algorithm. The minimization was set at an energy difference of less than 0.1 kcal/mol. The structures were further converted to the PDBQT format for docking.

3.2. Protein Pocket Analysis

The active sites of the receptor were predicted using CASTp (http://sts.bioe.uic.edu/castp/index.html?2pk9, accessed on 28 January 2021). The possible ligand-binding pockets that were solvent accessible, were ranked based on area and volume [37].

3.3. Molecular Docking and Interaction Analysis

AutoDock Vina 1.1.2 in PyRx 0.8 software (ver.0.8, Scripps Research, La Jolla, CA, USA) was used to predict the protein-ligand interactions of the triazole compounds against the SARS-CoV-2 main protease protein. Water compounds and attached ligands were eliminated from the protein structure prior to the docking experiments. The protein and ligand files were loaded to PyRx as macromolecules and ligands, which were then converted to PDBQT files for docking. These files were similar to pdb, with an inclusion of partial atomic charges (Q) and atom types (T) for each ligand. The binding pocket ranked first was selected (predicted from CASTp). Note that the other predicted pockets were relatively small and had lesser binding residues. The active sites of the receptor compounds were selected and were enclosed within a three-dimensional affinity grid box. The grid box was centered to cover the active site residues, with dimensions x = −13.83 Å, y = 12.30 Å, z = 72.67Å. The size of the grid wherein all the binding residues fit had the dimensions of x = 18.22 Å, y = 28.11 Å, z = 22.65 Å. This was followed by the molecular interaction process initiated via AutoDock Vina from PyRx [38]. The exhaustiveness of each of the three proteins was set at eight. Nine poses were predicted for each ligand with the spike protein. The binding energies of nine docked conformations of each ligand against the protein were recorded using Microsoft Excel (Office Version, Microsoft Corporation, Redmond, Washington, USA). Molecular docking was performed using the PyRx 0.8 AutoDock Vina module. The search space included the entire 3D structure chain A. Protein-ligand docking was initially visualized and analyzed by Chimera 1.15. The follow-up detailed analysis of amino acid and ligand interaction was performed with BIOVIA Discovery Studio Visualizer (BIOVIA, San Diego, CA, USA). The compounds with the best binding affinity values, targeting the COVID-19 main protease, were selected for further molecular dynamics simulation analysis.

3.4. Absorption, Distribution, Metabolism, Excretion, and Toxicity (ADMET) Analysis

Pharmacokinetic parameters related to the absorption, distribution, metabolism, excretion, and toxicity (ADMET) play a substantial role in the detection of novel drug candidates. To predict candidate molecules using in silico methods pkCSM (http://biosig.unimelb.edu.au/pkcsm/prediction, accessed on 28 February 2021), webtools were used. Parameters such as AMES toxicity, maximum tolerated dose (human), hERG I and hERG II inhibitory effects, oral rat acute and chronic toxicities, hepatotoxicity, skin sensitization, and T. pyriformis toxicity and fathead minnow toxicity were explored. In addition to these, molecular weight, hydrogen bond acceptor, hydrogen bond donor, number of rotatable bonds, topological polar surface area, octanol/water partition coefficient, aqueous solubility scale, blood-brain barrier permeability, CYP2D6 inhibitor hepatotoxicity, and number of violations of Lipinski's rule of five were also surveyed.

3.5. In Silico Antiviral Assay

A quantitative structure-activity relationship (QSAR) approach was used in AVCpred to predict the antiviral potential of the candidates through the AVCpred server (http://crdd.osdd.net/servers/avcpred/batch.php, accessed on 28 January 2021). This prediction was conducted based on the relationships connecting molecular descriptors and inhibition. In this method, we used the most promising compounds screened against: human immunodeficiency virus (HIV), hepatitis C virus (HCV), hepatitis B virus (HBV), human herpesvirus (HHV), and 26 other important viruses (listed in Supplementary Table S1), with experimentally validated percentage inhibition from ChEMBL, a large-scale bioactivity database for drug discovery. This was followed by descriptor calculation and selection of the best performing molecular descriptors. The latter were then used as input for a support vector machine (in regression mode) to develop QSAR models for different viruses, as well as a general model for other viruses. [39].

3.6. MD Simulation Studies

The five best protein-ligand complexes were chosen for MD simulation according to the lowest binding energy with the best docked pose. Additional binding interactions were used for molecular simulation studies. The simulation was carried out using the GROMACS 2020 package (University of Groningen, Groningen, Netherland), utilizing a charmm36 all-atom force field using empirical, semi-empirical and quantum mechanical energy functions for molecular systems. The topology and parameter files for the input ligand file were generated on the CGenff server (http://kenno.org/pro/cgenff/, accessed on 27 February 2021). A TIP3P water model was used to incorporate the solvent, adding counter ions to neutralize the system. The energy minimization process involved 50,000 steps for each steepest descent, followed by conjugant gradients. PBC condition was defined for x, y, and z directions, and simulations were performed at a physiological temperature of 300 K. The SHAKE algorithm was applied to constrain all bonding involved, hydrogen, and long-range electrostatic forces treated with PME (particle mesh Ewald). The system was then heated gradually at 300 K, using 100 ps in the canonical ensemble (NVT) MD with 2 fs time step. For the isothermal-isobaric ensemble (NPT) MD, the atoms were relaxed at 300 K and 1 atm using 100 ps with 2 fs time step. After equilibrating the system at desired temperature and pressure, the MD run for the system was carried out at 40 ns with time step of 2 fs at 20,000,000 steps. The coordinates and energies were saved at every 10 ps for analysis.

MD simulation trajectories were analyzed by using a trajectory analysis module integrated into the GROMACS 2020.01 simulation package, qtgrace, VMD, and Chimera software (University of California San Francisco, San Francisco, CA, USA). The trajectory files were first analyzed using GROMCAS tools: gmx rmsd, gmx gyrate, gmx sasa, gmx hbond, gmx covar, and gmx energy for extracting the graph of root-mean square deviation (RMSD), root-mean square fluctuations (RMSFs), radius of gyration (Rg), solvent accessible surface area (SASA), hydrogen bond, principal component, potential energy, kinetic energy, and enthalpy, with python3 free energy surface calculation and visualization. The .mdp files scripts for NVT, NPT, MD production and interaction energy were added in the Supplementary File as .mdp file Supplementary Script S1 to S4.

4. Conclusions

The present study explored the molecular interactions of ligands, Bemcentinib, Bisoctriazole, PYIITM, and NIPFC. These were analyzed as prospective drug candidates against the SARS-CoV-2 (M^{pro}) protein. The screened compounds showed excellent docking scores, excellent pharmacokinetic profiles, MD simulation data, and interaction energy profile. Furthermore, these compounds positively cohere with the predetermined amino acid residues present in the core palm region of the M^{pro} protein, thus inhibiting the processing of the polyproteins that are translated from viral RNA. The ADMET results revealed excellent bioavailability and enzymatic inhibitory effects. The four compounds under investigation in this paper are already approved for other medical applications. This paper demonstrated the first occasion that the inhibitory action of these compounds was simulated for use against the SARS-CoV-2 virus. The interaction energy estimation using GROMACS extension revealed that the selected inhibitors, Bemcentinib, Bisoctriazole, PYIITM, and NIPFC, possess extremely high interaction energy and molecular affinity. Therefore, we propose that the selected compounds might be used as lead compounds in COVID-19 therapy. The pharmacological profiling, docking analysis, MD simulation, MD trajectory, and interaction energy studies indicated that Bemcentinib, Bisoctriazole, PYIITM, and NIPFC could be used as possible drug candidates for inhibition against the SARS-CoV-2 M^{pro} protein to interrupt the essential role it plays in processing polyproteins translated from viral RNA. Based on the data presented in this paper, the compounds investigated in this study could be considered for further clinical studies and thereafter for potential treatment of COVID-19.

Supplementary Materials: The following are available online, Supplementary Table S1: List of viruses used for triazole based ligands antiviral activity screening; Supplementary Table S2: List of interacting residues participating in M^{pro} ligand pocket formation; Supplementary Table S3: List of best ligand molecules according to their binding affinity score during the docking process; Supplementary Table S4: Evaluation of Lipinski's rule of five with a drug-likeness score by Molsoft L.L.C.: Drug likeness and molecular property prediction of the selected molecules (best four ligands); Supplementary Table S5: Ligands already used as M^{pro} inhibitor, used as a reference with triazole ligands docking study; Supplementary Table S6: Triazole based organic ligands antiviral activity screening through web based antiviral compound prediction server; Supplementary Figure S1: 2D and 3D chemical structure of the best 4 triazole based organic ligands; Supplementary Figure S2: 2D chemical structure of the best 23 triazole based organic ligands; Supplementary Figure S3: Drug likeness evaluation of selected ligands using Molsoft L.L.C.: Drug likeness and molecular property prediction. Bemcentinib (DB12411) (A), Bisoctrizole (DB11262) (B), PYIITM (DB07213) (C), and NIPFC (DB07020) (D). Supplementary Script 1 NVT run; Supplementary Script 2 NPT run; Script 3 Supplementary MD run; Script 4 Supplementary Interaction energy run.

Author Contributions: All authors were involved in the data research, manuscript authorship, reviewed and editing of the final article. V.P.S.: conceptualization; methodology; software; visualization; data curation; performed most of the experiments, including designing the experiments, protein structure prediction, and MD simulation; and writing original draft. M.K.S.: writing original draft; data review; and editing. K.K.: project supervision; funding acquisition; manuscript revision; and editing. All authors have read and agreed to the published version of the manuscript.

Funding: This research was funded by the grant of Ministry of Health of the Czech Republic (NU20-03-00309); by the project "BIOCEV—Biotechnology and Biomedicine Centre of the Academy of Sciences and Charles University" (CZ.1.05/1.1.00/02.0109) from the European Regional Development Fund (www.biocev.eu accessed on 27 February 2021); and by the Institutional support of the Institute of Biotechnology of the Czech Academy of Sciences RVO (86652036).

Institutional Review Board Statement: Not applicable.

Informed Consent Statement: Not applicable.

Data Availability Statement: Not applicable.

Acknowledgments: We thanks to GROMACS team University of Groningen, Netherland for their free dynamic software, cgenff server, pkCSM (http://biosig.unimelb.edu.au/pkcsm/prediction, accessed on 27 February 2021) webtool server, RCSB (http://www.rcsb.org/pdb, accessed on 27 February 2021) and DrugBank 3.0 database (https://go.drugbank.com, accessed on 27 February 2021).

Conflicts of Interest: The authors declare no conflict of interest.

Sample Availability: Samples of the compounds are not available from the authors.

References

1. Yan, R.; Zhang, Y.; Li, Y.; Xia, L.; Guo, Y.; Zhou, Q. Structural basis for the recognition of SARS-CoV-2 by full-length human ACE2. *Science* **2020**, *367*, 1444. [CrossRef]
2. Ye, Z.-W.; Yuan, S.; Yuen, K.-S.; Fung, S.-Y.; Chan, C.-P.; Jin, D.-Y. Zoonotic origins of human coronaviruses. *Int. J. Biol. Sci.* **2020**, *16*, 1686–1697. [CrossRef] [PubMed]
3. Li, X.; Geng, M.; Peng, Y.; Meng, L.; Lu, S. Molecular immune pathogenesis and diagnosis of COVID-19. *J. Pharm. Anal.* **2020**, *10*, 102–108. [CrossRef]
4. Kumar, S.; Nyodu, R.; Maurya, V.K.; Saxena, S.K. Morphology, Genome Organization, Replication, and Pathogenesis of Severe Acute Respiratory Syndrome Coronavirus 2 (SARS-CoV-2). In *Coronavirus Disease 2019 (COVID-19)*; Springer: Berlin, Germany, 2020; pp. 23–31. [CrossRef]
5. Mousavizadeh, L.; Ghasemi, S. Genotype and phenotype of COVID-19: Their roles in pathogenesis. *J. Microbiol. Immunol. Infect.* **2021**, *54*, 159–163. [CrossRef]
6. Fehr, A.R.; Perlman, S. Coronaviruses: An Overview of Their Replication and Pathogenesis. In *Coronaviruses: Methods and Protocols*; Maier, H.J., Bickerton, E., Britton, P., Eds.; Springer: New York, NY, USA, 2015; pp. 1–23. [CrossRef]
7. Keretsu, S.; Bhujbal, S.P.; Cho, S.J. Rational approach toward COVID-19 main protease inhibitors via molecular docking, molecular dynamics simulation and free energy calculation. *Sci. Rep.* **2020**, *10*, 17716. [CrossRef] [PubMed]
8. Trezza, A.; Iovinelli, D.; Santucci, A.; Prischi, F.; Spiga, O. An integrated drug repurposing strategy for the rapid identification of potential SARS-CoV-2 viral inhibitors. *Sci. Rep.* **2020**, *10*, 13866. [CrossRef] [PubMed]
9. Sliwoski, G.; Kothiwale, S.; Meiler, J.; Lowe, E.W. Computational Methods in Drug Discovery. *Pharmacol. Rev.* **2014**, *66*, 334. [CrossRef] [PubMed]
10. Cava, C.; Bertoli, G.; Castiglioni, I. In Silico Discovery of Candidate Drugs against COVID-19. *Viruses* **2020**, *12*, 404. [CrossRef]
11. Wang, J. Fast Identification of Possible Drug Treatment of Coronavirus Disease-19 (COVID-19) through Computational Drug Repurposing Study. *J. Chem. Inf. Model.* **2020**, *60*, 3277–3286. [CrossRef]
12. Liang, J.; Pitsillou, E.; Karagiannis, C.; Darmawan, K.K.; Ng, K.; Hung, A.; Karagiannis, T.C. Interaction of the prototypical α-ketoamide inhibitor with the SARS-CoV-2 main protease active site in silico: Molecular dynamic simulations highlight the stability of the ligand-protein complex. *Comput. Biol. Chem.* **2020**, *87*, 107292. [CrossRef]
13. Gaudêncio, S.P.; Pereira, F. A Computer-Aided Drug Design Approach to Predict Marine Drug-Like Leads for SARS-CoV-2 Main Protease Inhibition. *Mar. Drugs* **2020**, *18*, 633. [CrossRef]
14. Sun, Y.; Chen, X.; Liu, H.; Liu, S.; Yu, H.; Wang, X.; Qin, Y.; Li, P. Preparation of New Sargassum fusiforme Polysaccharide Long-Chain Alkyl Group Nanomicelles and Their Antiviral Properties against ALV-J. *Molecules* **2021**, *26*, 3265. [CrossRef] [PubMed]
15. Zeng, M.; Zhang, Y. Colloidal nanoparticle inks for printing functional devices: Emerging trends and future prospects. *J. Mater. Chem. A* **2019**, *7*, 23301–23336. [CrossRef]

16. Zeng, M.; Chen, M.; Huang, D.; Lei, S.; Zhang, X.; Wang, L.; Cheng, Z. Engineered two-dimensional nanomaterials: An emerging paradigm for water purification and monitoring. *Mater. Horiz.* **2021**, *8*, 758–802. [CrossRef]
17. Witika, B.A.; Makoni, P.A.; Matafwali, S.K.; Mweetwa, L.L.; Shandele, G.C.; Walker, R.B. Enhancement of Biological and Pharmacological Properties of an Encapsulated Polyphenol: Curcumin. *Molecules* **2021**, *26*, 4244. [CrossRef]
18. Sánchez-López, E.; Gomes, D.; Esteruelas, G.; Bonilla, L.; Lopez-Machado, A.L.; Galindo, R.; Cano, A.; Espina, M.; Ettcheto, M.; Camins, A.; et al. Metal-Based Nanoparticles as Antimicrobial Agents: An Overview. *Nanomaterials* **2020**, *10*, 292. [CrossRef] [PubMed]
19. Sur, V.P.; Mazumdar, A.; Kopel, P.; Mukherjee, S.; Vítek, P.; Michalkova, H.; Vaculovičová, M.; Moulick, A. A Novel Ruthenium Based Coordination Compound Against Pathogenic Bacteria. *Int. J. Mol. Sci.* **2020**, *21*, 2656. [CrossRef] [PubMed]
20. Murcia, R.A.; Leal, S.M.; Roa, M.V.; Nagles, E.; Muñoz-Castro, A.; Hurtado, J.J. Development of Antibacterial and Antifungal Triazole Chromium(III) and Cobalt(II) Complexes: Synthesis and Biological Activity Evaluations. *Molecules* **2018**, *23*, 2013. [CrossRef]
21. Sumrra, S.H.; Kausar, S.; Raza, M.A.; Zubair, M.; Zafar, M.N.; Nadeem, M.A.; Mughal, E.U.; Chohan, Z.H.; Mushtaq, F.; Rashid, U. Metal based triazole compounds: Their synthesis, computational, antioxidant, enzyme inhibition and antimicrobial properties. *J. Mol. Struct.* **2018**, *1168*, 202–211. [CrossRef]
22. Karypidou, K.; Ribone, S.R.; Quevedo, M.A.; Persoons, L.; Pannecouque, C.; Helsen, C.; Claessens, F.; Dehaen, W. Synthesis, biological evaluation and molecular modeling of a novel series of fused 1,2,3-triazoles as potential anti-coronavirus agents. *Bioorganic Med. Chem. Lett.* **2018**, *28*, 3472–3476. [CrossRef]
23. Seck, I.; Nguemo, F. Triazole, imidazole, and thiazole-based compounds as potential agents against coronavirus. *Results Chem.* **2021**, *3*, 100132. [CrossRef]
24. Kumari, A.; Rajput, V.S.; Nagpal, P.; Kukrety, H.; Grover, S.; Grover, A. Dual inhibition of SARS-CoV-2 spike and main protease through a repurposed drug, rutin. *J. Biomol. Struct. Dyn.* **2020**, 1–13. [CrossRef]
25. Dai, W.; Zhang, B.; Jiang, X.-M.; Su, H.; Li, J.; Zhao, Y.; Xie, X.; Jin, Z.; Peng, J.; Liu, F.; et al. Structure-based design of antiviral drug candidates targeting the SARS-CoV-2 main protease. *Science* **2020**, *368*, 1331–1335. [CrossRef] [PubMed]
26. Chandra, A.; Chaudhary, M.; Qamar, I.; Singh, N.; Nain, V. In silico identification and validation of natural antiviral compounds as potential inhibitors of SARS-CoV-2 methyltransferase. *J. Biomol. Struct. Dyn.* **2021**, 1–11. [CrossRef]
27. Saha, S.; Nandi, R.; Vishwakarma, P.; Prakash, A.; Kumar, D. Discovering Potential RNA Dependent RNA Polymerase Inhibitors as Prospective Drugs Against COVID-19: An in silico Approach. *Front. Pharmacol.* **2021**, *12*. [CrossRef]
28. Tamura, S.; Yang, G.-M.; Yasueda, N.; Matsuura, Y.; Komoda, Y.; Murakami, N. Tellimagrandin I, HCV invasion inhibitor from Rosae Rugosae Flos. *Bioorganic Med. Chem. Lett.* **2010**, *20*, 1598–1600. [CrossRef]
29. Bhowmik, D.; Nandi, R.; Jagadeesan, R.; Kumar, N.; Prakash, A.; Kumar, D. Identification of potential inhibitors against SARS-CoV-2 by targeting proteins responsible for envelope formation and virion assembly using docking based virtual screening, and pharmacokinetics approaches. *Infect. Genet. Evol.* **2020**, *84*, 104451. [CrossRef]
30. Sinha, M.; Jagadeesan, R.; Kumar, N.; Saha, S.; Kothandan, G.; Kumar, D. In-silico studies on Myo inositol-1-phosphate synthase of Leishmania donovani in search of anti-leishmaniasis. *J. Biomol. Struct. Dyn.* **2020**, 1–14. [CrossRef]
31. Ambrosioni, J.; Blanco, J.L.; Reyes-Urueña, J.M.; Davies, M.-A.; Sued, O.; Marcos, M.A.; Martínez, E.; Bertagnolio, S.; Alcamí, J.; Miro, J.M.; et al. Overview of SARS-CoV-2 infection in adults living with HIV. *Lancet HIV* **2021**, *8*, e294–e305. [CrossRef]
32. Gutierrez, M.D.M.; Mur, I.; Mateo, M.G.; Vidal, F.; Domingo, P. Pharmacological considerations for the treatment of COVID-19 in people living with HIV (PLWH). *Expert Opin. Pharmacother.* **2021**, 1127–1141. [CrossRef]
33. Nygaard, M.; Kragelund, B.B.; Papaleo, E.; Lindorff-Larsen, K. An Efficient Method for Estimating the Hydrodynamic Radius of Disordered Protein Conformations. *Biophys. J.* **2017**, *113*, 550–557. [CrossRef]
34. Prakash, A.; Kumar, V.; Meena, N.K.; Lynn, A.M. Elucidation of the structural stability and dynamics of heterogeneous intermediate ensembles in unfolding pathway of the N-terminal domain of TDP-43. *RSC Adv.* **2018**, *8*, 19835–19845. [CrossRef]
35. Banerjee, P.; Bagchi, B. Dynamical control by water at a molecular level in protein dimer association and dissociation. *Proc. Natl. Acad. Sci. USA* **2020**, *117*, 2302. [CrossRef]
36. Teli, D.M.; Shah, M.B.; Chhabria, M.T. In silico Screening of Natural Compounds as Potential Inhibitors of SARS-CoV-2 Main Protease and Spike RBD: Targets for COVID-19. *Front. Mol. Biosci.* **2021**, *7*. [CrossRef] [PubMed]
37. Kumar, N.; Srivastava, R.; Prakash, A.; Lynn, A.M. Structure-based virtual screening, molecular dynamics simulation and MM-PBSA toward identifying the inhibitors for two-component regulatory system protein NarL of Mycobacterium Tuberculosis. *J. Biomol. Struct. Dyn.* **2020**, *38*, 3396–3410. [CrossRef] [PubMed]
38. Trott, O.; Olson, A.J. AutoDock Vina: Improving the speed and accuracy of docking with a new scoring function, efficient optimization, and multithreading. *J. Comput. Chem.* **2010**, *31*, 455–461. [CrossRef] [PubMed]
39. Qureshi, A.; Kaur, G.; Kumar, M. AVCpred: An integrated web server for prediction and design of antiviral compounds. *Chem. Biol. Drug Des.* **2017**, *89*, 74–83. [CrossRef] [PubMed]

molecules MDPI

Article

In Silico Exploration of Potential Natural Inhibitors against SARS-Cov-2 nsp10

Ibrahim H. Eissa [1,*], Mohamed M. Khalifa [1], Eslam B. Elkaeed [2], Elsayed E. Hafez [3], Aisha A. Alsfouk [4] and Ahmed M. Metwaly [5,*]

1 Pharmaceutical Medicinal Chemistry & Drug Design Department, Faculty of Pharmacy (Boys), Al-Azhar University, Cairo 11884, Egypt; mohamedKhalifa2321.el@azhar.edu.eg
2 Department of Pharmaceutical Sciences, College of Pharmacy, Almaarefa University, Riyadh 13713, Saudi Arabia; ikaeed@mcst.edu.sa
3 Department of Plant Protection and Biomolecular Diagnosis, ALCRI, City of Scientific Research and Technological Applications, New Borg El-Arab City 21934, Egypt; elsayed_hafez@yahoo.com
4 Department of Pharmaceutical Sciences, College of Pharmacy, Princess Nourah bint Abdulrahman University, Riyadh 11564, Saudi Arabia; aaalsfouk@pnu.edu.sa
5 Pharmacognosy and Medicinal Plants Department, Faculty of Pharmacy (Boys), Al-Azhar University, Cairo 11884, Egypt
* Correspondence: ibrahimeissa@azhar.edu.eg (I.H.E.); ametwaly@azhar.edu.eg (A.M.M.)

Abstract: In continuation of our previous effort, different *in silico* selection methods were applied to 310 naturally isolated metabolites that exhibited antiviral potentialities before. The applied selection methods aimed to pick the most relevant inhibitor of SARS-CoV-2 nsp10. At first, a structural similarity study against the co-crystallized ligand, S-Adenosyl Methionine (**SAM**), of SARS-CoV-2 nonstructural protein (nsp10) (PDB ID: 6W4H) was carried out. The similarity analysis culled 30 candidates. Secondly, a fingerprint study against **SAM** preferred compounds **44, 48, 85, 102, 105, 182, 220, 221, 282, 284, 285, 301**, and **302**. The docking studies picked **48, 182, 220, 221**, and **284**. While the ADMET analysis expected the likeness of the five candidates to be drugs, the toxicity study preferred compounds **48** and **182**. Finally, a density-functional theory (DFT) study suggested vidarabine (**182**) to be the most relevant SARS-Cov-2 nsp10 inhibitor.

Keywords: COVID-19; natural products; SARS-Cov-2 nsp10; structural similarity; fingerprint; molecular docking; ADMET; toxicity; DFT

Citation: Eissa, I.H.; Khalifa, M.M.; Elkaeed, E.B.; Hafez, E.E.; Alsfouk, A.A.; Metwaly, A.M. *In Silico* Exploration of Potential Natural Inhibitors against SARS-Cov-2 nsp10. *Molecules* **2021**, *26*, 6151. https://doi.org/10.3390/molecules26206151

Academic Editors: Giovanni N. Roviello and Giosuè Costa

Received: 9 September 2021
Accepted: 6 October 2021
Published: 12 October 2021

1. Introduction

More than 217 million humans around the world were confirmed to be infected with COVID-19 and another 4.5 million families lost one of their beloveds as stated by the WHO on 2 September 2021 [1]. In response, all scientists in the field of drug discovery should work unceasingly to discover a cure against the notorious virus.

Computer-assisted (based or aided) drug design is a well-established branch of drug design that covers various *in silico* computational and theoretical approaches. These approaches are essential contributors to the development of new bioactive agents [2–8]. Computer-assisted drug design has been applied in drug discovery [9–11], computational chemistry [12,13], toxicity prediction [14–16], ADMET assessment [17–19], molecular modeling [20], molecular design [21,22], and rational drug design [23–27]. All these techniques have great popularity and have been used in both academic fields in addition to the pharmaceutical industries [28]. This approach has been introduced successfully and recurrently as a powerful weapon in the global fight against COVID-19 [29–32].

The relationship between humans and nature dates back to the prehistoric ages. The latter supplied the former with food, tools of beauty, and treatment [33,34]. Plants [35,36] and lately microorganisms [37,38] have been extensively screened to explore their healing power. Scientists isolated the secondary metabolites produced by these natural sources and labeled them as the key element in bioactivity. These candidates belonged to various classes as isochromenes [39], α-pyrones [40], diterpenes [41,42], sesquiterpenes [43,44], steroids [45], flavonoids [46,47], alkaloids [48], and saponins [49,50].

SARS-CoV-2 is an enveloped positive-sensed RNA virus. The replication of SARS-CoV-2 depends on a group of 16 non-structural proteins. These proteins have the codes of nsp1–nsp16. Between them, the two proteins nsp10 and nsp16 make an essential protein complex [51]. That complex is responsible for the vital methylation reaction at the ribose 2′-O position of the penultimate nucleotide of the viral RNA cap [52]. Accordingly, if a molecule could bind with that enzyme and inhibit this essential step, the replication process will be stopped.

The targeting of SARS-CoV-2 nsp-16 with a library of 10 [53] and 265 [54] FDA-approved compounds was studied before. Likely, a group set of 22 natural compounds from some Indian plants was computationally screened against six non-structural-proteins of SARS-CoV-2 [55].

In this study, different computational (*in silico*) selection methods were applied to 310 candidates. The examined candidates were chosen through a deep database search according to three parameters. The first parameter was to be naturally isolated. The second was having exhibited antiviral potentiality before. Lastly, we considered that the culled compounds belong to different chemical classes and accordingly have various chemical structures. The applied computational techniques were a structural similarity study against **SAM** followed by a fingerprint study against the same target. The selected candidates were docked against nsp10 (PDB ID: 6W4H) to prefer **44**, **48**, **85**, **102**, **105**, **182**, **220**, **221**, **282**, **284**, **285**, **301**, and **302**. Then ADMET and toxicity studies further picked two candidates. Finally, a DFT study suggested the most relevant inhibitor of SARS-Cov-2 nsp10 (Figure 1).

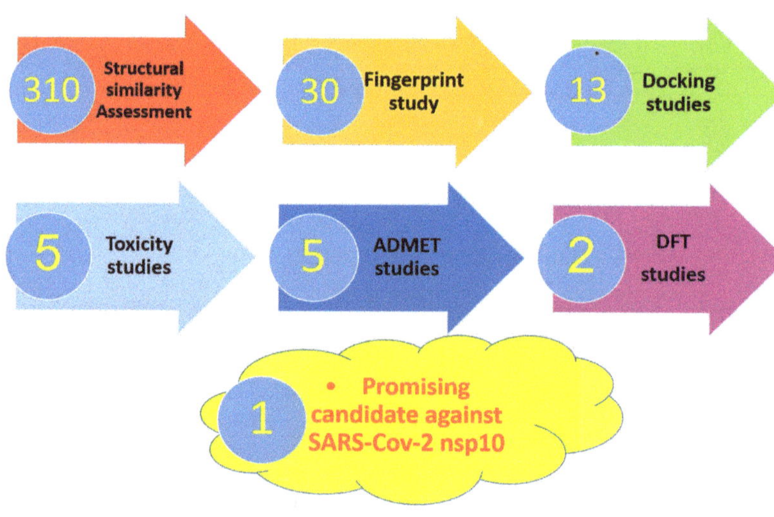

Figure 1. The applied *in silico* selection protocols.

2. Results and Discussion

2.1. *Molecular Similarity against* **SAM**

The basic principle of 2D Molecular similarity is that molecules with similar chemical structures are expected to have similar biological activities [56].

To measure the similarity of two objects, their general features have to be compared. On a molecular level, the molecular features or descriptors of any compound start from the general physicochemical properties and extend to more specific structural features such as partition coefficient (ALog p) [57], molecular weight (M. Wt) [58], hydrogen bond donors (HBA) [59], hydrogen bond acceptors (HBD) [60], number of rotatable bonds [61], number of rings, and also aromatic rings [62], in addition to molecular fractional polar surface area (MFPSA) [63].

All mentioned molecular properties were used in the applied similarity study between the natural candidate's set (Figure S1, Supplementary Materials) and the co-crystallized ligand (**SAM**) of SARS-CoV-2 nonstructural protein (nsp10) (PDB ID: 6W4H) using Discovery studio software. Thirty candidates (Figure 2) were chosen to be the most similar to **SAM**.

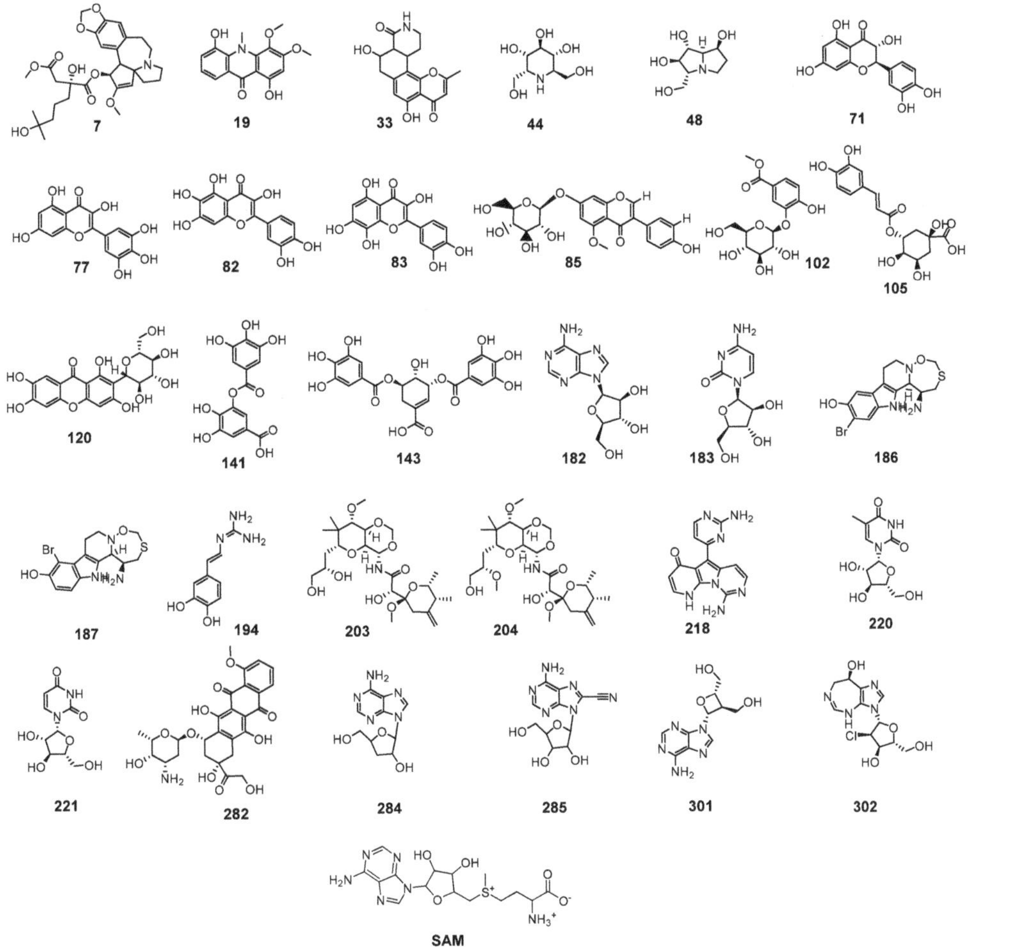

Figure 2. The most similar candidates with (**SAM**).

As shown in Figure 2, the similar candidates showed a high degree of structural similarity with **SAM**. In detail, most candidates have a sugar-like moiety as that of **SAM** as candidates **85**, **102**, **105**, **120**, **182**, **183**, **203**, **204**, **220**, **221**, **282**, **284**, **285**, **301**, and **302**. These moieties may serve as a good center for hydrogen bonding interaction with the target receptor. Furthermore, most candidates have hetero bicyclic structures as present in **SAM**. Besides, xanthine-like structures were defined in many similar candidates such as **182**, **284**, **285**, and **301**.

As shown in Figure 3, the candidate's set was divided into six smaller sets. From the first set to the fifth comprised 50 candidates while the sixth set was 60.

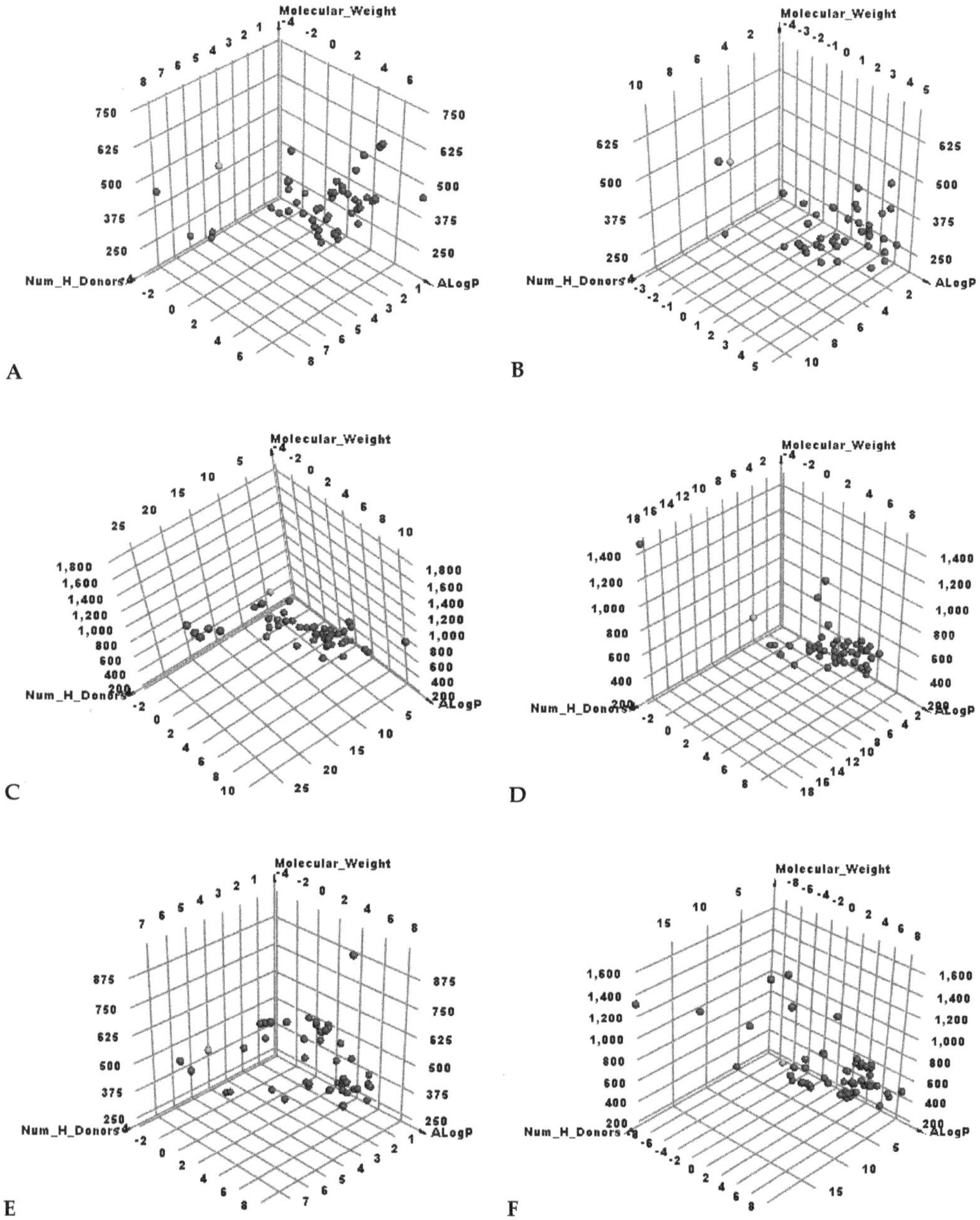

Figure 3. The results of similarity analysis of the test sets and **SAM**. Green = **SAM**, red = similar candidate, blue = not similar candidate. **(A)** Candidates 1–50, **(B)** 51–100, **(C)** 101–150, **(D)** 151–200, **(E)** 201–250, **(F)** 251–310.

Table 1 demonstrates the molecular properties of the similar candidates as well as **SAM**.

Table 1. Structural properties of the similar candidates with **SAM**.

Candidate	ALog p [1]	M. Wt [2]	HBA [3]	HBD [4]	Rotatable Bonds	Rings	Aromatic Rings	MFPSA [5]	Minimum Distance
7	0.857	546.629	9	3	11	5	1	0.223	1.272
19	2.643	301.294	6	2	2	3	2	0.261	1.441
33	0.674	315.321	5	3	0	4	1	0.333	1.472
44	−4.182	194.206	5	6	2	1	0	0.597	1.379
48	−3.556	190.217	4	5	1	2	0	0.466	1.454
71	1.479	304.252	7	5	1	3	2	0.467	1.491
77	1.388	318.235	8	6	1	3	2	0.526	1.477
82	1.388	318.235	8	6	1	3	2	0.526	1.477
83	1.388	318.235	8	6	1	3	2	0.526	1.477
85	0.436	446.404	10	5	5	4	2	0.373	1.093
102	−0.729	330.287	9	5	5	2	1	0.457	0.491
105	−1.814	353.301	9	5	5	2	1	0.489	0.375
120	−0.396	422.34	11	8	2	4	2	0.533	0.652
141	0.207	321.216	9	5	4	2	2	0.563	0.632
143	0.007	477.352	13	7	7	3	2	0.536	0.565
182	−1.881	267.241	8	4	2	3	2	0.539	0.489
183	−2.396	243.217	7	4	2	2	0	0.545	0.747
186	1.045	371.273	4	3	0	4	2	0.339	1.039
187	1.045	371.273	4	3	0	4	2	0.339	1.039
194	0.253	193.203	5	4	2	1	1	0.501	0.964
203	−0.499	503.583	10	4	8	3	0	0.268	1.113
204	−0.091	517.61	10	3	9	3	0	0.237	1.229
218	0.536	293.283	7	3	1	4	3	0.444	0.903
220	−2.005	258.228	6	4	2	2	0	0.479	0.877
221	−2.451	244.201	6	4	2	2	0	0.525	0.876
282	−1.049	544.527	11	6	5	5	2	0.403	0.534
284	−1.308	251.242	7	3	2	3	2	0.482	0.406
285	−1.595	292.251	9	4	2	3	2	0.57	0.432
301	−1.614	251.242	7	3	3	3	2	0.48	0.364
302	−1.526	302.714	7	4	2	3	1	0.401	0.510
SAM	−4.254	399.445	9	4	7	3	2	0.483	

[1] Partition coefficient, [2] Molecular weight, [3] Hydrogen bond acceptors, [4] Hydrogen bond donors, [5] Molecular fractional polar surface area.

2.2. Filter Using Fingerprints

The fingerprint is another similarity technique that depends on the 2D molecular structures of two different ligands in a binary format. This technique computes the presence and/or absence of several sub-structural fragments to calculate the degree of intermolecular structural similarity. This technique is utilized as a tool to detect the degree of similarity between a hit candidate and a lead one [64] The fingerprint approach examines the following parameters: charges [65], hybridization [66], H-bond acceptors, and donors [67], positive and negative ionizable moieties [68], halogens, and aromatic rings beside the ALogP category of candidates. The experiment was carried out using Discovery Studio.

The fingerprint's output depends on Tanimoto coefficient (SA/(SA + SB + SC)). SA is a symbol that represents the number of bits present in the reference molecule (**SAM**) and the examined candidate. On the other hand, SB and SC represent the number of bits in the examined candidate but not **SAM** and the number of bits in **SAM** but not the examined candidate, respectively. The Tanimoto coefficient gives values with a range of zero (no shared bits) to one (all bits the same).

The results revealed the significant fingerprint similarity of **44**, **48**, **85**, **102**, **105**, **182**, **220**, **221**, **282**, **284**, **285**, **301**, and **302** with **SAM** (Table 2).

The reported antiviral potentialities of the preferred metabolites were summarized in the Supplementary Materials.

2.3. Docking Studies

Molecular docking studies were achieved to study the binding modes, orientations, and affinities of the candidates **44**, **48**, **85**, **102**, **105**, **182**, **220**, **221**, **282**, **284**, **285**, **301**, and **302** inside the SARS-CoV-2 nonstructural protein (nsp10) (PDB ID: 6W4H, resolution: 1.80 Å) active site using MOE 14.0 software.

The docking process was validated through a redocking step of **SAM** against active pockets of SARS-CoV-2 nonstructural protein (nsp10). The suitability of the performed protocol was demonstrated by the small RMSD (0.60 Å) that was found between the docked pose and **SAM** (Figure 4).

Table 2. Fingerprint similarity between the tested candidates and **SAM**.

Comp.	Similarity	SA	SB	SC
SAM	1	237	0	0
44	0.503	159	79	78
48	0.423	110	23	127
85	0.423	200	236	37
102	0.497	149	63	88
105	0.529	165	75	72
182	0.717	160	−14	77
220	0.475	135	47	102
221	0.458	125	36	112
282	0.443	250	327	−13
284	0.685	150	−18	87
285	0.671	159	0	78
301	0.642	145	−11	92
302	0.552	139	15	98

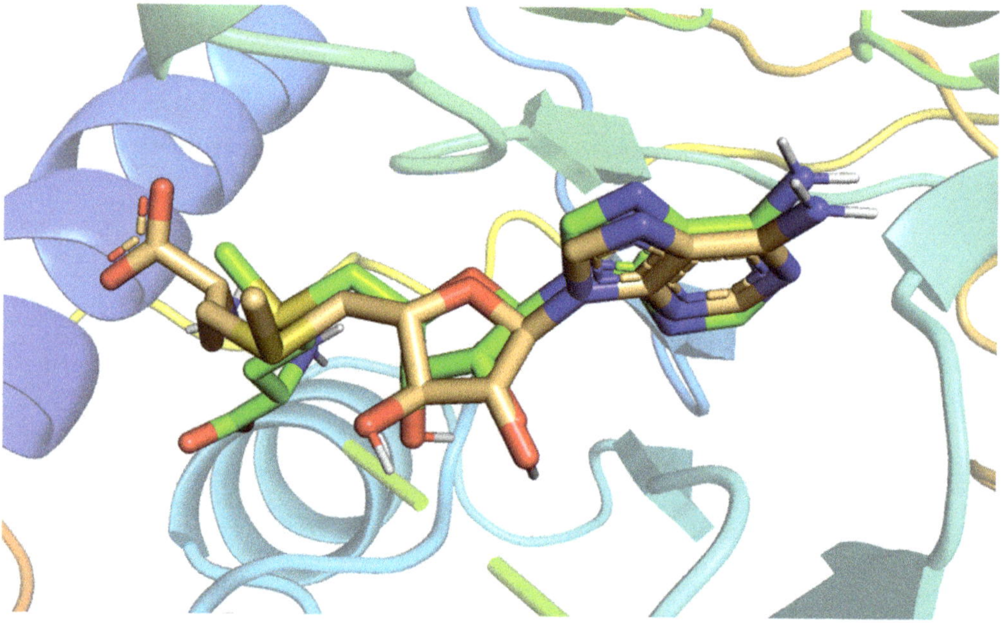

Figure 4. Superimposition of the co-crystallized ligand pose (**green**) and the docking pose (**wheat**).

The mode of binding of **SAM** inside COVID-19 nsp10 was illustrated in Figure 5. It was noticed that **SAM** interacted with the active site via the formation of six hydrogen bonds with Lys6844, Leu6898, Asn6899, Asp6912, Cys6913, and Tyr6930.

Among all studied metabolites, members **220**, **48**, **182**, **221**, and **284** exhibited the greatest binding free energies of docking (Table 3).

The methylpyrimidine-2,4-dione derivative (**220**) possessed a good potential affinity of −21.17 into the COVID-19 nsp10 active site. This high affinity is attributed to the formation of five hydrogen bond interactions. The pyrimidine moiety of candidate **220** was involved in two hydrogen-bonding interactions with Asp6912 and Cys6913. While the furan part interacted with the active site by three hydrogen bonds with Leu6898 and Tyr6930 (Figure 6).

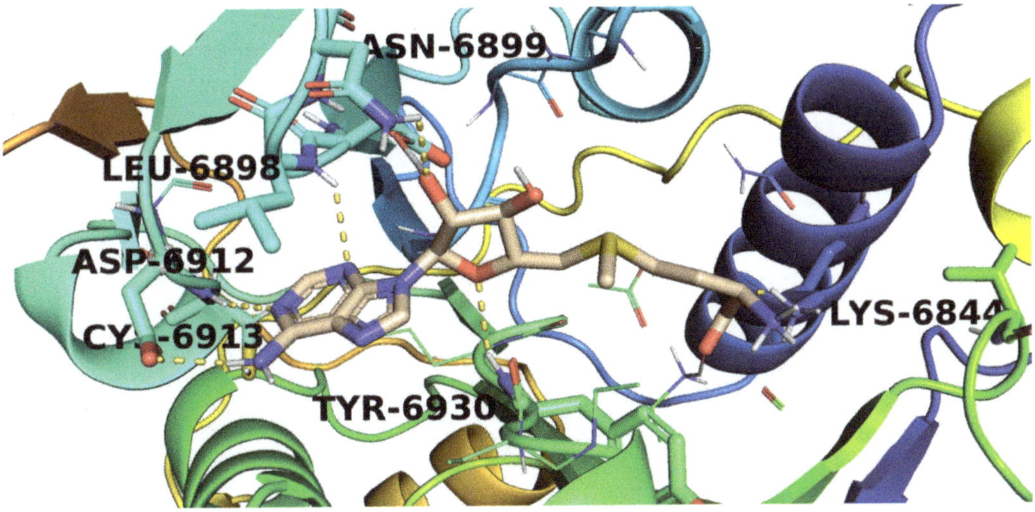

Figure 5. The proposed binding pattern of **SAM**.

Table 3. The calculated binding free energies of the examined candidates and **SAM** inside COVID-19 nsp10.

Comp.	ΔG [Kcal/mol]	Comp.	ΔG [Kcal/mol]
44	−18.65	**221**	−20.09
48	−21.15	**282**	−19.85
85	−19.32	**284**	−20.07
102	−18.98	**285**	−19.02
105	−20.01	**301**	−18.72
182	−21.10	**302**	−16.96
220	−21.17	**SAM**	−22.05

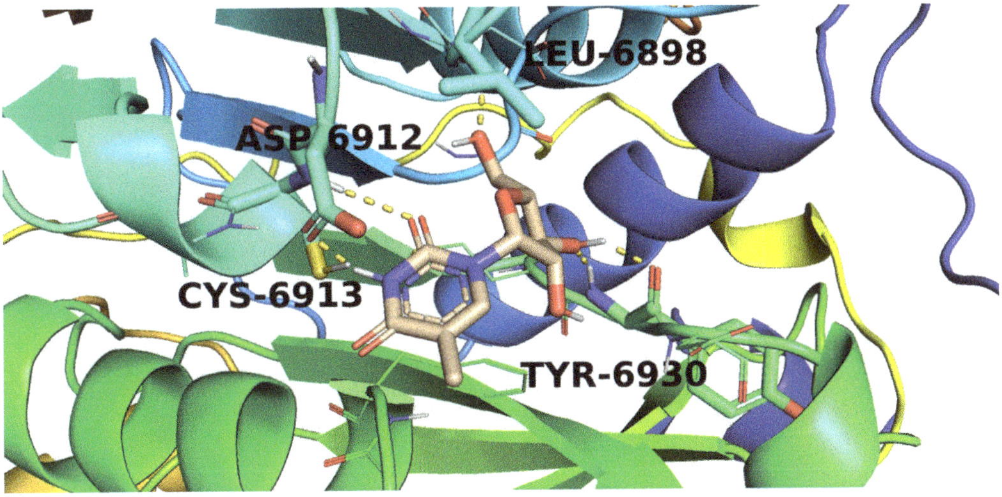

Figure 6. The proposed binding pattern of candidate **220**.

Candidate (**48**) exhibited a binding mode like that of **SAM** with the formation of four hydrogen bonds with Cys6913, Tyr6930, and Leu6898 (Figure 7).

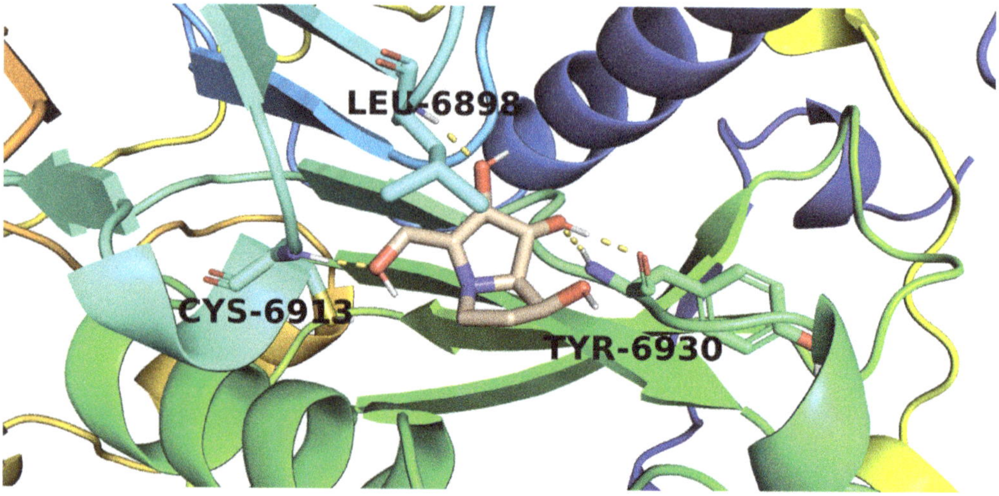

Figure 7. The proposed binding pattern of candidate **48**.

Investigation of the top docking poses of the 6-aminopurine member (**182**) showed that it interacted with the COVID-19 nsp10 active site through the formation of three hydrogen bond interactions. Its amino group was involved in a hydrogen bond with Asp6912 while one purine nitrogen atom formed a hydrogen bond with Cys6913. In addition, the furan oxygen interacted by a hydrogen bond with Tyr6930 (Figure 8).

The proposed binding pattern of the pyrimidinedione derivative (**221**) was illustrated in Figure 9. It interacted with the active site via the formation of five hydrogen bonds with Asn6899, Asp6897, Cys6913, and Tyr6930.

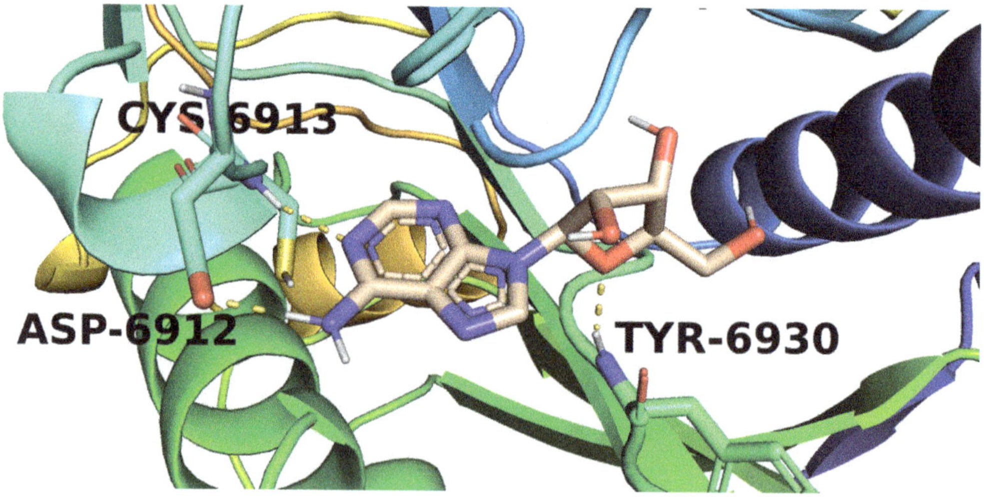

Figure 8. The proposed binding pattern of candidate **182**.

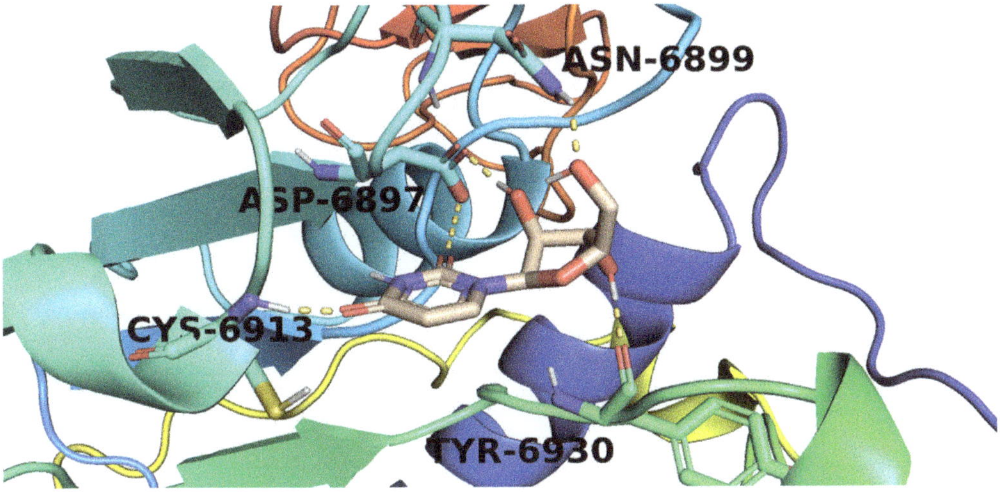

Figure 9. The proposed binding pattern of candidate **221**.

Figure 10 The proposed binding mode of candidate **284**. The purine moiety of **284** formed a hydrogen bond with Asp6912 while the attached amino group interacted with another hydrogen bond with Cys6913. The tetrahydrofuran-3-ol part formed two hydrogen bonds with Tyr6930 and Asn6899. Furthermore, the hydroxymethyl side chain was involved by a hydrogen bond with Gly6871.

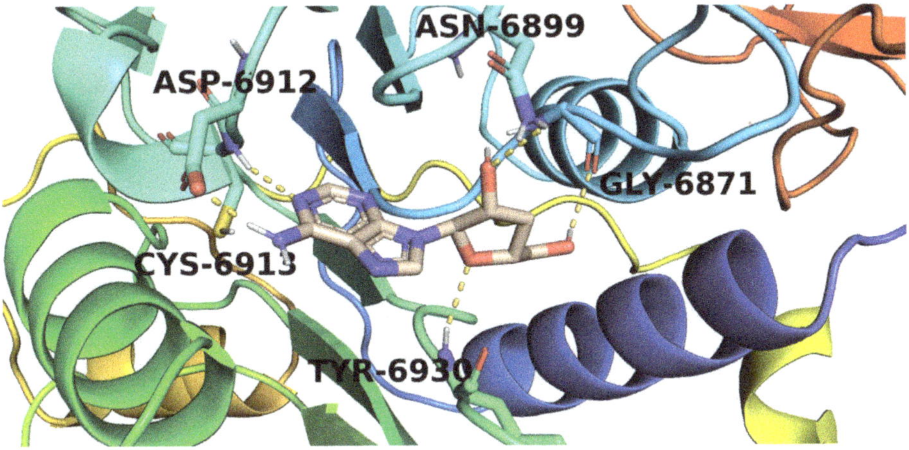

Figure 10. The proposed binding pattern of candidate **284**.

2.4. In Silico ADMET Analysis

Five parameters were examined for candidates **48**, **182**, **220**, **221**, and **284** using Discovery studio software. Acyclovir, the potent anti-viral drug, was used as a reference candidate. The results are illustrated in Figure 11.

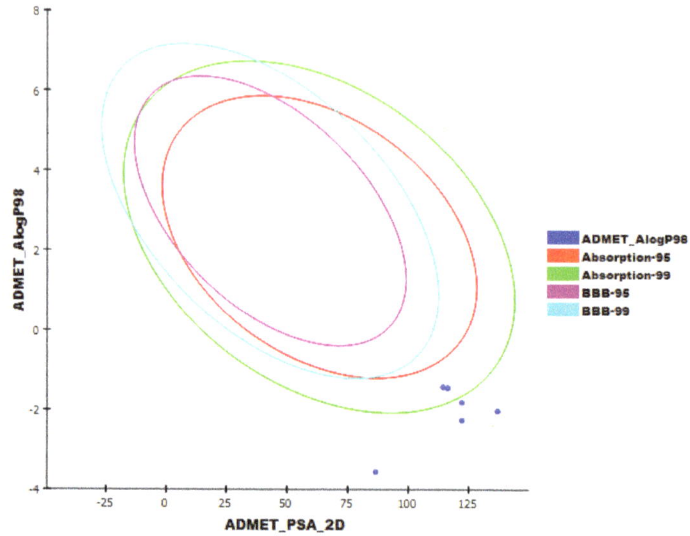

Figure 11. The expected ADMET study.

All the tested candidates have a very low chance to penetrate BBB. This indicates the high safety margin of such derivatives against the CNS. Additionally, all candidates exhibited an aqueous solubility character. For intestinal absorption, candidates **48**, **182**, **220**, and **221** were predicted to have poor to very poor levels, while candidate **284** was expected to have a moderate level. Furthermore, all candidates were predicted to be CYP2D6 non-inhibitors and can bind plasma protein by less than 90%. These results indicated that all the tested candidates have good pharmacokinetic properties and can be utilized for further investigations.

2.5. In Silico Toxicity Studies

Candidates **48**, **182**, **220**, **221**, and **284** were tested *in silico* for their proposed toxicity using Discovery studio software. In this test, seven toxicity models were utilized using ribavirin as a reference. The results are summarized in Table 4.

Table 4. Toxicity properties of candidates.

Comp.	FDA Rodent Carcinogenicity (Mouse-Female)	Carcinogenic Potency TD_{50} (Mouse) mg/kg Body Weight/Day	Rat Maximum Tolerated Dose (Feed) [a]	Rat Oral LD_{50} [a]	Rat Chronic LOAEL [a]	Ocular Irritancy	Skin Irritancy
48	Non-Carcinogen	9.295	0.191	0.778	0.018	Severe	Mild
182	Non-Carcinogen	4.245	0.175	1.119	0.010	Moderate	Mild
220	Single-Carcinogen	67.851	0.095	6.173	0.009	Moderate	Mild
221	Single-Carcinogen	55.437	0.094	4.343	0.006	Moderate	Mild
284	Multi-Carcinogen	6.402	0.155	1.213	0.004	Moderate	Mild
Ribavirin	Non-Carcinogen	13.111	0.154	0.750	0.013	Mild	Mild

[a] Unit = g/kg body weight.

FDA rodent carcinogenicity in female mice indicated that candidates **48** and **182** were non-carcinogenic, while candidates **220**, **221**, and **284** had some sort of carcinogenicity. Besides, candidates **48**, **182**, and **284** showed TD_{50} values of 9.295, 4.245, and 6.402 mg/kg body weight/day, respectively. Candidates **220** and **221** showed high carcinogenic potency TD_{50} values of 67.851 and 55.437 mg/kg body weight/day, respectively. Furthermore, candidates **48** and **182** showed high rat maximum tolerated dose values of 0.191 and 0.175 g/kg body weight, respectively. On the other hand, candidates **220** and **221** showed low rate maximum tolerated dose values of 0.095 and 0.094 g/kg body weight, respectively. Candidate **284** showed a comparable rat maximum tolerated dose value (0.155 g/kg body weight) with ribavirin (0.154 g/kg body weight). The tested candidates showed rat oral LD_{50} values ranging from 0.778 to 6.173 g/kg body weight, which were higher than the reference drug LD_{50} = 0.750 g/kg body weight. For the rat chronic LOAEL model, candidates **48** and **182** showed high values of 0.018 and 0.010 g/kg body weight, while candidates **220**, **221**, and **284** showed low values of 0.009, 0.006, and 0.004 g/kg body weight, respectively. All candidates were predicted to have mild to moderate irritant effects against ocular irritancy and skin irritancy models. Accordingly, candidates **48** and **182** had low toxicity profiles and were preferred for further studies.

2.6. DFT Studies

DFT parameters (Table 5) were studied for candidates 48 and 182 [69,70] against **SAM** as a reference using Discovery studio software (Table 5, Figures 12 and 13).

Table 5. Spatial distribution of molecular orbitals for candidates **48** and **182**.

Name	Total Energy *	Binding Energy *	HOMO Energy *	LUMO Energy *	Dipole Mag	Band Gap Energy *
48	−664.379	−4.841	−0.366	−0.156	1.391	0.210
182	−955.658	−6.102	−0.195	−0.068	1.396	0.128
SAM	−1675.931	−8.815	−0.270	−0.174	3.631	0.097

* Unit = kcal/mol for all descriptors except Dipole Mag.

2.6.1. Molecular Orbital Analysis

Candidates **48**, **182**, and **SAM** exhibited total energy values of −664.379, −955.658, and −1675.931 kcal/mol, respectively. The higher total energy of candidate **182** indicates a higher reactivity against the biological target. The two tested candidates, **48** and **182**, showed almost equal dipole moment values of 1.391 and 1.396, respectively. The Molecular Orbital (MO) analysis of EHOMO represents the energy of the highest occupied molecular orbital. On the other side, ELUMO represents the lowest unoccupied molecular orbital energies. The MO analysis is one of the essential parameters that is linked to the chemical reactivity and stability of a molecule. The HOMO spatial distributions of **SAM** are mainly presented on the 2-aminobutanoic acid moiety (the electron transfer zones), while its LUMO spatial distributions are located on the tetrahydrofuran-3,4-diol moiety (the electron acceptor zones). For candidate **48**, the HOMO spatial distributions are mainly located on the (2*R*,3*R*,4*R*)-2-(hydroxymethyl)pyrrolidine-3,4-diol moiety, while its LUMO spatial distributions are found on the (*S*)-pyrrolidin-3-ol moiety. For candidate **182**, the HOMO spatial distributions are mainly presented on the 9*H*-purin-6-amine moiety, while its LUMO spatial distributions are located on the (2*R*,3*S*,4*R*)-2-(hydroxymethyl)tetrahydrofuran-3,4-diol moiety. Furthermore, the gap energy of candidate **182** (0.128 kcal/mol) was less than that of candidate **48** (0.210 kcal/mol), confirming the high reactivity of candidate **182**. Consequently, candidate **182** may serve as a promising candidate for further studies.

2.6.2. Molecular Electrostatic Potential Maps (MEP)

MEP was used to specify the electrostatic potential of **48**, **182**, and **SAM** in a 3D form via the calculation of the partial charges, electronegativity, and chemical reactivity [71]. The electrostatic potential affects the binding of a drug with a specific protein and gives a deeper insight into drug–receptor interaction [72]. In MEP, the red color denotes the electronegative atoms, which can go through hydrogen bonding interactions as an acceptor. Additionally, the blue color denotes the electron-poor atoms that can form a donor in hydrogen bonding. The green to yellow color denotes the neutral atoms, which can form hydrophobic interactions [73].

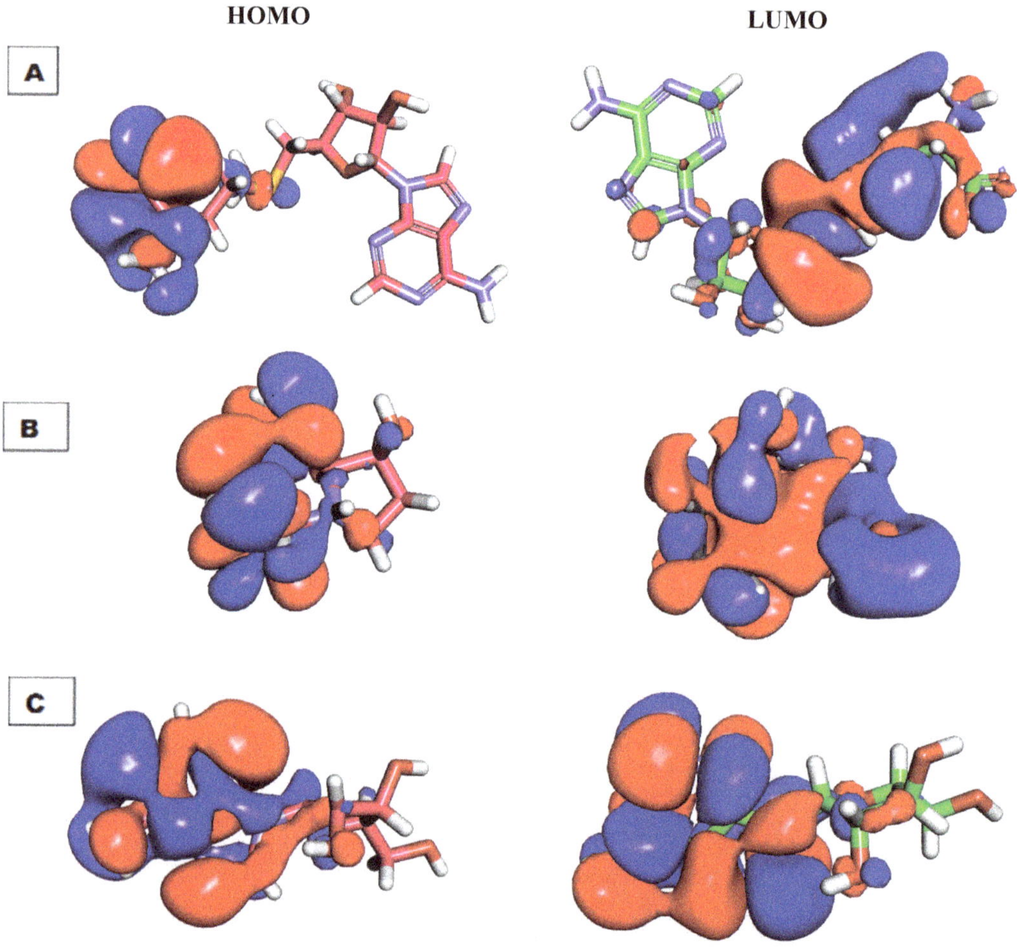

Figure 12. Spatial distribution of molecular orbitals for (**A**) **SAM**, (**B**) candidate **48**, and (**C**) candidate **182**.

The MEPs of **SAM**, **48**, and **182**, were illustrated in Figure 13A, B, and C, respectively. Investigating these figures indicated that **SAM** has eight red patches that are suitable for hydrogen bonding acceptors and are considered favorable sites for the electrophilic attack. Also, it comprises six blue patches that are suitable for hydrogen bond donors (the most favorable sites for the nucleophilic attack). Candidate **182** has six red patches and five blue patches. In addition, there is a yellow patch on the 9*H*-purine nucleus indicating a high possibility for hydrophobic interaction. These findings are highly like that of **SAM**. The MEP of candidate **48** is slightly different from **SAM**. In detail, it has four red patches and four blue patches. These results indicated that candidate **182** has a greater similarity with **SAM** than candidate **48**. Because of that, candidate **182** was singled out.

The antiviral activities of the preferred candidate, vidarabine (**182**), were reported against several viruses in different reports. It was active against herpes simplex encephalitis and neonatal herpes simplex infection [74,75], HBV [76], varicella-zoster virus [77], human polyomavirus [78], adenovirus [79], and Epstein–Barr virus infection [80].

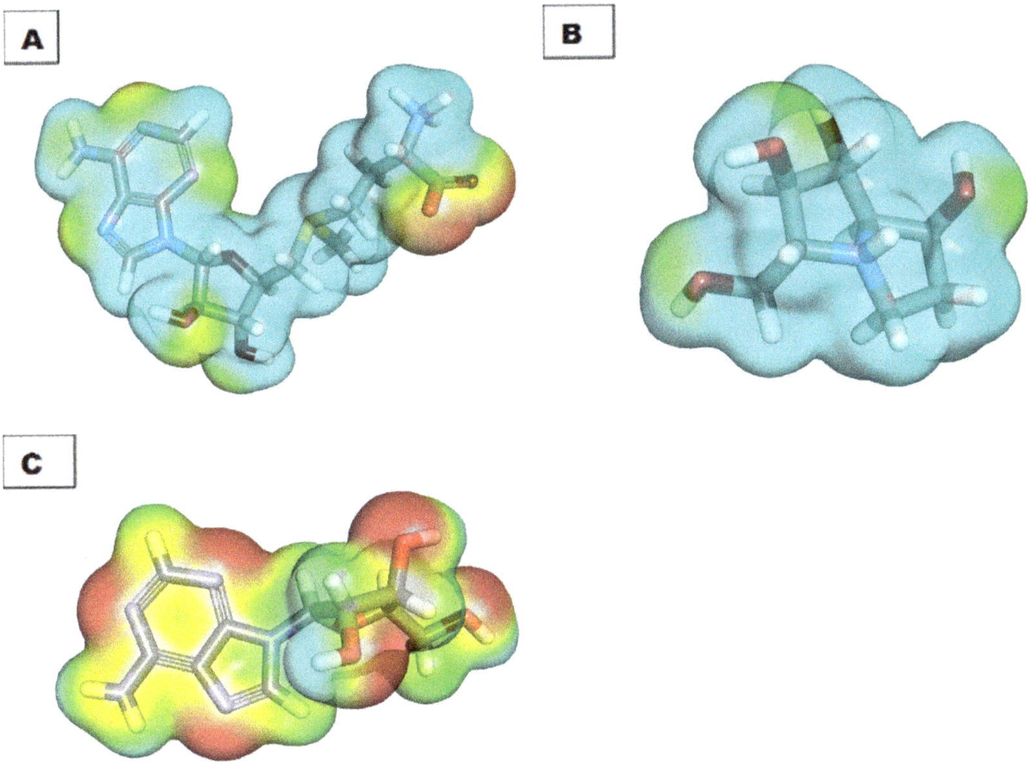

Figure 13. Molecular electrostatic potential map of (**A**) **SAM**, (**B**) candidate **48**, and (**C**) candidate **182**.

3. Method

3.1. Molecular Similarity Detection

Achieved by Discovery studio software (see method part in Supplementary Materials).

3.2. Pharmacophoric Study

Achieved by Discovery studio software (see method part in Supplementary Materials).

3.3. Docking Studies

Docking studies were achieved by MOE.14 software (see method part in Supplementary Materials).

3.4. ADMET Analysis

Achieved by Discovery studio 4.0 (see method part in Supplementary Materials).

3.5. Toxicity Studies

Achieved by Discovery studio software [81–83] (see method part in Supplementary Materials).

3.6. DFT Studies

Achieved by Discovery studio software [84] (see method part in Supplementary Materials).

4. Conclusions

Vidarabine (**182**) was suggested to be the most relevant SARS-Cov-2 nsp10 inhibitor among 310 naturally isolated metabolites that exhibited antiviral potentialities before. This suggestion was based on different computational (*in silico*) selection methods that included molecular similarity assessment, molecular fingerprint, docking studies, toxicity, ADMET, and DFT. The selected candidate showed various antiviral activities before. Further in vitro and in vivo biological studies have to be conducted to confirm the effect of **182** against SARS-Cov-2 nsp10 and its potential as an anti-COVID-19 drug.

Supplementary Materials: The following are available online, Figure S1: Chemical structures of the examined natural antiviral compounds, Table S1: Detailed toxicity report, in addition to the method (Molecular Similarity, Pharmacophore, Docking studies, ADMET studies, Toxicity studies and DFT studies).

Author Contributions: Conceptualization, A.M.M. and I.H.E.; methodology I.H.E., E.E.H.; software, M.M.K., E.B.E. and I.H.E.; writing—review and editing, A.M.M., E.B.E., E.E.H., A.A.A. and I.H.E. supervision, A.M.M. and I.H.E.; project administration, A.M.M. and I.H.E.; funding acquisition, E.B.E. All authors have read and agreed to the published version of the manuscript.

Funding: The authors extend their appreciation to the Research center at Almaarefa University for funding this work.

Institutional Review Board Statement: Not applicable.

Informed Consent Statement: Not applicable.

Conflicts of Interest: The authors declare no conflict of interest.

Sample Availability: Samples of the compounds are not available from the authors.

References

1. WHO. WHO Coronavirus (COVID-19) Dashboard. Available online: https://covid19.who.int/ (accessed on 2 September 2021).
2. Prieto-Martínez, F.D.; López-López, E.; Juárez-Mercado, K.E.; Medina-Franco, J.L. Computational drug design methods—Current and future perspectives. In *In Silico Drug Design*; Elsevier: Amsterdam, The Netherlands, 2019; pp. 19–44.
3. Nasser, A.A.; Eissa, I.H.; Oun, M.R.; El-Zahabi, M.A.; Taghour, M.S.; Belal, A.; Saleh, A.M.; Mehany, A.B.; Luesch, H.; Mostafa, A.E.; et al. Discovery of new pyrimidine-5-carbonitrile derivatives as anticancer agents targeting EGFR WT and EGFR T790M. *Org. Biomol. Chem.* **2020**, *18*, 7608–7634. [CrossRef] [PubMed]
4. Abbass, E.M.; Khalil, A.K.; Mohamed, M.M.; Eissa, I.H.; El-Naggar, A.M. Design, efficient synthesis, docking studies, and anticancer evaluation of new quinoxalines as potential intercalative Topo II inhibitors and apoptosis inducers. *Bioorg. Chem.* **2020**, *104*, 104255. [CrossRef]
5. Alanazi, M.M.; Mahdy, H.A.; Alsaif, N.A.; Obaidullah, A.J.; Alkahtani, H.M.; Al-Mehizia, A.A.; Alsubaie, S.M.; Dahab, M.A.; Eissa, I.H. New bis ([1,2,4][1,2,4][1,2,4][1,2,4][1,2,4][1,2,4][1,2,4][1,2,4][1,2,4][1,2,4][1,2,4][1,2,4][1,2,4][1,2,4][1,2,4][1,2,4][1,2,4][1,2,4][1,2,4] triazolo)[4,3-*a*:3′,4′-*c*] quinoxaline derivatives as VEGFR-2 inhibitors and apoptosis inducers: Design, synthesis, in silico studies, and anticancer evaluation. *Bioorg. Chem.* **2021**, *112*, 104949. [PubMed]
6. El-Helby, A.-G.A.; Sakr, H.; Ayyad, R.R.; Mahdy, H.A.; Khalifa, M.M.; Belal, A.; Rashed, M.; El-Sharkawy, A.; Metwaly, A.M.; Elhendawy, M.A.; et al. Design, synthesis, molecular modeling, in vivo studies and anticancer activity evaluation of new phthalazine derivatives as potential DNA intercalators and topoisomerase II inhibitors. *Bioorg. Chem.* **2020**, *103*, 104233. [CrossRef]
7. Eissa, I.H.; Ibrahim, M.K.; Metwaly, A.M.; Belal, A.; Mehany, A.B.; Abdelhady, A.A.; Elhendawy, M.A.; Radwan, M.M.; ElSohly, M.A.; Mahdy, H.A. Design, molecular docking, in vitro, and in vivo studies of new quinazolin-4 (3H)-ones as VEGFR-2 inhibitors with potential activity against hepatocellular carcinoma. *Bioorg. Chem.* **2021**, *107*, 104532. [CrossRef]
8. Abo-Ashour, M.F.; Eldehna, W.M.; Nocentini, A.; Bonardi, A.; Bua, S.; Ibrahim, H.S.; Elaasser, M.M.; Kryštof, V.; Jorda, R.; Gratteri, P.; et al. 3-Hydrazinoisatin-based benzenesulfonamides as novel carbonic anhydrase inhibitors endowed with anticancer activity: Synthesis, in vitro biological evaluation and in silico insights. *Eur. J. Med. Chem.* **2019**, *184*, 111768. [CrossRef]
9. Marrone, T.J.; Briggs, A.; James, M.; McCammon, J.A. Structure-based drug design: Computational advances. *Annu. Rev. Pharmacol. Toxicol.* **1997**, *37*, 71–90. [CrossRef]
10. Li, N.; Wang, Y.; Li, W.; Li, H.; Yang, L.; Wang, J.; Mahdy, H.A.; Mehany, A.; Jaiash, D.A.; Santali, E.Y.; et al. Screening of Some Sulfonamide and Sulfonylurea Derivatives as Anti-Alzheimer's Agents Targeting BACE1 and PPARγ. *J. Chem.* **2020**, *2020*, 1631243. [CrossRef]

11. Abdel-Aziz, H.A.; Eldehna, W.M.; Fares, M.; Al-Rashood, S.T.; Al-Rashood, K.A.; Abdel-Aziz, M.M.; Soliman, D.H. Synthesis, biological evaluation and 2D-QSAR study of halophenyl bis-hydrazones as antimicrobial and antitubercular agents. *Int. J. Mol. Sci.* **2015**, *16*, 8719–8743. [CrossRef] [PubMed]
12. Kairys, V.; Baranauskiene, L.; Kazlauskiene, M.; Matulis, D.; Kazlauskas, E. Binding affinity in drug design: Experimental and computational techniques. *Expert Opin. Drug Discov.* **2019**, *14*, 755–768. [CrossRef]
13. Al-Warhi, T.; El Kerdawy, A.M.; Aljaeed, N.; Ismael, O.E.; Ayyad, R.R.; Eldehna, W.M.; Abdel-Aziz, H.A.; Al-Ansary, G.H. Synthesis, biological evaluation and in silico studies of certain oxindole–indole conjugates as anticancer CDK inhibitors. *Molecules* **2020**, *25*, 2031. [CrossRef]
14. El-Metwally, S.A.; Abou-El-Regal, M.M.; Eissa, I.H.; Mehany, A.B.; Mahdy, H.A.; Elkady, H.; Elwan, A.; Elkaeed, E.B. Discovery of thieno [2,3-d] pyrimidine-based derivatives as potent VEGFR-2 kinase inhibitors and anti-cancer agents. *Bioorg. Chem.* **2021**, *112*, 104947. [CrossRef]
15. Alanazi, M.M.; Eissa, I.H.; Alsaif, N.A.; Obaidullah, A.J.; Alanazi, W.A.; Alasmari, A.F.; Albassam, H.; Elkady, H.; Elwan, A. Design, synthesis, docking, ADMET studies, and anticancer evaluation of new 3-methylquinoxaline derivatives as VEGFR-2 inhibitors and apoptosis inducers. *J. Enzym. Inhib. Med. Chem.* **2021**, *36*, 1760–1782. [CrossRef]
16. Alanazi, M.M.; Alaa, E.; Alsaif, N.A.; Obaidullah, A.J.; Alkahtani, H.M.; Al-Mehizia, A.A.; Alsubaie, S.M.; Taghour, M.S.; Eissa, I.H. Discovery of new 3-methylquinoxalines as potential anti-cancer agents and apoptosis inducers targeting VEGFR-2: Design, synthesis, and in silico studies. *J. Enzym. Inhib. Med. Chem.* **2021**, *36*, 1732–1750. [CrossRef]
17. Alsaif, N.A.; Taghour, M.S.; Alanazi, M.M.; Obaidullah, A.J.; Al-Mehizia, A.A.; Alanazi, M.M.; Aldawas, S.; Elwan, A.; Elkady, H. Discovery of new VEGFR-2 inhibitors based on bis ([1,2,4]triazolo)[4,3-*a*: 3′,4′-*c*] quinoxaline derivatives as anticancer agents and apoptosis inducers. *J. Enzym. Inhib. Med. Chem.* **2021**, *36*, 1093–1114. [CrossRef]
18. Alsaif, N.A.; Dahab, M.A.; Alanazi, M.M.; Obaidullah, A.J.; Al-Mehizia, A.A.; Alanazi, M.M.; Aldawas, S.; Mahdy, H.A.; Elkady, H. New quinoxaline derivatives as VEGFR-2 inhibitors with anticancer and apoptotic activity: Design, molecular modeling, and synthesis. *Bioorg. Chem.* **2021**, *110*, 104807. [CrossRef]
19. El-Adl, K.; Ibrahim, M.-K.; Alesawy, M.S.; Eissa, I.H. [1,2,4]Triazolo [4,3-*c*]quinazoline and bis([1,2,4][1,2,4][1,2,4][1,2,4][1,2,4][1,2,4][1,2,4][1,2,4][1,2,4][1,2,4][1,2,4][1,2,4][1,2,4]triazolo)[4,3-*a*:4′,3′-*c*] quinazoline derived DNA intercalators: Design, synthesis, in silico ADMET profile, molecular docking and anti-proliferative evaluation studies. *Bioorg. Med. Chem.* **2021**, *30*, 115958.
20. March-Vila, E.; Pinzi, L.; Sturm, N.; Tinivella, A.; Engkvist, O.; Chen, H.; Rastelli, G. On the integration of in silico drug design methods for drug repurposing. *Front. Pharmacol.* **2017**, *8*, 298. [CrossRef]
21. Zhang, W.; Pei, J.; Lai, L. Computational multitarget drug design. *J. Chem. Inf. Model.* **2017**, *57*, 403–412. [CrossRef]
22. Youssef, M.I.; Zhou, Y.; Eissa, I.H.; Wang, Y.; Zhang, J.; Jiang, L.; Hu, W.; Qi, J.; Chen, Z. Tetradecyl 2,3-dihydroxybenzoate alleviates oligodendrocyte damage following chronic cerebral hypoperfusion through IGF-1 receptor. *Neurochem. Int.* **2020**, *138*, 104749. [CrossRef]
23. Zhong, F.; Xing, J.; Li, X.; Liu, X.; Fu, Z.; Xiong, Z.; Lu, D.; Wu, X.; Zhao, J.; Tan, X. Artificial intelligence in drug design. *Sci. China Life Sci.* **2018**, *61*, 1191–1204. [CrossRef]
24. Hagras, M.; El Deeb, M.A.; Elzahabi, H.S.; Elkaeed, E.B.; Mehany, A.B.; Eissa, I.H. Discovery of new quinolines as potent colchicine binding site inhibitors: Design, synthesis, docking studies, and anti-proliferative evaluation. *J. Enzym. Inhib. Med. Chem.* **2021**, *36*, 640–658. [CrossRef]
25. Eissa, I.H.; Dahab, M.A.; Ibrahim, M.K.; Alsaif, N.A.; Alanazi, A.; Eissa, S.I.; Mehany, A.B.; Beauchemin, A.M. Design and discovery of new antiproliferative 1,2,4-triazin-3 (2H)-ones as tubulin polymerization inhibitors targeting colchicine binding site. *Bioorg. Chem.* **2021**, *112*, 104965. [CrossRef]
26. Eissa, I.H.; El-Helby, A.-G.A.; Mahdy, H.A.; Khalifa, M.M.; Elnagar, H.A.; Mehany, A.B.; Metwaly, A.M.; Elhendawy, M.A.; Radwan, M.M.; ElSohly, M.A.; et al. Discovery of new quinazolin-4 (3H)-ones as VEGFR-2 inhibitors: Design, synthesis, and anti-proliferative evaluation. *Bioorg. Chem.* **2020**, *105*, 104380. [CrossRef]
27. El-Adl, K.; El-Helby, A.-G.A.; Ayyad, R.R.; Mahdy, H.A.; Khalifa, M.M.; Elnagar, H.A.; Mehany, A.B.; Metwaly, A.M.; Elhendawy, M.A.; Radwan, M.M.; et al. Design, synthesis, and anti-proliferative evaluation of new quinazolin-4 (3H)-ones as potential VEGFR-2 inhibitors. *Bioorg. Med. Chem.* **2021**, *29*, 115872. [CrossRef]
28. Hopfinger, A. Computer-assisted drug design. *J. Med. Chem.* **1985**, *28*, 1133–1139. [CrossRef]
29. Jalmakhanbetova, R.I.; Suleimen, Y.M.; Oyama, M.; Elkaeed, E.B.; Eissa, I.; Suleimen, R.N.; Metwaly, A.M.; Ishmuratova, M.Y. Isolation and In Silico Anti-COVID-19 Main Protease (Mpro) Activities of Flavonoids and a Sesquiterpene Lactone from Artemisia sublessingiana. *J. Chem.* **2021**, *2021*, 13. [CrossRef]
30. Al-Karmalawy, A.A.; Dahab, M.A.; Metwaly, A.M.; Elhady, S.S.; Elkaeed, E.B.; Eissa, I.H.; Darwish, K.M. Molecular Docking and Dynamics Simulation Revealed the Potential Inhibitory Activity of ACEIs Against SARS-CoV-2 Targeting the hACE2 Receptor. *Front. Chem.* **2021**, *9*, 661230. [CrossRef]
31. Alesawy, M.S.; Abdallah, A.E.; Taghour, M.S.; Elkaeed, E.B.; Eissa, I.H.; Metwaly, A.M. In Silico Studies of Some Isoflavonoids as Potential Candidates against COVID-19 Targeting Human ACE2 (hACE2) and Viral Main Protease (Mpro). *Molecules* **2021**, *26*, 2806. [CrossRef]
32. El-Demerdash, A.; Metwaly, A.M.; Hassan, A.; El-Aziz, A.; Mohamed, T.; Elkaeed, E.B.; Eissa, I.H.; Arafa, R.K.; Stockand, J.D. Comprehensive virtual screening of the antiviral potentialities of marine polycyclic guanidine alkaloids against SARS-CoV-2 (COVID-19). *Biomolecules* **2021**, *11*, 460. [CrossRef]

33. Metwaly, A.M.; Ghoneim, M.M.; Eissa, I.H.; Elsehemy, I.A.; Mostafa, A.E.; Hegazy, M.M.; Afifi, W.M.; Dou, D. Traditional ancient Egyptian medicine: A review. *Saudi J. Biol. Sci.* **2021**, *28*, 5823–5832. [CrossRef]
34. Han, X.; Yang, Y.; Metwaly, A.M.; Xue, Y.; Shi, Y.; Dou, D. The Chinese herbal formulae (Yitangkang) exerts an antidiabetic effect through the regulation of substance metabolism and energy metabolism in type 2 diabetic rats. *J. Ethnopharmacol.* **2019**, *239*, 111942. [CrossRef]
35. Metwaly, A.M.; Zhu, L.; Huang, L.; Dou, D. Black ginseng and its saponins: Preparation, phytochemistry and pharmacological effects. *Molecules* **2019**, *24*, 1856. [CrossRef]
36. Wang, Y.-M.; Ran, X.-K.; Riaz, M.; Yu, M.; Cai, Q.; Dou, D.-Q.; Metwaly, A.M.; Kang, T.-G.; Cai, D.-C. Chemical constituents of stems and leaves of *Tagetespatula* L. and its fingerprint. *Molecules* **2019**, *24*, 3911. [CrossRef]
37. Metwaly, A. Comparative biological evaluation of four endophytic fungi isolated from nigella sativa seeds. *Al-Azhar J. Pharm. Sci.* **2019**, *59*, 123–136. [CrossRef]
38. Metwaly, A.M.; Wanas, A.S.; Radwan, M.M.; Ross, S.A.; ElSohly, M.A. New α-Pyrone derivatives from the endophytic fungus *Embellisia* sp. *Med. Chem. Res.* **2017**, *26*, 1796–1800. [CrossRef]
39. Metwaly, A.M.; Kadry, H.A.; Atef, A.; Mohammad, A.-E.I.; Ma, G.; Cutler, S.J.; Ross, S.A. Nigrosphaerin A a new isochromene derivative from the endophytic fungus *Nigrospora sphaerica*. *Phytochem. Lett.* **2014**, *7*, 1–5. [CrossRef]
40. Metwaly, A.M.; Fronczek, F.R.; Ma, G.; Kadry, H.A.; Atef, A.; Mohammad, A.-E.I.; Cutler, S.J.; Ross, S.A. Antileukemic α-pyrone derivatives from the endophytic fungus *Alternaria phragmospora*. *Tetrahedron Lett.* **2014**, *55*, 3478–3481. [CrossRef]
41. Imieje, V.O.; Zaki, A.A.; Metwaly, A.M.; Mostafa, A.E.; Elkaeed, E.B.; Falodun, A. Comprehensive In Silico Screening of the Antiviral Potentialities of a New Humulene Glucoside from *Asteriscus hierochunticus* against SARS-CoV-2. *J. Chem.* **2021**, *2021*, 5541876. [CrossRef]
42. Zhanzhaxina, A.; Suleimen, Y.; Metwaly, A.M.; Eissa, I.H.; Elkaeed, E.B.; Suleimen, R.; Ishmuratova, M.; Akatan, K.; Luyten, W. In Vitro and In Silico Cytotoxic and Antibacterial Activities of a Diterpene from *Cousinia alata* Schrenk. *J. Chem.* **2021**, *2021*, 5542455. [CrossRef]
43. Imieje, V.O.; Zaki, A.A.; Metwaly, A.M.; Eissa, I.H.; Elkaeed, E.B.; Ali, Z.; Khan, I.A.; Falodun, A. Antileishmanial Derivatives of Humulene from *Asteriscus hierochunticus* with in silico Tubulin Inhibition Potential. *Rec. Nat. Prod.* **2021**, *16*, 150–171.
44. Jalmakhanbetova, R.; Elkaeed, E.B.; Eissa, I.H.; Metwaly, A.M.; Suleimen, Y.M. Synthesis and Molecular Docking of Some Grossgemin Amino Derivatives as Tubulin Inhibitors Targeting Colchicine Binding Site. *J. Chem.* **2021**, *2021*, 5586515. [CrossRef]
45. Suleimen, Y.M.; Metwaly, A.M.; Mostafa, A.E.; Elkaeed, E.B.; Liu, H.-W.; Basnet, B.B.; Suleimen, R.N.; Ishmuratova, M.Y.; Turdybekov, K.M.; Van He ke, K. Isolation, Crystal Structure, and In Silico Aromatase Inhibition Activity of Ergosta-5,22-dien-3β-ol from the Fungus *Gyromitra esculenta*. *J. Chem.* **2021**, *2021*, 5529786. [CrossRef]
46. Ghoneim, M.M.; Afifi, W.M.; Ibrahim, M.; Elagawany, M.; Khayat, M.T.; Aboutaleb, M.H.; Metwaly, A.M. Biological evaluation and molecular docking study of metabolites from *Salvadora persica* L. Growing in Egypt. *Pharmacogn. Mag.* **2019**, *15*, 232.
47. Liu, L.; Luo, S.; Yu, M.; Metwaly, A.M.; Ran, X.; Ma, C.; Dou, D.; Cai, D. Chemical Constituents of Tagetes patula and Their Neuroprotecting Action. *Nat. Prod. Commun.* **2020**, *15*, 1934578X20974507.
48. Metwaly, A.M.; Ghoneim, M.M.; Musa, A. Two new antileishmanial diketopiperazine alkaloids from the endophytic fungus *Trichosporum* sp. *Derpharmachemica* **2015**, *7*, 322–327.
49. Yassin, A.M.; El-Deeb, N.M.; Metwaly, A.M.; El Fawal, G.F.; Radwan, M.M.; Hafez, E.E. Induction of apoptosis in human cancer cells through extrinsic and intrinsic pathways by *Balanites aegyptiaca* furostanol saponins and saponin-coated silvernanoparticles. *Appl. Biochem. Biotechnol.* **2017**, *182*, 1675–1693. [CrossRef]
50. Sharaf, M.H.; El-Sherbiny, G.M.; Moghannem, S.A.; Abdelmonem, M.; Elsehemy, I.A.; Metwaly, A.M.; Kalaba, M.H. New combination approaches to combat methicillin-resistant *Staphylococcus aureus* (MRSA). *Sci. Rep.* **2021**, *11*, 4240. [CrossRef] [PubMed]
51. Lin, S.; Chen, H.; Chen, Z.; Yang, F.; Ye, F.; Zheng, Y.; Yang, J.; Lin, X.; Sun, H.; Wang, L.; et al. Crystal structure of SARS-CoV-2 nsp10 bound to nsp14-ExoN domain reveals an exoribonuclease with both structural and functional integrity. *Nucleic Acids Res.* **2021**, *49*, 5382–5392. [CrossRef]
52. Lin, S.; Chen, H.; Ye, F.; Chen, Z.; Yang, F.; Zheng, Y.; Cao, Y.; Qiao, J.; Yang, S.; Lu, G. Crystal structure of SARS-CoV-2 nsp10/nsp16 2′-*O*-methylase and its implication on antiviral drug design. *Signal Transduct. Target. Ther.* **2020**, *5*, 131. [CrossRef] [PubMed]
53. Tazikeh-Lemeski, E.; Moradi, S.; Raoufi, R.; Shahlaei, M.; Janlou, M.A.M.; Zolghadri, S. Targeting SARS-COV-2 non-structural protein 16: A virtual drug repurposing study. *J. Biomol. Struct. Dyn.* **2020**, *39*, 4633–4646. [CrossRef]
54. Yadav, R.; Parihar, R.D.; Dhiman, U.; Dhamija, P.; Kumar, S. Docking of fda approved drugs targeting nsp-16, n-protein and main protease of sars-cov-2 as dual inhibitors. *Biointerface Res. Appl. Chem.* **2020**, *11*, 9848–9861.
55. Parida, P.K.; Paul, D.; Chakravorty, D. Nature's therapy for COVID-19: Targeting the vital non-structural proteins (NSP) from SARS-CoV-2 with phytochemicals from Indian medicinal plants. *Phytomed. Plus* **2021**, *1*, 100002. [CrossRef]
56. Nasser, M.; Salim, N.; Hamza, H.; Saeed, F.; Rabiu, I. Improved deep learning based method for molecular similarity searching using stack of deep belief networks. *Molecules* **2021**, *26*, 128. [CrossRef]
57. Turchi, M.; Cai, Q.; Lian, G. An evaluation of in-silico methods for predicting solute partition in multiphase complex fluids–A case study of octanol/water partition coefficient. *Chem. Eng. Sci.* **2019**, *197*, 150–158. [CrossRef]

58. Sullivan, K.M.; Enoch, S.J.; Ezendam, J.; Sewald, K.; Roggen, E.L.; Cochrane, S. An adverse outcome pathway for sensitization of the respiratory tract by low-molecular-weight chemicals: Building evidence to support the utility of in vitro and in silico methods in a regulatory context. *Applied In Vitro Toxicology* **2017**, *3*, 213–226. [CrossRef]
59. Altamash, T.; Amhamed, A.; Aparicio, S.; Atilhan, M. Effect of hydrogen bond donors and acceptors on CO_2 absorption by deep eutectic solvents. *Processes* **2020**, *8*, 1533. [CrossRef]
60. Wan, Y.; Tian, Y.; Wang, W.; Gu, S.; Ju, X.; Liu, G. In silico studies of diarylpyridine derivatives as novel HIV-1 NNRTIs using docking-based 3D-QSAR, molecular dynamics, and pharmacophore modeling approaches. *RSC Adv.* **2018**, *8*, 40529–40543. [CrossRef]
61. Escamilla-Gutiérrez, A.; Ribas-Aparicio, R.M.; Córdova-Espinoza, M.G.; Castelán-Vega, J.A. In silico strategies for modeling RNA aptamers and predicting binding sites of their molecular targets. *Nucleosides Nucleotides Nucleic Acids* **2021**, *40*, 798–807. [CrossRef]
62. Kaushik, A.C.; Kumar, A.; Bharadwaj, S.; Chaudhary, R.; Sahi, S. Ligand-Based Approach for In-silico Drug Designing. In *Bioinformatics Techniques for Drug Discovery*; Springer: Berlin, Germany, 2018; pp. 11–19.
63. Zhang, H.; Ren, J.-X.; Ma, J.-X.; Ding, L. Development of an in silico prediction model for chemical-induced urinary tract toxicity by using naïve Bayes classifier. *Mol. Divers.* **2019**, *23*, 381–392. [CrossRef]
64. Willett, P. Similarity-based virtual screening using 2D fingerprints. *Drug Discov. Today* **2006**, *11*, 1046–1053. [CrossRef]
65. Ieritano, C.; Campbell, J.L.; Hopkins, W.S. Predicting differential ion mobility behaviour in silico using machine learning. *Analyst* **2021**, *146*, 4737–4743. [CrossRef]
66. Taha, M.; Ismail, N.H.; Ali, M.; Rashid, U.; Imran, S.; Uddin, N.; Khan, K.M. Molecular hybridization conceded exceptionally potent quinolinyl-oxadiazole hybrids through phenyl linked thiosemicarbazide antileishmanial scaffolds: In silico validation and SAR studies. *Bioorg. Chem.* **2017**, *71*, 192–200. [CrossRef]
67. Chu, H.; He, Q.-X.; Wang, J.; Hu, Y.; Wang, Y.-Q.; Lin, Z.-H. In silico design of novel benzohydroxamate-based compounds as inhibitors of histone deacetylase 6 based on 3D-QSAR, molecular docking, and molecular dynamics simulations. *New J. Chem.* **2020**, *44*, 21201–21210. [CrossRef]
68. Opo, F.A.D.M.; Rahman, M.M.; Ahammad, F.; Ahmed, I.; Bhuiyan, M.A.; Asiri, A.M. Structure based pharmacophore modeling, virtual screening, molecular docking and ADMET approaches for identification of natural anti-cancer agents targeting XIAP protein. *Sci. Rep.* **2021**, *11*, 4049. [CrossRef]
69. Fleming, I. *Frontier Orbitals and Organic Chemical Reactions*; Wiley: Hoboken, NJ, USA, 1977.
70. El-Nahass, M.; Kamel, M.; El-Deeb, A.; Atta, A.; Huthaily, S. Ab initio HF, DFT and experimental (FT-IR) investigation of vibrational spectroscopy of PN, N-dimethylaminobenzylidenemalononitrile (DBM). *Spectrochim. Acta Part. A Mol. Biomol. Spectrosc.* **2011**, *79*, 443–450. [CrossRef]
71. Suhasini, M.; Sailatha, E.; Gunasekaran, S.; Ramkumaar, G. Vibrational and electronic investigations, thermodynamic parameters, HOMO and LUMO analysis on Lornoxicam by density functional theory. *J. Mol. Struct.* **2015**, *1100*, 116–128. [CrossRef]
72. Bitencourt-Ferreira, G.; de Azevedo Junior, W.F. Electrostatic Potential Energy in Protein-Drug Complexes. *Curr. Med. Chem.* **2021**, *28*, 4954–4971. [CrossRef] [PubMed]
73. Matin, M.M.; Hasan, M.S.; Uzzaman, M.; Bhuiyan, M.M.H.; Kibria, S.M.; Hossain, M.E.; Roshid, M.H. Synthesis, spectroscopic characterization, molecular docking, and ADMET studies of mannopyranoside esters as antimicrobial agents. *J. Mol. Struct.* **2020**, *1222*, 128821. [CrossRef]
74. Whitley, R.J.; Alford, C.A.; Hirsch, M.S.; Schooley, R.T.; Luby, J.P.; Aoki, F.Y.; Hanley, D.; Nahmias, A.J.; Soong, S.-J.; NIAID Collaborative Antiviral Study Group. Vidarabine versus acyclovir therapy in herpes simplex encephalitis. *N. Engl. J. Med.* **1986**, *314*, 144–149. [CrossRef]
75. Whitley, R.J.; Nahmias, A.J.; Soong, S.-J.; Galasso, G.G.; Fleming, C.L.; Alford, C.A.; Connor, J.; Bryson, Y.; Linnemann, C. Vidarabine therapy of neonatal herpes simplex virus infection. *Pediatrics* **1980**, *66*, 495–501. [PubMed]
76. Pollard, R.B.; Smith, J.L.; Neal, E.A.; Gregory, P.B.; Merigan, T.C.; Robinson, W.S. Effect of vidarabine on chronic hepatitis B virus infection. *JAMA* **1978**, *239*, 1648–1650. [CrossRef] [PubMed]
77. Miwa, N.; Kurosaki, K.; Yoshida, Y.; Kurokawa, M.; Saito, S.; Shiraki, K. Comparative efficacy of acyclovir and vidarabine on the replication of varicella-zoster virus. *Antivir. Res.* **2005**, *65*, 49–55. [CrossRef] [PubMed]
78. Chapman, C.; Flower, A.; Durrant, S. The use of vidarabine in the treatment of human polyomavirus associated acute haemorrhagic cystitis. *Bone Marrow Transplant.* **1991**, *7*, 481–483. [PubMed]
79. Kurosaki, K.; Miwa, N.; Yoshida, Y.; Kurokawa, M.; Kurimoto, M.; Endo, S.; Shiraki, K. Therapeutic basis of vidarabine on adenovirus-induced haemorrhagic cystitis. *Antivir. Chem. Chemother.* **2004**, *15*, 281–285. [CrossRef]
80. Kimura, H.; Morita, M.; Tsuge, I.; Hoshino, Y.; Tanaka, N.; Ito, Y.; Morishima, T. Vidarabine therapy for severe chronic active Epstein–Barr virus infection. *J. Pediatric Hematol. Oncol.* **2001**, *23*, 294–299. [CrossRef]
81. Yousef, R.; Sakr, H.; Eissa, I.; Mehany, A.; Metwaly, A.; Elhendawy, M.A.; Radwan, M.; ElSohly, M.A.; Abulkhair, H.S.; El-Adl, K. New quinoxaline-2 (1H)-ones as potential VEGFR-2 inhibitors: Design, synthesis, molecular docking, ADMET profile and anti-proliferative evaluations. *New J. Chem.* **2021**, *45*, 16949–16964. [CrossRef]
82. Amer, H.H.; Alotaibi, S.H.; Trawneh, A.H.; Metwaly, A.M.; Eissa, I.H. Anticancer activity, spectroscopic and molecular docking of some new synthesized sugar hydrazones, Arylidene and α-Aminophosphonate derivatives. *Arab. J. Chem.* **2021**, *14*, 103348. [CrossRef]

83. Alesawy, M.S.; Al-Karmalawy, A.A.; Elkaeed, E.B.; Alswah, M.; Belal, A.; Taghour, M.S.; Eissa, I.H. Design and discovery of new 1,2,4-triazolo [4,3-*c*] quinazolines as potential DNA intercalators and topoisomerase II inhibitors. *Arch. Pharm.* **2021**, *354*, 2000237. [CrossRef]
84. Parmar, D.R.; Soni, J.Y.; Guduru, R.; Rayani, R.H.; Kusurkar, R.V.; Vala, A.G.; Talukdar, S.N.; Eissa, I.H.; Metwaly, A.M.; Khalil, A.; et al. Discovery of new anticancer thiourea-azetidine hybrids: Design, synthesis, in vitro antiproliferative, SAR, in silico molecular docking against VEGFR-2, ADMET, toxicity, and DFT studies. *Bioorg. Chem.* **2021**, *115*, 105206. [CrossRef]

molecules

MDPI

Review

Catechins: Therapeutic Perspectives in COVID-19-Associated Acute Kidney Injury

Lúcio Ricardo Leite Diniz [1], Hatem A. Elshabrawy [2], Marilia Trindade de Santana Souza [3], Allana Brunna Sucupira Duarte [4], Sabarno Datta [5] and Damião Pergentino de Sousa [4,*]

1 College of Nordeste da Bahia, Coronel João Sá 48590-000, BA, Brazil; luciodiniz@yahoo.com.br
2 Department of Molecular and Cellular Biology, College of Osteopathic Medicine, Sam Houston State University, Conroe, TX 77304, USA; hatem.elshabrawy@shsu.edu
3 Department of Pharmacy, Federal University of Sergipe, São Cristóvão 49100-000, SE, Brazil; biomari@hotmail.com
4 Department of Pharmaceutical Sciences, Federal University of Paraíba, João Pessoa 58051-970, PB, Brazil; allanabrunna@gmail.com
5 College of Osteopathic Medicine, Sam Houston State University, Conroe, TX 77304, USA; sxd071@shsu.edu
* Correspondence: damiao_desousa@yahoo.com.br; Tel.: +55-83-3216-7347

Abstract: Data obtained from several intensive care units around the world have provided substantial evidence of the strong association between impairment of the renal function and in-hospital deaths of critically ill COVID-19 patients, especially those with comorbidities and requiring renal replacement therapy (RRT). Acute kidney injury (AKI) is a common renal disorder of various etiologies characterized by a sudden and sustained decrease of renal function. Studies have shown that 5–46% of COVID-19 patients develop AKI during hospital stay, and the mortality of those patients may reach up to 100% depending on various factors, such as organ failures and RRT requirement. Catechins are natural products that have multiple pharmacological activities, including anti-coronavirus and reno-protective activities against kidney injury induced by nephrotoxic agents, obstructive nephropathies and AKI accompanying metabolic and cardiovascular disorders. Therefore, in this review, we discuss the anti-SARS-CoV-2 and reno-protective effects of catechins from a mechanistic perspective. We believe that catechins may serve as promising therapeutics in COVID-19-associated AKI due to their well-recognized anti-SARS-CoV-2, and antioxidant and anti-inflammatory properties that mediate their reno-protective activities.

Keywords: SARS-CoV-2; coronavirus; renal disease; nephrotoxicity; reno-protective effect; green tea; epigallocatechin-3-gallate (EGCG); flavonoids; natural products; medicinal plant

Citation: Diniz, L.R.L.; Elshabrawy, H.A.; Souza, M.T.d.S.; Duarte, A.B.S.; Datta, S.; de Sousa, D.P. Catechins: Therapeutic Perspectives in COVID-19-Associated Acute Kidney Injury. *Molecules* **2021**, *26*, 5951. https://doi.org/10.3390/molecules26195951

Academic Editors: Giovanni N. Roviello and Caterina Vicidomini

Received: 16 August 2021
Accepted: 22 September 2021
Published: 30 September 2021

1. Introduction

Coronaviruses (CoVs) are positive single-stranded (+ss) RNA viruses that are known to infect several mammalian and avian species [1,2]. Seven human CoVs (HCoVs) have been reported as the causative agents of respiratory diseases, of which only three are associated with severe respiratory illnesses with high fatality, whereas others cause mild symptoms [3,4]. Severe Acute Respiratory Syndrome-CoV (SARS-CoV) was first discovered following a disease outbreak that occurred in China in 2002–2003 [5]. Middle East Respiratory Syndrome-CoV (MERS-CoV) was then discovered in 2012 after reports of severe respiratory disease cases in Saudi Arabia [6]. In December 2019, SARS-CoV-2 was identified as the novel causative agent of Coronavirus Disease 2019 (COVID-19) that originated in Wuhan, China, and later became a pandemic, resulting in millions of deaths worldwide [4]. SARS-CoV-2 is transmitted primarily via respiratory droplets, after which it infects and replicates in lung epithelial cells, causing respiratory symptoms observed in COVID-19 patients [7].

The SARS-CoV-2 uses its receptor binding domain (RBD), within the S1 domain of spike (S) protein on the viral surface, for binding to the angiotensin converting enzyme 2

(ACE2) receptor and infecting lung cells [8]. After binding to ACE2, S protein is processed by cell membrane-bound serine protease TMPRSS2 or cathepsin L in late endosomes into S1 and S2 subunits [9,10]. A fusion peptide in the S2 domain then triggers the fusion of the viral envelope with cellular membranes, releasing the viral RNA into the cytoplasm [9]. Similar to SARS-CoV, which shares 80% genomic sequence identity with SARS-CoV-2, there is a cap at the 5′ end of the SARS-CoV-2 genome and a poly-A tail at the 3′ end [11,12]. Once the SARS-CoV-2 genome is in the cytoplasm, the 5′ end is translated into two polyproteins, pp1a and pp1ab [12]. These two polyproteins are processed by two proteases, main protease (Mpro or 3CLpro) and papain-like protease (PLpro), which are part of polyproteins. Processing of pp1a and pp1ab produces 16 nonstructural proteins (NSP1–16) with different roles in viral replication, including RdRp (NSP12) and others which have enzymatic activity, such as NSP15 (endoribonuclease) and NSP13 (helicase) [12]. The rest of the genome is translated into structural and nonstructural proteins. The structural proteins, S, membrane (M), envelope (E) and nucleocapsid (N) proteins assemble with the viral RNA, following genome replication, to form new viral particles which are released from infected cells to initiate infection of neighboring healthy cells [12].

COVID-19 patients have shown an unpredictable and variable clinical outcome that ranges from asymptomatic presentation to multiple organ failure and death [13–15]. Although pulmonary complications have been the main clinical presentation of COVID-19, there is increasing evidence that acute kidney injury (AKI) is strongly associated with the in-hospital high mortality of COVID-19 patients, especially those with underlying comorbidities and/or who require renal replacement therapy (RRT) [16–18]. AKI can lead to serious complications, such as hydro-electrolytic disturbance, alterations in blood pressure control, impairment of acid–base homeostasis, rise of serum concentration of toxic metabolites and increased risk of drug overdose [19,20]. Moreover, AKI contributes to the production of diverse inflammatory mediators and vasoactive agents that play key roles in pathological mechanisms of several diseases [21].

Although at the beginning of the COVID-19 pandemic renal involvement in severe COVID-19 was insignificant, clinical reports have shown that AKI is associated with the severity of the disease and in-hospital deaths of COVID-19 patients [16,22,23]. The incidence of AKI in hospitalized COVID-19 patients can vary from 5% to 46%, which most likely require RRT. It has been estimated that approximately 20% of COVID-19 patients, admitted to intensive care units (ICUs), would require RRT on average 15 days after disease onset [24–26]. The prolonged hospitalization and poor prognosis of COVID-19 patients with AKI who require RRT are not observed, at least in the same magnitude, in COVID-19 patients with other commodities or renal disease. For example, SARS-CoV-2-infected patients on peritoneal dialysis or with chronic kidney disease (CKD) not requiring RRT showed no significant difference in incidence and mortality compared to the general population, while COVID-19 patients with dialysis-dependent CKD showed greater risk of in-hospital deaths [27,28]. Therefore, patients on dialysis with earlier-stage renal disease requiring RRT have become a vulnerable group to COVID-19-related morbidity and mortality, and it is extremely necessary to find an effective treatment for impairment of renal functions and damages promoted by SARV-CoV-2 infection.

In this context, reno-protective compounds that showed therapeutic activity in different experimental models of AKI and antiviral activity against SARS-CoV-2 would be ideal in COVID-19-associated AKI [29].

Catechins are natural polyphenolic compounds found in a wide variety of fruits, vegetables and plant-based food and beverages. Green tea extract is a recognized rich dietary source of catechins, containing a substantial amount of catechin, (-)-epicatechin (EC), (-)-epicatechin-3-gallate (ECG), (-)-epigallocatechin (EGC) and (-)-epigallocatechin-3-gallate (EGCG) [30,31]. A broad range of pharmacological activities have been reported for catechins, including antiviral and nephroprotective activities, which are strongly associated with their anti-inflammatory and antioxidant properties [32–37].

In the present study, we review the anti-SARS-CoV-2 activity of catechins, safety and effectiveness of catechins in AKI induced by diverse nephrotoxic stimuli, and provide evidence that catechins could be used as antiviral and reno-protective agents to prevent COVID-19-induced AKI.

2. Materials and Methods

We performed a literature search of catechins, coronaviruses and acute kidney injury. The literature search, performed in the PubMed database, included studies published in English until March 2021 and used the following keywords: catechins and COVID-19, catechins and coronaviruses, catechins and SARS-CoV-2, catechins and acute kidney injury, catechins and acute renal failure, catechins and SARS-COV-2-induced acute kidney injury, catechins and SARS-COV-2-induced acute renal injury and catechins and COVID-19-induced acute renal injury. We selected studies which investigated the anti-SARS-CoV-2 activities of catechins as well as the reno-protective effects of catechins and isomers in experimental or clinical AKI in accordance with KDIGO stages 1, 2 or 3 definitions. Studies on the reno-protective activities of crude extracts or beverages as well as combinations of catechins with other bioactive drugs were not included. Reported data of reno-protective effects attributed to catechins assessed by in vitro assays or performed in experimental models of chronic renal injury were not selected.

3. Catechins: General Pharmacological Properties

Catechins are natural polyphenolic compounds, belonging to the flavanols group of flavonoids, found in a wide variety of fruits, vegetables and plant-based food and beverages, such as fresh tea leaves, red wines, black grapes, cocoa beans, apricots and others [31,34]. Green tea extract is a recognized rich dietary source of catechins, containing a substantial amount of catechin, (-)-epicatechin (EC), (-)-epicatechin-3-gallate (ECG), (-)-epigallocatechin (EGC) and (-)-epigallocatechin-3-gallate (EGCG). EGCG is the most active and abundant polyphenol in green tea, accounting for approximately 50% of green tea polyphenols [30–38].

Pharmacokinetic studies have shown that, after oral administration, catechins and their metabolites are absorbed by the gastrointestinal tract and may undergo three types of biotransformation in humans: methylation, glucuronidation and sulfation in the liver and intestinal tissues [39–41]. Following absorption in the small intestine, the peak plasma concentrations (nmol/L) of EC, EGC and EGCG might vary from 1.5 to 2.5 h after intraduodenal administration of 20 and 30 mg/kg BW catechins, and return to baseline values between 2.5 and 4 h after administration, respectively [39]. The catechin metabolites are conjugated in the liver and excreted in the urine. The urinary excretion levels of catechins vary from 0% to 28.5% in human volunteers [40,41]. The bioavailability of catechins depend on several factors, such as enterohepatic recirculation, used dose, interaction with food components and flavan-3-ol stereochemistry. Different catechins were ranked as (-)-EC > (+)-EC = (+)-catechin > (-)-catechin on the basis of plasma concentrations and urinary excretion of the aglycones [38–43]. Recently, Andreu-Fernandez et al. [44] reported that the oral administration of EGCG in the form of green tea extract, in a single dose of 250 mg after overnight fasting, resulted in the highest peak concentrations (Cmax), area under the curve (AUC) values of 0–360 and average concentrations (Cav) both in men (5.95 ng/mL/kg, 3.86 ± 4.11 μg/mL/kg/6 h, 2.96 ng/mL/kg) and women (6.66 ng/mL/kg, 3.33 ± 1.08 μg/mL/kg/6 h, 3.66 ng/mL). Moreover, the study reported that t1/2 after oral ECGC administration were 192 ± 66 and 133 ± 28 min in men and women, respectively. A Mediterranean diet breakfast was shown to reduce the bioavailability of EGCG (more than 100% in males and 30% in females) [44].

Over the last few years, studies have demonstrated a broad range of pharmacological activities of catechins, including neuroprotective, anticarcinogenic, antihypertensive, antibacterial, antiviral and anti-inflammatory activities [45–47]. The majority of therapeutical indications attributed to catechins are strongly associated with their anti-inflammatory and antioxidant properties [48–54]. Catechins possess direct antioxidant effects, through free radical scavenging and metal ion chelation, as well as indirect antioxidant effects by induction of antioxidant enzymes, inhibition of pro-oxidant enzymes and production of the phase II detoxification enzymes [50,51]. Interestingly, studies have shown that catechins may produce a dose-dependent pro-oxidant effect, due to activation of the nuclear factor-erythroid factor 2-related factor 2 (Nrf2) pathway, leading to production of ROS [52,53].

Studies have provided evidence that pure catechins or plants with high concentrations of catechins exert anti-inflammatory effects in inflammatory diseases [48,49]. Mechanistic studies revealed that catechins decrease the production of pro-inflammatory cytokines, such as TNF-α, IL-1β, IL-6 and INF-γ, reduce the expression of adhesion molecules, inhibit infiltration and proliferation of immune cells and decrease the activity of inflammatory-related enzymes, such as iNOS and COX-2. Catechins' therapeutic effects are mediated by impairing multiple inflammation-related and oxidative stress-related pathways that involve JNK/NF-κB, MAPKs, Nrf2 and STAT1/3 transcription factors [49,54]. Figure 1 illustrates the catechins discussed in this study.

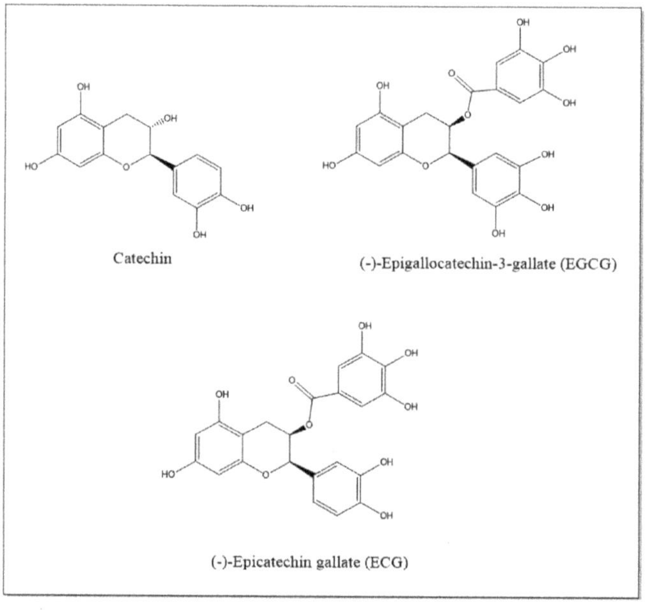

Figure 1. Chemical structure of reno-protective catechins.

4. Anti-SARS-CoV-2 Activity of Catechins

Given the urgent need to develop effective antiviral drugs against SARS-CoV-2, many research groups have tested several drug candidates, including plant-derived compounds, for anti-SARS-CoV-2 activity. Polyphenols, such as catechins, have been well-known for their antiviral activity against several RNA viruses, including coronaviruses [55–60]. A study has shown that epigallocatechin-3-gallate (EGCG) inhibited the SARS-CoV main protease (Mpro or 3CLpro) with an IC_{50} of 73 μM [59]. Another study has shown that increasing the concentration of (-)-catechin gallate (CAG) gallate potently inhibited SARS-CoV N protein and its association with RNA, with an IC_{50} of 0.05 μg/mL [61]. Based on their effectiveness against SARS-CoV, several studies have tested catechins for anti-SARS-CoV-2 activities (Figure 2). Molecular docking studies have shown that EGCG, epicatechin-3-gallate (ECG) and CAG bind strongly to SARS-CoV-2-Mpro's amino acid residues, His41 and Cys145, that are important for the enzymes' catalytic activity [62–64].

The Mpro-catechins complexes were found to be of high stability, which indicates that they can be further developed into potent Mpro inhibitors and SARS-CoV-2 antivirals. The high-affinity bindings of ECG and EGCG to Mpro were confirmed by an in vitro study which showed that ECG and EGCG inhibited recombinant Mpro activity with IC_{50} values of 5.21 ± 0.5 and 7.51 ± 0.21, respectively [65]. Furthermore, CAG and (-)-gallocatechin-3-gallate (GCG) inhibited Mpro with IC_{50} values of 2.98 ± 0.21 and 6.38 ± 0.5, respectively. EGCG was also reported by another group as an inhibitor of Mpro [66].

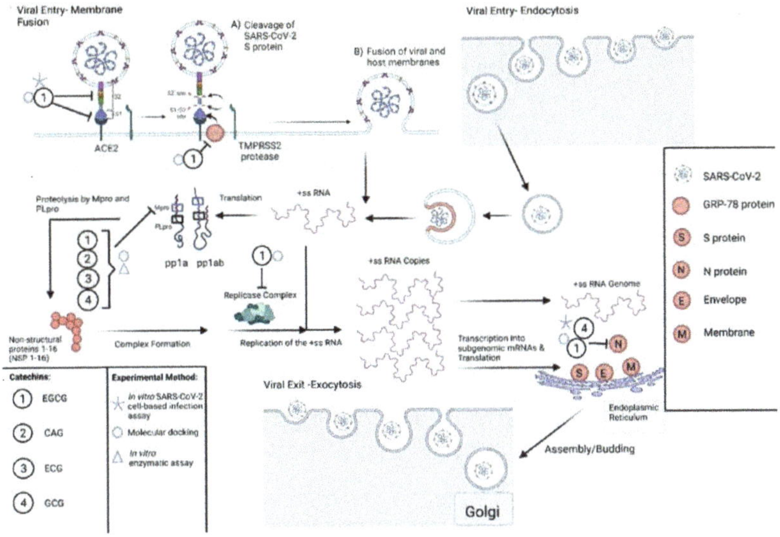

Figure 2. Catechins exert anti-SARS-CoV-2 activities by targeting different steps of the SARS-CoV-2 lifecycle. Catechins, such as EGCG, bind to SARS-CoV-2 S protein and inhibit its binding to the ACE2 receptor. EGCG also binds to GRP-78, which potentially blocks its binding to S protein, which may inhibit viral entry. EGCG, ECG, CAG and GCG inhibit Mpro of SARS-CoV-2, which blocks viral replication. Molecular docking studies have shown that EGCG binds to RNA-dependent RNA polymerase and other proteins of the replicase complex (NSP6 and NSP15) which may block viral replication. Furthermore, EGCG and GCG bind to and inhibit association of N protein with the RNA genome blocking viral assembly. This figure was created with BioRender.com (accessed on 25 July 2021).

A molecular docking study of 18 plant constituents to SARS-CoV-2 proteins showed that EGCG bound with higher affinity than the antiviral drugs chloroquine and remdesivir to all tested SARS-CoV-2 protein targets, including Mpro, S protein, S2 subunit of the S protein, RBD-ACE2 complex and NSP15 endoribonuclease [67]. The previous study was in agreement with another molecular docking study which showed high binding affinity of EGCG and catechin to SARS-CoV-2 proteins such as Mpro, RNA-dependent RNA polymerase, PLpro, RBD of S protein, NSP6, N protein, ACE2 receptor and ACE2 receptor bound to RBD [68–71]. The binding of EGCG to SARS-CoV-2-S protein was further confirmed by a molecular docking study which showed high-affinity binding to S protein [72], and another study which showed high-affinity binding not only to the S protein of the wild-type strain but to the UK mutant strain S protein [73]. The ability of EGCG to target S protein and its potential to impair its binding to the ACE2 receptor indicates that it could be further developed as an entry inhibitor for SARS-CoV-2. Furthermore, all docking studies provide evidence that EGCG and other catechins target multiple SARS-CoV-2 viral proteins, which make catechins potentially effective antivirals against SARS-CoV-2 and emerging variants since it would not be feasible for the virus to mutate all viral targets to avoid inhibition. Glucose-Regulated Protein-78 (GRP78) is an ER protein which plays a major role in the unfolded protein response (UPR), ensuring proper folding of proteins and reducing the amounts of unfolded proteins [74]. Studies have shown that GRP-78 binds to SARS-CoV-2-S protein, and this interaction is critical for viral entry [75]. Moreover, EGCG was shown to bind to and inhibit GRP-78 and thus could prevent its binding to SARS-CoV-2-S protein [74,76]. These findings suggest that EGCG could be investigated as a potential viral entry inhibitor for SARS-CoV-2 by disrupting SARS-CoV-2-S protein–GRP-78 binding.

The previous research findings were complemented by in vitro SARS-CoV-2 infection assays using Vero E6 cells, which showed that the catechin mixture, extracted from green tea, inhibited viral replication and reduced viral titer in a dose-dependent manner [77]. Another study identified EGCG as an entry inhibitor not only for SARS-CoV-2 but also for SARS-CoV and MERS-CoV [78]. Furthermore, 1 mM of EGCG reduced the infectivity of SARS-CoV-2 by binding to S protein and impairing its interaction with the ACE2 receptor [79]. The previous study findings are consistent with molecular docking studies which showed high-affinity binding of EGCG to SARS-CoV-2-S protein [67]. Moreover, EGCG was reported as a NSP15 inhibitor and effectively blocked SARS-CoV-2 viral replication, with an IC_{50} = 0.2 M [80]. In addition to EGCG, GCG was also reported as an inhibitor for SARS-CoV-2 replication by targeting N protein, with an IC_{50} = 44.4 M and a selectivity index (SI) of 3.5 [81]. Interestingly, a mixture of catechins consisting of 32.8% EGCG, 15.2% ECG, 13.2% EC, 10.8% EGC, 10.4% GC and 4.4% catechin impaired SARS-CoV-2 replication in Vero E6 cells [82]. Moreover, the same study demonstrated the in vivo efficacy of the mixture in inducing cell-mediated immunity by increasing the frequency of CD8+ T cells and mitigating lipopolysaccharide-induced cytokine storm in mice. The in vivo activity of catechins implies that it can be used not only to block viral replication but to alleviate symptoms, due to SARS-COV-2-induced cytokine storm, and boost immunity against SARS-CoV-2.

5. Reno-Protective Effect of Catechins in the Acute Kidney Injury

The reno-protective effects of catechins (Figure 3) have been investigated in experimental models of AKI over the past few years [32,33]. As illustrated in Table 1, several studies have shown that catechins and their isomers significantly reduced the impairment of renal function and structural damages caused by nephrotoxic drugs, obstructive nephropathy and AKI accompanying metabolic and cardiovascular disorders [83–89].

Pretreatment of animals with catechin (40 mg/kg, p.o.), twice a day for 4 days, prevented deterioration of kidney function and preserved renal morphology following injection of 50% glycerol solution (8 mL/kg, i.m.) and ferric nitrilotriacetate (8 mg iron/kg, i.p.) [83,90]. In a traditional model of AKI caused by a nephrotoxic dose of gentamicin (100 mg/kg/day, i.p.), the treatment with catechin hydrate (50 mg/kg), once daily for 14 days, inhibited the significant increases of BUN and SCr and protected glomeruli and tubules against gentamicin-induced damage [91]. In a nephrotoxicity model induced by cisplatin, it has been shown that 15 mg/kg of EGCG significantly inhibited the elevation of BUN and SCr induced by cisplatin (7 mg/kg) in rats [84]. In another study, 100 mg/kg of EGCG reduced SCr and BUN as well as kidney structural damages such as tubular dilatation, cast formation, granulovaculoar degeneration and tubular cell necrosis induced by cisplatin (20 mg/kg, i.p.) [92]. Similar reno-protective effects were reported by i.p. pretreatment with 5 mg/kg of ECG for 10 days, which protected against renal dysfunction and tubular necrosis induced by cisplatin (8 mg/kg) [86]. Furthermore, Gao et al. [93] showed that IV administration of EGCG (10 and 20 mg/kg) was effective in protection against contrast-induced nephropathy (CIN). The prophylactic or therapeutic treatment with EGCG was effective in CIN, as determined by low SCr and BUN levels, and a reduction in renal damage [93]. Soussi et al. [88] also showed that 5 mg/kg of EGCG protected against ammonium metavanadate (AMV)-induced glomerular hypertrophy and tubular dilatation in male Wistar rats.

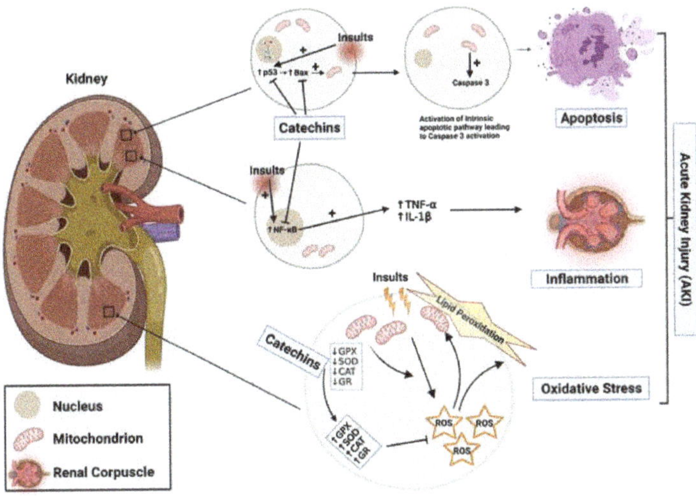

Figure 3. Catechins protect against AKI through different mechanisms. Catechins exert their reno-protective effects by inhibiting apoptosis (inhibition of p53 induction and Bax expression), reducing inflammation by decreasing accumulation of NF-kB in the nucleus, which lowers production of pro-inflammatory cytokines (TNF- α and IL-1 β), and attenuating oxidative stress and lipid peroxidation by restoring antioxidant enzyme activities, which detoxify reactive oxygen species (ROS). This figure was created with BioRender.com (accessed on 1 August 2021).

In obstructive nephropathy models of AKI, catechins significantly reduced renal structural and functional abnormalities, whereas in an ischemia/reperfusion (I/R) model of AKI, pretreatment with catechin (40 mg/kg, p.o.), twice daily for 4 days and 2 h prior to renal ischemia, markedly attenuated renal dysfunction and morphological alterations induced by I/R nephropathy [94]. Lv et al. showed that i.p. treatment with 50 mg/kg of EGCG protected the kidneys and prevented histological changes induced by clamping the left renal artery for 45 min, followed by 24 h reperfusion and contralateral nephrectomy in rats [95]. It was also found that i.p. administration of 5 mg/kg of EGCG for 14 days alleviated glomerular and tubular injury, inhibited renal tubulointerstitial fibrosis and reduced tubular cell apoptosis in a mouse model of unilateral ureteral obstruction (UUO) [96]. Moreover, i.p. pretreatment with 50 mg/kg of EGCG promoted a marked decrease of mRNA levels of TNF-α and the two markers of kidney damage, KIM-1 and NGAL, in the obstructed kidney of UUO Wistar rats. However, there was no significant effect of EGCG on renal blood flow, GFR, urine volume and urinary sodium excretion [86]. Considering the severity and the timeline of AKI onset induced by the UUO model, we believe that the use of higher doses of EGCG could have a more profound therapeutic effect [97,98].

Table 1. Reno-protective effects of catechins in animal models of acute kidney injury (AKI).

Acute Kidney Injury Induced by Nephrotoxic Drugs			
Experimental Model	**Reno-Protective Effect**	**Mechanism**	**Reference**
Rhabdomyolysis-induced AKI (50% glycerol-8 mL/kg, i.m.)	Catechin (40 mg/kg) inhibited the increase of BUN and SCr	Reduced the lipid peroxidation and increased glutathione levels. Restored the activity of renal antioxidant enzymes (SOD, CAT and GR).	[90]

Table 1. *Cont.*

Experimental Model	Reno-Protective Effect	Mechanism	Reference
Acute Kidney Injury Induced by Nephrotoxic Drugs			
Fe-NTA-induced AKI (8 mg iron/kg, i.p.)	Catechin (40 mg/kg) inhibited the increase of BUN and SCr	Reduced the lipid peroxidation and increased glutathione levels. Restored the activities of renal antioxidant enzymes (SOD, CAT and GR).	[83]
Gentamicin-induced AKI (100 mg/kg/day, i.p, for 14 days)	Catechin hydrate (50 mg/kg) inhibited the increase of BUN and SCr	Restored levels of renal glutathione.	[91]
Cisplatin-induced AKI (7, 8 and 20 mg/kg, i.p.)	EGCG (50 mg/kg) inhibited the increase of BUN and SCr	Restored the Se, Zn and Cu ion concentration in renal tissue. Restored the activities of renal SOD, GPX and CAT. Reduced lipid peroxidation.	[84]
	EGCG (100 mg/kg) improved cisplatin-induced tubular dilatation, cast formation, granulovaculoar degeneration and tubular cell necrosis	Restored the activities of renal antioxidant enzymes (MnSOD and GPX). Reduced production of TNF-α and IL-1β. Decreased accumulation of NF-κB in nucleus, and reduced p53 activation and apoptotic cell death.	[92]
	ECG (5 mg/kg) inhibited the increase of BUN and SCr	Reduced the lipid peroxidation. Restored the activities of renal antioxidant enzymes (SOD and CAT). Increased GSH. Reduced TNF-α and IL-6. Attenuated the activation of MAPK pathway by decrease phosphorylation of ERK1/2, JNK and p38 in renal tissues.	[87]
Contrast-induced nephropathy (CIN)	EGCG (10 mg/kg) normalized SCr and BUN levels, and improved renal histopathological scoring	Reduced the lipid peroxidation. Restored the activities of renal antioxidant enzyme (SOD) and reduced IL-1β via up-regulation of HO-1.	[93]
Ammonium metavanadate-induced AKI (5 mg/kg, i.p.)	EGCG (5 mg/kg) inhibited oxidative stress	Restored the activities of renal antioxidant enzymes (CAT, SOD and GPx). Reduced lipid peroxidation	[88]
Acute kidney injury induced by obstructive nephropathy			
Unilateral ureteral obstruction (UUO)	Catechin (2.5, 5 and 10 mg/kg) inhibited oxidative stress	Increased GSH and ROS. Increased mRNA and protein expression of Nrf2 and γ-GCS.	[99]
	EGCG (5 mg/kg) alleviated glomerular and tubular injury and attenuated renal interstitial fibrosis in UUO mice	Decreased macrophage infiltration and reduced production of inflammatory cytokines. Decreased expression of kidney damage markers (KIM-1 and NGAL) via NF-κB and Nrf2 nuclear translocation. Promoted HO-1 production	[96,97]
Renal ischemia-reperfusion injury	Catechin (40 mg/kg) inhibited the increase of BUN and SCr	Reduced lipid peroxidation, increased glutathione levels and restored activities of renal antioxidant enzymes (SOD, CAT and GR).	[89]
	EGCG (50 mg/kg) inhibited the increase of BUN and SCr	Reduced expression of TNF-α, IL-1β, IL-6, Bax and levels of cleaved caspase 3.	[90]
Acute kidney injury accompanying other morbidities			
Streptozotocin-induced diabetic nepropathy	Catechins (35 mg/day) prevented functional and morphological deterioration of kidneys, reduced albuminuria and increased creatinine clearance	Catechins reduced lipid peroxidation.	[32,33]
Salt-induced hypertension and renal injury in Dahl salt-sensitive rats (8%)	EGCG (50 mg/kg) reduced of urinary volume, urine protein and renal fibrosis, and increased CCr	Reduced the lipid peroxidation. Decreased the numbers of infiltrating macrophages and T cells. Induced the apoptosis of NRK-49F cells.	[99]
Cardiopulmonary by-pass-induced AKI	EGCG (1 mmol/L) inhibited the increase of SCr	Reduced oxidative stress and kidney damage as demonstrated by lower expression of KIM-1 and less production of 8-hydroxy-20 -deoxyguanosine	[85]

AKI is also a common comorbidity in cardiovascular and metabolic disorders, such as hypertension and diabetes. It was observed that administration of 5 mg/kg of catechins in the drinking water for 12 weeks prevented the progression of kidney damage in streptozotocin-induced diabetic rats [32,33]. In addition, Funamoto et al. investigated the reno-protective effect of oral pretreatment with EGCG for 2 weeks before cardiopulmonary bypass (CPB) in diabetic rats. The study reported that 30 min of CPB induced renal damage and that pretreatment with EGCG attenuated tubular injury and reduced KIM-1 expression [85]. Recently, Luo et al. reported that the oral administration of 50 mg/kg of EGCG, twice daily for 6 weeks, decreased blood pressure, lowered the 24 h urine protein levels and creatinine clearance and reduced the severity of renal fibrosis in Dahl rats with salt-sensitive hypertension [99]. The previous findings suggest that ECGC protected the kidneys, leading to a decrease in blood pressure. Treatment with EGCG also decreased serum cystatin C levels (an early marker for acute kidney injury), and urinary N-acetyl-β-d-glucosaminidase, NAG (a tubular injury marker), in high-fat diet-induced kidney injury [97]. Since AKI is strongly associated with high mortality of COVID-19 patients with underlying comorbidities, including diabetes and hypertension [98,99], the reno-protective effects of catechins are very promising and should be investigated in COVID-19-induced AKI.

In general, the mechanisms underlying nephrotoxicity are: (a) kidney-specific mitochondrial oxidative stress, caused by altered activities of mitochondrial electron transport chain enzyme complexes and accompanied by impaired antioxidant defenses, (b) inflammation, as evidenced by elevated levels of inflammatory mediators, such as TNF-α, IL-1, increased activities of NF-κB and p53 induction, and (c) apoptotic cell death, as indicated by high caspase-3 activity and DNA fragmentation in diverse clinical and experimental studies of AKI [100,101]. Oxidative stress is an important risk factor for AKI, as evidenced by a decrease in antioxidant enzymes' activities, an increase in lipid peroxidation and elevated levels of reactive oxygen species (ROS) and reactive nitrogen species (RNS) in AKI induced by nephrotoxic drugs and obstructive nephropathies [52–55]. In view of this, the improvement of renal function following catechins treatment is frequently accompanied by reduced levels of thiobarbituric acid reactive substances (TBARS, a marker of lipid peroxidation) and restoration of renal antioxidant enzymes. This indicates that catechins exert reno-protective effects, probably by radical scavenging and antioxidant activities [52–54]. It has been shown that (+)-catechin, catechin hydrate and ECGC attenuate lipid peroxidation and preserve activities of antioxidant enzymes in several AKI experimental models, including kidney injury induced by cisplatin, gentamicin, ammonium metavanadate, Fe-NTA, streptozotocin and unilateral urethra obstruction and renal ischemia/reperfusion models [32,33,83,88,89,91,94].

In line with previous antioxidant activities of catechins, a study has shown that treatment of mice with EGCG markedly attenuated cisplatin-induced mitochondrial oxidative/nitrative stress [84,87,92]. According to Zhou et al., the antioxidant activities of catechins might be partly via activation of the Nrf2 signaling pathway because ECG increased mRNA and protein levels of Nrf2 and γ-glutamylcysteine synthetase (γ-GCS) in obstructive AKI induced by renal ischemia and reperfusion [89]. Increased levels of heme oxygenase-1 (HO-1) were reported in contrast-induced nephropathy (CIN), while treatment with EGCG further increased HO-1 levels accompanied by an increase in Nrf2. Interestingly, the blockade of HO-1 with protoporphyrin IX zinc (II) (ZnPP) reversed the protective effect of EGCG on CIN. Its ability to increase the activity of antioxidant enzymes and to reduce inflammation and oxidative stress indicates that HO-1 is the upstream molecule that regulates the EGCG therapeutic effects [93]. Additionally, Funamoto et al. reported that reduced expression of KIM-1 following EGCG treatment was accompanied by lower production of 8-hydroxy-20-deoxyguanosine, indicating reduced oxidant stress [85].

The reno-protective effects of catechins are also due to their specific targeting of inflammatory and apoptotic pathways that are well-described in the genesis and progression of renal injuries. Pretreatment with EGCG promoted a significant decrease of the renal expression of proinflammatory cytokines TNF-α, IL-1β and IL-6 and reduced the numbers of macrophages and T cells infiltrating the renal tissue in salt-induced renal injury in Dahl salt-sensitive rats [99]. Moreover, EGCG decreased macrophage infiltration and inflammatory cytokine production, and attenuated renal interstitial fibrosis, in UUO animals, through regulation of NF-κB, Nrf2 and TGF-β/Smad signaling pathways [48,49,54,92,93]. Moreover, treatment with catechins reduced the expression of proapoptotic proteins Bax and caspase-3, increased the expression of the antiapoptotic protein, Bcl-2, and attenuated the activation of the MAPK pathway in renal tissues of rat models of AKI [94–97].

6. COVID-19 and Acute Kidney Injury (AKI)

AKI is defined according to RIFLE (Risk, Injury, Failure, Loss, End-Stage Kidney Disease), Acute Kidney Network (AKIN) and Kidney Disease Improving Global Outcomes (KDIGO). Briefly, RIFLE uses the Glomerular Filtration Rate (GFR) measurement based on levels of creatinine in serum (SCr) and urine, and urine output over 7 days. Risk, Injury and Failure stages were determined by increases in serum creatinine (SCr) $\geq$ 1.5-, 2- and 3-fold from a known baseline, respectively [102–104]. AKIN diagnosis of AKI is driven by observations of minor increases in SCr over a shorter period of time (48 h). AKIN stage 1 is defined as an increase in SCr by $\geq$0.3 mg/dL ($\geq$26.4 μmol/L) over 48 h, whereas an absolute rise in SCr to $\geq$2- and 3-fold above baseline are criteria for diagnosis and classification of AKIN stages 2 and 3, respectively [19,105]. The current definition of AKI according to KDIGO is similar to AKIN, but the timeframe is extended from 48 h to 7 days. The decrease of urinary output to less than 0.5 mL kg^{-1} h^{-1} for 6 h was also similar to the RIFLE and AKIN definitions [19,106].

AKI has been shown to develop in COVID-19 patients 5 to 9 days after hospital admission and mostly in patients with severe COVID-19 disease [107,108]. In general, elevated levels of SCr and blood urea nitrogen (BUN) and reduced GFR were reported in COVID-19 patients following hospital admission, whereas an increase in SCr accompanied by proteinuria or hematuria was observed in 7–63% and 26.7% of cases, respectively [20,109–111]. Hypokalemia with increased kaliuresis were also observed in COVID-19 patients, which could be resulting from elevated angiotensin II levels, alterations of tubular reabsorption of potassium ions by drugs such as diuretics and/or SARS-CoV-2-induced diarrhea [112–114].

The postmortem histopathological analysis of renal tissues of patients who died of COVID-19 showed significant glomerular and tubular lesions [115,116]. Diffuse acute proximal tubular injury and detachment of podocytes containing numerous spherical particles typical of viral inclusion bodies were observed in these tissues [117–119]. In addition, diffuse erythrocyte aggregation and obstruction of glomerular and peritubular capillaries were evident. In a few cases, platelets, thrombi or fibrinoid necrosis were detected within the glomerular capillary loops [117,118]. All the previous data have suggested glomerular ischemia, endothelial cell injury and coagulation activation in COVID-19 patients [114]. Glomerular and tubular damages were also reported in other viral infections. Membranous glomerulopathy, glomerulosclerosis, membranoproliferative glomerulonephritis, interstitial nephritis and necrotizing tubulointerstitial nephritis are common clinical manifestations caused by HIV, HCV, HBV and adenoviral infections [120–122]. Histological examination demonstrated acute distal tubular necrosis and the presence of viral particles in epithelial cells, as well as in Bowman's capsule, which differ significantly from pathological changes reported in bacterial infection [122–124].

The precise mechanisms of AKI development in COVID-19 patients are not yet fully understood. However, the mechanisms might be divided into specific and unspecific mechanisms [23,24,117]. Among the specific mechanisms are the direct invasion of renal parenchyma by SARS-CoV-2 and the imbalance of the renin-angiotensin-aldosterone system (RAAS) with generation of inflammatory mediators, oxidative stress and microthrombosis [125,126]. The viral invasion of renal tissue has been described as an important mechanism for the development of virus-associated kidney injury [126,127]. The detection of viral particles in urine and the absence of CD4+ and CD8+ T cells in kidney tissue for a long time after initial viral infection suggest that immune cells are not recruited to infected renal tissue as a viral strategy to establish persistent infection [128–130]. RAAS has also been shown to play a key role in renal dysfunctional and AKI [131]. RAAS is comprised of the classical axis that consists of angiotensin-converting enzyme (ACE), angiotensin II (Ang II) and the angiotensin type 1 receptor (AT1R), and the alternative axis, which is composed of ACE2, Ang 1–7 and the Mas receptor [132–134]. The two RAAS axes act as counter-regulatory systems. Increases of ACE expression and activity lead to intense production of Ang II, which by binding to AT1R results in vasoconstriction, inflammation and pro-proliferative effects [132,134]. The reno-protective effects of ACE inhibitors and AT1R blockers in diabetic nephropathy and renal ischemia have suggested that the classical RAS axis is involved in the pathogenesis and progression of nephropathies [135,136]. On the other hand, several studies have shown that Ang 1–7, via Mas receptor activation, were reno-protective by counteracting AT1R-mediated Ang II renal damage [133,134,137,138]. It has been suggested that SARS-CoV-2 promoted both inhibition of ACE2 activity and lysosomal degradation of membrane-bound ACE2. Downregulation of ACE2 in SARS-CoV-2 infection leads to accumulation of Ang II, resulting in increased inflammation, fibrotic and vasoconstrictor effects [139–142]. Recently, Yang et al. demonstrated that kidney injury molecule-1 (KIM1), a molecule dramatically upregulated upon kidney injury, binds with the receptor-binding domain (RBD) of SARS-CoV-2 and MERS-CoV, facilitating their attachment to the cell membrane, with the immunoglobulin variable Ig-like (Ig V) domain of KIM1, which in turn suggest KIM1 as a novel receptor for SARS-CoV-2 and other coronaviruses. According to the authors, KIM1 may mediate and exacerbate the renal infection of SARS-CoV-2 in a 'vicious cycle', and KIM1 could be further explored as a therapeutic target [140].

The unspecific mechanisms of COVID-19-induced AKI are similar to those described in kidney impairment induced by different etiologies, such as renal ischemia/reperfusion, nephrotoxic compounds and bacterial infections [26,141]. Glomerular and tubular damages are believed to occur secondary to ischemia with the redistribution of blood flow from renal medulla to the cortex, the deterioration of microcirculatory oxygenation and the generation of local inflammatory mediators, pro-fibrotic agents and reactive oxygen species (ROS) [142,143]. A strong cytokines storm, characterized by overproduction of type I (IFN-α/β) and type II IFNs (IFN-γ), IL-1β, IL-6, IL-2 and IL-4, followed by a sudden decrease of virus loads in the mesangial cells and vascular endothelial cells, was described following Duck Hepatitis A virus (DHAV) infection [120–123,144]. The potential role of cytokine storm in COVID-19-induced kidney damage has also been described [24,124]. In addition, immune reactions or immune complex deposition during viral infections may play an important role in severe coagulopathy, endothelial damage and increased vascular permeability, resulting in glomerular membrane proliferation [24,120–124].

7. Conclusions

Catechins are phytochemicals present in several natural foods and medicinal plants with high therapeutic potential against various pathologies, especially in inflammatory diseases [29–35]. According to studies discussed in this review, these natural products exhibit significant nephroprotective effects in blocking or attenuating renal dysfunctions and glomerulus/tubular lesions caused by varied nephrotoxic origins [86–97]. Even though more studies are needed to elucidate the mechanisms underlying nephroprotective effects of catechins, the renoprotection promoted by catechins appears to be strongly associated with their antioxidant, anti-inflammatory and anti-apoptotic activities into renal tissues [100,101]. Among the catechins discussed, ECGC is a very promising nephroprotective and anti-SARS-CoV-2 compound, especially due to the antiviral activities discussed earlier in this review [67–69]. Together, the experimental and clinical data in the present review support the indication of catechins as promising molecules in the treatment of COVID-19-associated AKI [83–99]. In addition, catechins can also be used as a prototype for the synthesis of more active analogs for the treatment of AKI.

Author Contributions: Writing—original draft preparation, L.R.L.D., M.T.d.S.S. and H.A.E.; writing—review and editing, H.A.E., S.D. and A.B.S.D.; conceptualization, methodology, review and editing, and supervision, D.P.d.S. All authors have read and agreed to the published version of the manuscript.

Funding: This research received no external funding.

Institutional Review Board Statement: Not applicable.

Informed Consent Statement: Not applicable.

Data Availability Statement: Not applicable.

Acknowledgments: This research was supported by the National Council for Scientific and Technological Development (CNPq) and the Coordination for the Improvement of Higher Education Personnel (CAPES).

Conflicts of Interest: No potential conflict of interest was reported by the authors.

References

1. Woo, P.C.; Huang, Y.; Lau, S.K.; Yuen, K.Y. Coronavirus genomics and bioinformatics analysis. *Viruses* **2010**, *2*, 1804–1820. [CrossRef]
2. Woo, P.C.; Lau, S.K.; Lam, C.S.; Lau, C.C.; Tsang, A.K.; Lau, J.H.; Bai, R.; Teng, J.L.; Tsang, C.C.; Wang, M.; et al. Discovery of Seven Novel Mammalian and Avian Coronaviruses in the Genus *Deltacoronavirus* Supports Bat Coronaviruses as the Gene Source of *Alphacoronavirus* and *Betacoronavirus* and Avian Coronaviruses as the Gene Source of *Gammacoronavirus* and *Deltacoronavirus*. *J. Virol.* **2012**, *86*, 3995–4008. [CrossRef]
3. Lim, Y.X.; Ng, Y.L.; Tam, J.P.; Liu, D.X. Human Coronaviruses: A Review of Virus-Host Interactions. *Diseases* **2016**, *4*, 26. [CrossRef]
4. Zhu, N.; Zhang, D.; Wang, W.; Li, X.; Yang, B.; Song, J.; Zhao, X.; Huang, B.; Shi, W.; Lu, R.; et al. A Novel Coronavirus from Patients with Pneumonia in China, 2019. *N. Engl. J. Med.* **2020**, *382*, 727–733. [CrossRef]
5. Drosten, C.; Günther, S.; Preiser, W.; Werf, S.; Brodt, H.R.; Becker, S.; Rabenau, H.; Panning, M.; Kolesnikova, L.; Fouchier, R.A.; et al. Identification of a novel coronavirus in patients with severe acute respiratory syndrome. *N. Engl. J. Med.* **2003**, *348*, 1967–1976. [CrossRef]
6. Zaki, A.M.; van Boheemen, S.; Bestebroer, T.M.; Osterhaus, A.D.; Fouchier, R.A. Isolation of a novel coronavirus from a man with pneumonia in Saudi Arabia. *N. Engl. J. Med.* **2012**, *367*, 1814–1820. [CrossRef]
7. Li, B.; Li, X.; Wang, Y.; Han, Y.; Wang, Y.; Wang, C.; Zhang, G.; Jin, J.; Jia, H.; Fan, F.; et al. Diagnostic value and key features of computed tomography in Coronavirus Disease 2019. *Emerg. Microbes Infect.* **2020**, *9*, 787–793. [CrossRef] [PubMed]
8. Wan, Y.; Shang, J.; Graham, R.; Baric, R.S.; Li, F. Receptor Recognition by the Novel Coronavirus from Wuhan: An Analysis Based on Decade-Long Structural Studies of SARS Coronavirus. *J. Virol.* **2020**, *94*, e00127-20. [CrossRef] [PubMed]
9. Hoffmann, M.; Kleine-Weber, H.; Schroeder, S.; Krüger, N.; Herrler, T.; Erichsen, S.; Schiergens, T.S.; Herrler, G.; Wu, N.H.; Nitsche, A.; et al. SARS-CoV-2 Cell Entry Depends on ACE2 and TMPRSS2 and Is Blocked by a Clinically Proven Protease Inhibitor. *Cell* **2020**, *181*, 271–280. [CrossRef] [PubMed]
10. Ou, X.; Liu, Y.; Lei, X.; Li, P.; Mi, D.; Ren, L.; Guo, L.; Guo, R.; Chen, T.; Hu, J.; et al. Characterization of spike glycoprotein of SARS-CoV-2 on virus entry and its immune cross-reactivity with SARS-CoV. *Nat. Commun.* **2020**, *11*, 1620. [CrossRef] [PubMed]
11. Lu, R.; Zhao, X.; Li, J.; Niu, P.; Yang, B.; Wu, H.; Wang, W.; Song, H.; Huang, B.; Zhu, N.; et al. Genomic characterisation and epidemiology of 2019 novel coronavirus: Implications for virus origins and receptor binding. *Lancet* **2020**, *395*, 565–574. [CrossRef]
12. Wu, A.; Peng, Y.; Huang, B.; Ding, X.; Wang, X.; Niu, P.; Meng, J.; Zhu, Z.; Zhang, Z.; Wang, J.; et al. Genome Composition and Divergence of the Novel Coronavirus (2019-nCoV) Originating in China. *Cell Host Microbe* **2020**, *27*, 325–328. [CrossRef] [PubMed]
13. Hu, B.; Guo, H.; Zhou, P.; Shi, Z.L. Characteristics of SARS-CoV-2 and COVID-19. *Nat. Ver. Microbiol.* **2021**, *19*, 141–154. [CrossRef] [PubMed]
14. Shang, Y.; Xu, C.; Jiang, F.; Huang, R.; Li, Y.; Zhou, Y.; Dai, H. Clinical characteristics and changes of chest CT features in 307 patients with common COVID-19 pneumonia infected SARS-CoV-2: A multicenter study in Jiangsu, China. *Int. J. Infect. Dis.* **2020**, *96*, 157–162. [CrossRef] [PubMed]
15. Yao, R.; Martin, C.B.; Haase, V.S.; Tse, B.C.; Nishino, M.; Gheorghe, C.; Balli, K. Initial clinical characteristics of gravid severe acute respiratory syndrome coronavirus 2-positive patients and the risk of progression to severe coronavirus disease 2019. *Am. J. Obstet. Gynecol. MFM* **2021**, *3*, 100365. [CrossRef] [PubMed]
16. Ouyang, L.; Gong, Y.; Zhu, Y.; Gong, J. Association of acute kidney injury with the severity and mortality of SARS-CoV-2 infection: A meta-analysis. *Am. J. Emerg. Med.* **2021**, *43*, 149–157. [CrossRef] [PubMed]
17. Tocoglu, A.; Dheir, H.; Bektas, M.; Acikgoz, S.B.; Karabay, O.; Sipahi, S. Predictors of Mortality in Patients with COVID-19 Infection-associated Acute Kidney Injury. *J. Coll. Phys. Surg. Pak.* **2021**, *30*, S60–S65. [CrossRef]
18. Xu, H.; Garcia-Ptacek, S.; Annetorp, M.; Bruchfeld, A.; Cederholm, T.; Johnson, P.; Kivipelto, M.; Metzner, C.; Religa, D.; Eriksdotter, M. Acute kidney injury and mortality risk in older adults with COVID-19. *J. Nephrol.* **2021**, *34*, 295–304. [CrossRef]
19. Kellum, J.A.; Lameire, N. The definition of acute kidney injury. *Lancet* **2018**, *391*, 202–203. [CrossRef]

20. Ugwuowo, U.; Yamamoto, Y.; Arora, T.; Saran, I.; Partridge, C.; Biswas, A.; Martin, M.; Moledina, D.G.; Greenberg, J.H.; Simonov, M.; et al. Real-Time Prediction of Acute Kidney Injury in Hospitalized Adults: Implementation and Proof of Concept. *Am. J. Kidney Dis.* **2020**, *76*, 806–814.e1. [CrossRef]
21. Barton, A.L.; Williams, S.B.M.; Dickinson, S.J.; Parry, R.G.; Pollard, A. Acute Kidney Injury in Primary Care: A Review of Patient Follow-Up, Mortality, and Hospital Admissions following the Introduction of an AKI Alert System. *Nephron.* **2020**, *144*, 498–505. [CrossRef]
22. Chen, Y.T.; Shao, S.C.; Lai, E.C.; Hung, M.J.; Chen, Y.C. Mortality rate of acute kidney injury in SARS, MERS, and COVID-19 infection: A systematic review and meta-analysis. *Crit. Care* **2020**, *24*, 439. [CrossRef]
23. Piñeiro, G.J.; Molina-Andújar, A.; Hermida, E.; Blasco, M.; Quintana, L.F.; Rojas, G.M.; Mercadal, J.; Castro, P.; Sandoval, E.; Andrea, R.; et al. Severe acute kidney injury in critically ill COVID-19 patients. *J. Nephrol.* **2021**, *34*, 285–293. [CrossRef] [PubMed]
24. Pan, X.W.; Xu, D.; Zhang, H.; Zhou, W.; Wang, L.H.; Cui, X.G. Identification of a potential mechanism of acute kidney injury during the COVID-19 outbreak: A study based on single-cell transcriptome analysis. *Intensive Care Med.* **2020**, *46*, 1114–1116. [CrossRef] [PubMed]
25. Zhou, X.; Dong, P.; Pan, J.; Wang, H.; Xu, Z.; Chen, B. Renal replacement therapy modality in critically ill patients with acute kidney injury—A network meta-analysis of randomized controlled trials. *J. Crit. Care* **2021**, *64*, 82–90. [CrossRef] [PubMed]
26. Kolhe, N.V.; Fluck, R.J.; Selby, N.M.; Taal, M.W. Acute kidney injury associated with COVID-19: A retrospective cohort study. *PLoS Med.* **2020**, *17*, e1003406. [CrossRef] [PubMed]
27. Jiang, H.J.; Tang, H.; Xiong, F.; Chen, W.L.; Tian, J.B.; Sun, J.; Dong, J.W.; Wang, X.H.; Jin, X.F.; Ding, Y.Q.; et al. COVID-19 in Peritoneal Dialysis Patients. *Clin. J. Am. Soc. Nephrol.* **2020**, *16*, 121–123. [CrossRef]
28. Yang, D.; Xiao, Y.; Chen, J.; Chen, Y.; Luo, P.; Liu, Q.; Yang, C.; Xiong, M.; Zhang, Y.; Liu, X.; et al. COVID-19 and chronic renal disease: Clinical characteristics and prognosis. *QJM* **2020**, *113*, 799–805. [CrossRef]
29. Sawhney, S.; Tan, Z.; Black, C.; Marks, A.; Mclernon, D.J.; Ronksley, P.; James, M.T. Validation of Risk Prediction Models to Inform Clinical Decisions after Acute Kidney Injury. *Am. J. Kidney Dis.* **2021**, *78*, 28–37. [CrossRef]
30. Ramdani, D.; Chaudhry, A.S.; Seal, C.J. Chemical composition, plant secondary metabolites, and minerals of green and black teas and the effect of different tea-to-water ratios during their extraction on the composition of their spent leaves as potential additives for ruminants. *J. Agric. Food Chem.* **2013**, *61*, 4961–4970. [CrossRef]
31. Shii, T.; Tanaka, T.; Watarumi, S.; Matsuo, Y.; Miyata, Y.; Tamaya, K.; Tamaru, S.; Tanaka, K.; Matsui, T.; Kouno, I. Polyphenol composition of a functional fermented tea obtained by tea-rolling processing of green tea and loquat leaves. *J. Agric. Food Chem.* **2011**, *59*, 7253–7260. [CrossRef]
32. Chennasamudram, S.P.; Kudugunti, S.; Boreddy, P.R.; Moridani, M.Y.; Vasylyeva, T.L. Renoprotective effects of (+)-catechin in streptozotocin-induced diabetic rat model. *Nutr. Res.* **2012**, *32*, 347–356. [CrossRef]
33. Hase, M.; Babazono, T.; Karibe, S.; Kinae, N.; Iwamoto, Y. Renoprotective effects of tea catechin in streptozotocin-induced diabetic rats. *Int. Urol. Nephrol.* **2006**, *38*, 693–699. [CrossRef]
34. Chen, D.; Chen, G.; Sun, Y.; Zeng, X.; Ye, H. Physiological genetics, chemical composition, health benefits and toxicology of tea (*Camellia sinensis* L.) flower: A review. *Food Res. Int.* **2020**, *137*, 109584. [CrossRef] [PubMed]
35. Kochman, J.; Jakubczyk, K.; Antoniewicz, J.; Mruk, H.; Janda, K. Health Benefits and Chemical Composition of Matcha Green Tea: A Review. *Molecules* **2020**, *26*, 85. [CrossRef]
36. Lu, H.; Meng, X.; Li, C.; Sang, S.; Patten, C.; Sheng, S.; Hong, J.; Bai, N.; Winnik, B.; Ho, C.T.; et al. Glucuronides of tea catechins: Enzymology of biosynthesis and biological activities. *Drug Metab. Dispos.* **2003**, *31*, 452–461. [CrossRef] [PubMed]
37. Lambert, J.D.; Sang, S.; Yang, C.S. Biotransformation of green tea polyphenols and the biological activities of those metabolites. *Mol. Pharm.* **2007**, *4*, 819–825. [CrossRef] [PubMed]
38. Li, F.; Wang, Y.; Li, D.; Chen, Y.; Qiao, X.; Fardous, R.; Lewandowski, A.; Liu, J.; Chan, T.H.; Dou, Q.P. Perspectives on the recent developments with green tea polyphenols in drug discovery. *Expert Opin. Drug Discov.* **2018**, *13*, 643–660. [CrossRef] [PubMed]
39. Das, N.P. Studies on flavonoid metabolism. Absorption and metabolism of (+)-catechin in man. *Biochem. Pharmacol.* **1971**, *20*, 3435–3445. [CrossRef]
40. Lu, H.; Meng, X.; Yang, C.S. Enzymology of methylation of tea catechins and inhibition of catechol-O-methyltransferase by (-)-epigallocatechin gallate. *Drug Metab. Dispos.* **2003**, *31*, 572–579. [CrossRef]
41. Moridani, M.Y.; Scobie, H.; Salehi, P.; O'Brien, P.J. Catechin metabolism: Glutathione conjugate formation catalyzed by tyrosinase, peroxidase, and cytochrome p450. *Chem. Res. Toxicol.* **2001**, *14*, 841–848. [CrossRef]
42. Wein, S.; Beyer, B.; Gohlke, A.; Blank, R.; Metges, C.C.; Wolffram, S. Systemic Absorption of Catechins after Intraruminal or Intraduodenal Application of a Green Tea Extract in Cows. *PLoS ONE* **2016**, *11*, e0159428. [CrossRef]
43. Lee, M.J.; Wang, Z.Y.; Li, H.; Chen, L.; Sun, Y.; Gobbo, S.; Balentine, D.A.; Yang, C.S. Analysis of plasma and urinary tea polyphenols in human subjects. *Cancer Epidemiol. Biomark. Prev.* **1995**, *4*, 393–399.
44. Andreu-Fernández, V.; Almeida Toledano, L.; Pizarro, N.; Navarro-Tapia, E.; Gómez-Roig, M.D.; de la Torre, R.; García-Algar, Ó. Bioavailability of Epigallocatechin Gallate Administered with Different Nutritional Strategies in Healthy Volunteers. *Antioxidants* **2020**, *9*, 440. [CrossRef] [PubMed]
45. Jaffri, J.M.; Mohamed, S.; Rohimi, N.; Ahmad, I.N.; Noordin, M.M.; Manap, Y.A. Antihypertensive and cardiovascular effects of catechin-rich oil palm (*Elaeis guineensis*) leaf extract in nitric oxide-deficient rats. *J. Med. Food* **2011**, *14*, 775–783. [CrossRef] [PubMed]

46. Ruan, H.; Yang, Y.; Zhu, X.; Wang, X.; Chen, R. Neuroprotective effects of (+/−)-catechin against 1-methyl-4-phenyl-1,2,3,6-tetrahydropyridine (MPTP)-induced dopaminergic neurotoxicity in mice. *Neurosci. Lett.* **2009**, *450*, 152–157. [CrossRef]
47. Yang, H.Y.; Yang, S.C.; Chao, J.C.; Chen, J.R. Beneficial effects of catechin-rich green tea and inulin on the body composition of overweight adults. *Br. J. Nutr.* **2012**, *107*, 749–754. [CrossRef] [PubMed]
48. Kim, E.J.; Seo, J.B.; Yu, J.S.; Lee, S.; Lim, J.S.; Choi, J.U.; Lee, C.M.; Rashan, L.; Kim, K.H.; Cho, Y.C. Anti-Inflammatory Effects of a Polyphenol, Catechin-7,4′-O-Digallate, from Woodfordia uniflora by Regulating NF-κB Signaling Pathway in Mouse Macrophages. *Pharmaceutics* **2021**, *13*, 408. [CrossRef] [PubMed]
49. Nakano, E.; Kamei, D.; Murase, R.; Taki, I.; Karasawa, K.; Fukuhara, K.; Iwai, S. Anti-inflammatory effects of new catechin derivatives in a hapten-induced mouse contact dermatitis model. *Eur. J. Pharmacol.* **2019**, *845*, 40–47. [CrossRef]
50. Medina, I.; Gallardo, J.M.; Gonzalez, M.J.; Lois, S.; Hedges, N. Effect of molecular structure of phenolic families as hydroxycinnamic acids and catechins on their antioxidant effectiveness in minced fish muscle. *J. Agric. Food Chem.* **2007**, *55*, 3889–3895. [CrossRef]
51. Simos, Y.V.; Verginadis, I.I.; Toliopoulos, I.K.; Velalopoulou, A.P.; Karagounis, I.V.; Karkabounas, S.C.; Evangelou, A.M. Effects of catechin and epicatechin on superoxide dismutase and glutathione peroxidase activity, in vivo. *Redox Rep.* **2012**, *17*, 181–186. [CrossRef]
52. Xu, Y.Q.; Gao, Y.; Granato, D. Effects of epigallocatechin gallate, epigallocatechin and epicatechin gallate on the chemical and cell-based antioxidant activity, sensory properties, and cytotoxicity of a catechin-free model beverage. *Food Chem.* **2021**, *339*, 128060. [CrossRef] [PubMed]
53. Lu, N.; Chen, P.; Yang, Q.; Peng, Y.Y. Anti- and pro-oxidant effects of (+)-catechin on hemoglobin-induced protein oxidative damage. *Toxicol. In Vitro* **2011**, *25*, 833–838. [CrossRef] [PubMed]
54. Wang, Y.; Wang, B.; Du, F.; Su, X.; Sun, G.; Zhou, G.; Bian, X.; Liu, N. Epigallocatechin-3-Gallate Attenuates Oxidative Stress and Inflammation in Obstructive Nephropathy via NF-κB and Nrf2/HO-1 Signalling Pathway Regulation. *Basic Clin. Pharmacol. Toxicol.* **2015**, *117*, 164–172. [CrossRef] [PubMed]
55. Pusztai, R.; Béládi, I.; Bakai, M.; Mucsi, I.; Kukán, E. Study on the effect of flavonoids and related substances. I. The effect of quercetin on different viruses. *Acta Microbiol. Acad. Sci. Hung.* **1966**, *13*, 113–118.
56. Béládi, I.; Pusztai, R.; Mucsi, I.; Bakay, M.; Gábor, M. Activity of some flavonoids against viruses. *Ann. N. Y. Acad. Sci.* **1977**, *284*, 358–364. [CrossRef]
57. Kaul, T.N.; Middleton, E.; Ogra, P.L. Antiviral effect of flavonoids on human viruses. *J. Med. Virol.* **1985**, *15*, 71–79. [CrossRef]
58. Zhang, W.; Qiao, H.; Lv, Y.; Wang, J.; Chen, X.; Hou, Y.; Tan, R.; Li, E. Apigenin inhibits enterovirus-71 infection by disrupting viral RNA association with trans-acting factors. *PLoS ONE* **2014**, *9*, e110429. [CrossRef]
59. Nguyen, T.T.; Woo, H.J.; Kang, H.K.; Nguyen, V.D.; Kim, Y.M.; Kim, D.W.; Ahn, S.A.; Xia, Y.; Kim, D. Flavonoid-mediated inhibition of SARS coronavirus 3C-like protease expressed in *Pichia pastoris*. *Biotechnol. Lett.* **2012**, *34*, 831–838. [CrossRef]
60. Choi, H.J.; Kim, J.H.; Lee, C.H.; Ahn, Y.J.; Song, J.H.; Baek, S.H.; Kwon, D.H. Antiviral activity of quercetin 7-rhamnoside against porcine epidemic diarrhea virus. *Antivir. Res.* **2009**, *81*, 7–81. [CrossRef]
61. Roh, C. A facile inhibitor screening of SARS coronavirus N protein using nanoparticle-based RNA oligonucleotide. *Int. J. Nanomed.* **2012**, *7*, 2173–2179. [CrossRef]
62. Ghosh, R.; Chakraborty, A.; Biswas, A.; Chowdhuri, S. Evaluation of green tea polyphenols as novel corona virus (SARS CoV-2) main protease (Mpro) inhibitors—an in silico docking and molecular dynamics simulation study. *J. Biomol. Struct. Dyn.* **2021**, *39*, 4362–4374. [CrossRef]
63. Halder, P.; Pal, U.; Paladhi, P.; Dutta, S.; Paul, P.; Pal, S.; Das, D.; Ganguly, A.; IshitaDutta; SayarneelMandal; et al. Evaluation of potency of the selected bioactive molecules from Indian medicinal plants with MPro of SARS-CoV-2 through in silico analysis. *J. Ayurveda Integr. Med.* **2021**. [CrossRef]
64. Mahmud, S.; Biswas, S.; Paul, G.K.; Mita, M.A.; Promi, M.M.; Afrose, S.; Hasan, M.R.; Zaman, S.; Uddin, M.S.; Dhama, K.; et al. Plant-Based Phytochemical Screening by Targeting Main Protease of SARS-CoV-2 to Design Effective Potent Inhibitors. *Biology* **2021**, *10*, 589. [CrossRef]
65. Zhu, Y.; Xie, D.Y. Docking Characterization and in vitro Inhibitory Activity of Flavan-3-ols and Dimeric Proanthocyanidins against the Main Protease Activity of SARS-Cov-2. *Front. Plant Sci.* **2020**, *11*, 601316. [CrossRef]
66. Du, A.; Zheng, R.; Disoma, C.; Li, S.; Chen, Z.; Li, S.; Liu, P.; Zhou, Y.; Shen, Y.; Liu, S.; et al. Epigallocatechin-3-gallate, an active ingredient of Traditional Chinese Medicines, inhibits the 3CLpro activity of SARS-CoV-2. *Int. J. Biol. Macromol.* **2021**, *176*, 1–12. [CrossRef]
67. Mhatre, S.; Srivastava, T.; Naik, S.; Patravale, V. Antiviral activity of green tea and black tea polyphenols in prophylaxis and treatment of COVID-19: A review. *Phytomedicine* **2021**, *85*, 153286. [CrossRef]
68. Mhatre, S.; Naik, S.; Patravale, V. A molecular docking study of EGCG and theaflavin digallate with the druggable targets of SARS-CoV-2. *Comput. Biol. Med.* **2021**, *129*, 104137. [CrossRef]
69. Mishra, C.B.; Pandey, P.; Sharma, R.D.; Malik, M.Z.; Mongre, R.K.; Lynn, A.M.; Prasad, R.; Jeon, R.; Prakash, A. Identifying the natural polyphenol catechin as a multi-targeted agent against SARS-CoV-2 for the plausible therapy of COVID-19: An integrated computational approach. *Brief. Bioinform.* **2021**, *22*, 1346–1360. [CrossRef]
70. Sakshi, C.; Harikrishnan, A.; Jayaraman, S.; Choudhury, A.R.; Veena, V. Predictive medicinal metabolites from *Momordica dioica* against comorbidity related proteins of SARS-CoV-2 infections. *J. Biomol. Struct. Dyn.* **2021**, *11*, 1–14. [CrossRef]

71. Jena, A.B.; Kanungo, N.; Nayak, V.; Chainy, G.B.N.; Dandapat, J. Author Correction: Catechin and curcumin interact with S protein of SARS-CoV2 and ACE2 of human cell membrane: Insights from computational studies. *Sci. Rep.* **2021**, *11*, 8482. [CrossRef]
72. Maiti, S.; Banerjee, A. Epigallocatechin gallate and theaflavin gallate interaction in SARS-CoV-2 spike-protein central channel with reference to the hydroxychloroquine interaction: Bioinformatics and molecular docking study. *Drug Dev. Res.* **2021**, *82*, 86–96. [CrossRef]
73. Mhatre, S.; Gurav, N.; Shah, M.; Patravale, V. Entry-inhibitory role of catechins against SARS-CoV-2 and its UK variant. *Comput. Biol. Med.* **2021**, *135*, 104560. [CrossRef]
74. Ibrahim, I.M.; Abdelmalek, D.H.; Elfiky, A.A. GRP78: A cell's response to stress. *Life Sci.* **2019**, *226*, 156–163. [CrossRef]
75. Carlos, A.J.; Ha, D.P.; Yeh, D.W.; Van Krieken, R.; Tseng, C.C.; Zhang, P.; Gill, P.; Machida, K.; Lee, A.S. The chaperone GRP78 is a host auxiliary factor for SARS-CoV-2 and GRP78 depleting antibody blocks viral entry and infection. *J. Biol. Chem.* **2021**, *296*, 100759. [CrossRef]
76. Bhattacharjee, R.; Devi, A.; Mishra, S. Molecular docking and molecular dynamics studies reveal structural basis of inhibition and selectivity of inhibitors EGCG and OSU-03012 toward glucose regulated protein-78 (GRP78) overexpressed in glioblastoma. *J. Mol. Model.* **2015**, *21*, 272. [CrossRef]
77. Nishimura, H.; Okamoto, M.; Dapat, I.; Katumi, M.; Oshitani, H. Inactivation of SARS-CoV-2 by Catechins from Green Tea. *Jpn. J. Infect. Dis.* **2021**. [CrossRef]
78. Henss, L.; Auste, A.; Schürmann, C.; Schmidt, C.; von Rhein, C.; Mühlebach, M.D.; Schnierle, B.S. The green tea catechin epigallocatechin gallate inhibits SARS-CoV-2 infection. *J. Gen. Virol.* **2021**, *102*, 001574. [CrossRef]
79. Ohgitani, E.; Shin-Ya, M.; Ichitani, M.; Kobayashi, M.; Takihara, T.; Kawamoto, M.; Kinugasa, H.; Mazda, O. Significant Inactivation of SARS-CoV-2 In Vitro by a Green Tea Catechin, a Catechin-Derivative, and Black Tea Galloylated Theaflavins. *Molecules* **2021**, *26*, 3572. [CrossRef]
80. Hong, S.; Seo, S.H.; Woo, S.J.; Kwon, Y.; Song, M.; Ha, N.C. Epigallocatechin Gallate Inhibits the Uridylate-Specific Endoribonuclease Nsp15 and Efficiently Neutralizes the SARS-CoV-2 Strain. *J. Agric. Food Chem.* **2021**, *69*, 5948–5954. [CrossRef]
81. Zhao, M.; Yu, Y.; Sun, L.M.; Xing, J.Q.; Li, T.; Zhu, Y.; Wang, M.; Yu, Y.; Xue, W.; Xia, T.; et al. GCG inhibits SARS-CoV-2 replication by disrupting the liquid phase condensation of its nucleocapsid protein. *Nat. Commun.* **2021**, *12*, 2114. [CrossRef] [PubMed]
82. Yang, C.C.; Wu, C.J.; Chien, C.Y.; Chien, C.T. Green Tea Polyphenol Catechins Inhibit Coronavirus Replication and Potentiate the Adaptive Immunity and Autophagy-Dependent Protective Mechanism to Improve Acute Lung Injury in Mice. *Antioxidants* **2021**, *10*, 928. [CrossRef] [PubMed]
83. Chopra, K.; Singh, D.; Chander, V. Nephrotoxicity and its prevention by catechin in ferric nitrilotriacetate promoted oxidative stress in rats. *Hum. Exp. Toxicol.* **2004**, *23*, 137–143. [CrossRef] [PubMed]
84. Fatima, S.; Al-Mohaimeed, N.; Arjumand, S.; Banu, N.; Al-Jameil, N.; Al-Shaikh, Y. Effect of pre- and post-combined multidoses of epigallocatechin gallate and coenzyme Q10 on cisplatin-induced oxidative stress in rat kidney. *J. Biochem. Mol. Toxicol.* **2015**, *29*, 91–97. [CrossRef] [PubMed]
85. Funamoto, M.; Masumoto, H.; Takaori, K.; Taki, T.; Setozaki, S.; Yamazaki, K.; Minakata, K.; Ikeda, T.; Hyon, S.H.; Sakata, R. Green Tea Polyphenol Prevents Diabetic Rats from Acute Kidney Injury after Cardiopulmonary Bypass. *Ann. Thorac. Surg.* **2016**, *101*, 1507–1513. [CrossRef]
86. Hammad, F.T.; Lubbad, L. The effect of epigallocatechin-3-gallate on the renal dysfunction in the obstructed kidney in the rat. *Int. J. Physiol. Pathophysiol. Pharmacol.* **2017**, *9*, 119–126.
87. Malik, S.; Suchal, K.; Bhatia, J.; Gamad, N.; Dinda, A.K.; Gupta, Y.K.; Arya, D.S. Molecular mechanisms underlying attenuation of cisplatin-induced acute kidney injury by epicatechin gallate. *Lab. Investig.* **2016**, *96*, 853–861. [CrossRef] [PubMed]
88. Soussi, A.; Abdennabi, R.; Ghorbel, F.; Murat, J.C.; El Feki, A.F. Ameliorated Effects of (−)-Epigallocatechin Gallate against Toxicity Induced by Vanadium in the Kidneys of Wistar Rats. *Biol. Trace Elem. Res.* **2017**, *180*, 239–245. [CrossRef]
89. Zhou, P.; Yu, J.F.; Zhao, C.G.; Sui, F.X.; Teng, X.; Wu, Y.B. Therapeutic potential of EGCG on acute renal damage in a rat model of obstructive nephropathy. *Mol. Med. Rep.* **2013**, *7*, 1096–1102. [CrossRef]
90. Chander, V.; Singh, D.; Chopra, K. Catechin, a natural antioxidant protects against rhabdomyolysis-induced myoglobinuric acute renal failure. *Pharmacol. Res.* **2003**, *48*, 503–509. [CrossRef]
91. Sardana, A.; Kalra, S.; Khanna, D.; Balakumar, P. Nephroprotective effect of catechin on gentamicin-induced experimental nephrotoxicity. *Clin. Exp. Nephrol.* **2015**, *19*, 178–184. [CrossRef]
92. Pan, H.; Chen, J.; Shen, K.; Wang, X.; Wang, P.; Fu, G.; Meng, H.; Wang, Y.; Jin, B. Mitochondrial modulation by Epigallocatechin 3-Gallate ameliorates cisplatin induced renal injury through decreasing oxidative/nitrative stress, inflammation and NF-kB in mice. *PLoS ONE* **2015**, *10*, e0124775. [CrossRef]
93. Gao, Z.; Han, Y.; Hu, Y.; Wu, X.; Wang, Y.; Zhang, X.; Fu, J.; Zou, X.; Zhang, J.; Chen, X.; et al. Targeting HO-1 by Epigallocatechin-3-Gallate Reduces Contrast-Induced Renal Injury via Anti-Oxidative Stress and Anti-Inflammation Pathways. *PLoS ONE* **2016**, *11*, e0149032. [CrossRef] [PubMed]
94. Singh, D.; Chander, V.; Chopra, K. Protective effect of catechin on ischemia-reperfusion-induced renal injury in rats. *Pharmacol. Rep.* **2005**, *57*, 70–76.
95. Lv, J.; Feng, M.; Zhang, L.; Wan, X.; Zeng, Y.C.; Liang, P.F.; Xu, A.P. Protective effect of epigallocatechin gallate, a major constituent of green tea, against renal ischemia-reperfusion injury in rats. *Int. Urol. Nephrol.* **2015**, *47*, 1429–1435. [CrossRef] [PubMed]

96. Wang, Y.; Wang, B.; Du, F.; Su, X.; Sun, G.; Zhou, G.; Bian, X.; Liu, N. Epigallocatechin-3-gallate attenuates unilateral ureteral obstruction-induced renal interstitial fibrosis in mice. *J. Histochem. Cytochem.* **2015**, *63*, 270–279. [CrossRef] [PubMed]
97. Wang, Y.; Liu, N.; Bian, X.; Sun, G.; Du, F.; Wang, B.; Su, X.; Li, D. Epigallocatechin-3-gallate reduces tubular cell apoptosis in mice with ureteral obstruction. *J. Surg. Res.* **2015**, *197*, 145–154. [CrossRef] [PubMed]
98. Park, J.; Park, R.; Jang, M.; Park, Y.I. Therapeutic Potential of EGCG, a Green Tea Polyphenol, for Treatment of Coronavirus Diseases. *Life* **2021**, *11*, 197. [CrossRef]
99. Luo, D.; Xu, J.; Chen, X.; Zhu, X.; Liu, S.; Li, J.; Xu, X.; Ma, X.; Zhao, J.; Ji, X. Author Correction: (−)-Epigallocatechin-3-gallate (EGCG) attenuates salt-induced hypertension and renal injury in Dahl salt-sensitive rats. *Sci. Rep.* **2020**, *10586*. [CrossRef]
100. García-Ortuño, L.E.; Bobadilla, N.A. Integrative View of the Mechanisms that Induce Acute Kidney Injury and its Transition to Chronic Kidney Disease. *Rev. Investig. Clin.* **2018**, *70*, 261–268. [CrossRef]
101. Guzzi, F.; Cirillo, L.; Roperto, R.M.; Romagnani, P.; Lazzeri, E. Molecular Mechanisms of the Acute Kidney Injury to Chronic Kidney Disease Transition: An Updated View. *Int. J. Mol. Sci.* **2019**, *20*, 4941. [CrossRef] [PubMed]
102. Teo, S.H.; Endre, Z.H. Biomarkers in acute kidney injury (AKI). *Best. Pract. Res. Clin. Anaesthesiol.* **2017**, *31*, 331–344. [CrossRef]
103. Tsai, T.Y.; Chien, H.; Tsai, F.C.; Pan, H.C.; Yang, H.Y.; Lee, S.Y.; Hsu, H.H.; Fang, J.T.; Yang, C.W.; Chen, Y.C. Comparison of RIFLE, AKIN, and KDIGO classifications for assessing prognosis of patients on extracorporeal membrane oxygenation. *J. Formos. Med. Assoc.* **2017**, *116*, 844–851. [CrossRef]
104. Ülger, F.; Pehlivanlar Küçük, M.; Küçük, A.O.; İlkaya, N.K.; Murat, N.; Bilgiç, B.; Abanoz, H. Evaluation of acute kidney injury (AKI) with RIFLE, AKIN, CK, and KDIGO in critically ill trauma patients. *Eur. J. Trauma Emerg. Surg.* **2018**, *44*, 597–605. [CrossRef]
105. Huber, W.; Schneider, J.; Lahmer, T.; Küchle, C.; Jungwirth, B.; Schmid, R.M.; Schmid, S. Validation of RIFLE, AKIN, and a modified AKIN definition ("backward classification") of acute kidney injury in a general ICU: Analysis of a 1-year period. *Medicine* **2018**, *97*, e12465. [CrossRef]
106. Chen, Y. Current definition of acute kidney injury actually identifies a heterogenous group of patients with elevated serum creatinine and reduced urine output. *Crit. Care* **2020**, *24*, 247. [CrossRef]
107. Ghosn, M.; Attallah, N.; Badr, M.; Abdallah, K.; De Oliveira, B.; Nadeem, A.; Varghese, Y.; Munde, D.; Salam, S.; Abduljawad, B.; et al. Severe Acute Kidney Injury in Critically Ill Patients with COVID-19 Admitted to ICU: Incidence, Risk Factors, and Outcomes. *J. Clin. Med.* **2021**, *10*, 1217. [CrossRef]
108. Joseph, A.; Zafrani, L.; Mabrouki, A.; Azoulay, E.; Darmon, M. Acute kidney injury in patients with SARS-CoV-2 infection. *Ann. Intensive Care* **2020**, *10*, 117. [CrossRef]
109. Gabarre, P.; Dumas, G.; Dupont, T.; Darmon, M.; Azoulay, E.; Zafrani, L. Acute kidney injury in critically ill patients with COVID-19. *Intensive Care Med.* **2020**, *46*, 1339–1348. [CrossRef] [PubMed]
110. Jafari-Oori, M.; Fiorentino, M.; Castellano, G.; Ebadi, A.; Rahimi-Bashar, F.; Guest, P.C.; Vahedian-Azimi, A.; Sahebkar, A. Acute Kidney Injury and Covid-19: A Scoping Review and Meta-Analysis. *Adv. Exp. Med. Biol.* **2021**, *1321*, 309–324. [CrossRef] [PubMed]
111. Lowe, R.; Ferrari, M.; Nasim-Mohi, M.; Jackson, A.; Beecham, R.; Veighey, K.; Cusack, R.; Richardson, D.; Grocott, M.; Levett, D.; et al. Clinical characteristics and outcome of critically ill COVID-19 patients with acute kidney injury: A single centre cohort study. *BMC Nephrol.* **2021**, *22*, 92. [CrossRef] [PubMed]
112. Henry, B.M.; de Oliveira, M.H.S.; Cheruiyot, I.; Benoit, J.L.; Cooper, D.S.; Lippi, G.; Le Cras, T.D.; Benoit, S.W. Circulating level of Angiopoietin-2 is associated with acute kidney injury in coronavirus disease 2019 (COVID-19). *Angiogenesis* **2021**, *24*, 403–406. [CrossRef] [PubMed]
113. Hirsch, J.S.; Ng, J.H.; Ross, D.W.; Sharma, P.; Shah, H.H.; Barnett, R.L.; Hazzan, A.D.; Fishbane, S.; Jhaveri, K.D. Acute kidney injury in patients hospitalized with COVID-19. *Kidney Int.* **2020**, *98*, 209–218. [CrossRef] [PubMed]
114. Sharma, P.; Ng, J.H.; Bijol, V.; Jhaveri, K.D.; Wanchoo, R. Pathology of COVID-19-associated acute kidney injury. *Clin. Kidney J.* **2021**, *14*, i30–i39. [CrossRef]
115. Rivero, J.; Merino-López, M.; Olmedo, R.; Garrido-Roldan, R.; Moguel, B.; Rojas, G.; Chavez-Morales, A.; Alvarez-Maldonado, P.; Duarte-Molina, P.; Castaño-Guerra, R.; et al. Association between Postmortem Kidney Biopsy Findings and Acute Kidney Injury from Patients with SARS-CoV-2 (COVID-19). *Clin. J. Am. Soc. Nephrol.* **2021**, *16*, 685–693. [CrossRef] [PubMed]
116. Rossi, G.M.; Delsante, M.; Pilato, F.P.; Gnetti, L.; Gabrielli, L.; Rossini, G.; Re, M.C.; Cenacchi, G.; Affanni, P.; Colucci, M.E.; et al. Kidney Biopsy Findings in a Critically Ill COVID-19 Patient with Dialysis-Dependent Acute Kidney Injury: A Case against "SARS-CoV-2 Nephropathy". *Kidney Int. Rep.* **2020**, *5*, 1100–1105. [CrossRef]
117. Golmai, P.; Larsen, C.P.; DeVita, M.V.; Wahl, S.J.; Weins, A.; Rennke, H.G.; Bijol, V.; Rosenstock, J.L. Histopathologic and Ultrastructural Findings in Postmortem Kidney Biopsy Material in 12 Patients with AKI and COVID-19. *J. Am. Soc. Nephrol.* **2020**, *31*, 1944–1947. [CrossRef]
118. Kaya, İ. New Insight for Kidney Biopsy Series in Cases with COVID-19-related Acute Kidney Injury. *J. Coll. Phys. Surg. Pak.* **2021**, *30*, S48–S49. [CrossRef]
119. Santoriello, D.; Khairallah, P.; Bomback, A.S.; Xu, K.; Kudose, S.; Batal, I.; Barasch, J.; Radhakrishnan, J.; D'Agati, V.; Markowitz, G. Postmortem Kidney Pathology Findings in Patients with COVID-19. *J. Am. Soc. Nephrol.* **2020**, *31*, 2158–2167. [CrossRef]
120. Fujinaga, S.; Hara, T. Acute Kidney Injury Following Plastic Bronchitis Associated with Influenza B Virus in a Child with Nephrotic Syndrome. *Indian Pediatr.* **2015**, *52*, 523–525. [CrossRef]

121. Godinho, I.; Nogueira, E.; Jorge, S.; Alves, A.T.; Gomes da Costa, A. Membranoproliferative glomerulonephritis and interstitial nephritis in the setting of Epstein-Barr virus-related hemophagocytic syndrome. *Clin. Nephrol.* **2018**, *89*, 474–479. [CrossRef]
122. Nadasdy, T.; Hebert, L.A. Infection-related glomerulonephritis: Understanding mechanisms. *Semin. Nephrol.* **2011**, *31*, 369–375. [CrossRef] [PubMed]
123. Gupta, A.; Quigg, R.J. Glomerular Diseases Associated with Hepatitis B and C. *Adv. Chronic Kidney Dis.* **2015**, *22*, 343–351. [CrossRef] [PubMed]
124. Prasad, N.; Patel, M.R. Infection-Induced Kidney Diseases. *Front. Med.* **2018**, *5*, 327. [CrossRef] [PubMed]
125. Siew, E.D.; Parr, S.K.; Abdel-Kader, K.; Perkins, A.M.; Greevy, R.A., Jr.; Vincz, A.J.; Denton, J.; Wilson, O.D.; Hung, A.M.; Ikizler, T.A.; et al. Renin-angiotensin aldosterone inhibitor use at hospital discharge among patients with moderate to severe acute kidney injury and its association with recurrent acute kidney injury and mortality. *Kidney Int.* **2021**, *99*, 1202–1212. [CrossRef] [PubMed]
126. Soleimani, M. Acute Kidney Injury in SARS-CoV-2 Infection: Direct Effect of Virus on Kidney Proximal Tubule Cells. *Int. J. Mol. Sci.* **2020**, *21*, 3275. [CrossRef]
127. Ou, X.; Mao, S.; Jiang, Y.; Zhang, S.; Ke, C.; Ma, G.; Cheng, A.; Wang, M.; Zhu, D.; Chen, S.; et al. Viral-host interaction in kidney reveals strategies to escape host immunity and persistently shed virus to the urine. *Oncotarget* **2017**, *8*, 7336–7349. [CrossRef]
128. Agarwal, A.; Dong, Z.; Harris, R.; Murray, P.; Parikh, S.M.; Rosner, M.H.; Kellum, J.A.; Ronco, C. Cellular and Molecular Mechanisms of AKI. *J. Am. Soc. Nephrol.* **2016**, *27*, 1288–1299. [CrossRef]
129. Kawakita, C.; Kinomura, M.; Otaka, N.; Kitagawa, M.; Sugiyama, H.; Kusano, N.; Mizuno, M.; Wada, J. HIV-associated Immune Complex Kidney Disease with C3-dominant Deposition Induced by HIV Infection after Treatment of IgA Nephropathy. *Intern. Med.* **2019**, *58*, 3001–3007. [CrossRef]
130. Michal, J.L.; Rab, S.; Patel, M.; Kyle, A.W.; Miller, L.S.; Easley, K.A.; Kalapila, A.G. Incidence of Acute Kidney Injury in Patients Coinfected with HIV and C Virus Receiving Tenofovir Disoproxil Fumarate and Ledipasvir/Sofosbuvir in a Real-World, Urban, Ryan White Clinic. *AIDS Res. Hum. Retrovir.* **2018**, *34*, 690–698. [CrossRef]
131. Simões, E.; Silva, A.C.; Flynn, J.T. The renin-angiotensin-aldosterone system in 2011: Role in hypertension and chronic kidney disease. *Pediatr. Nephrol.* **2012**, *27*, 1835–1845. [CrossRef]
132. Gunalan, S.; Somarathinam, K.; Bhattacharya, J.; Srinivasan, S.; Jaimohan, S.M.; Manoharan, R.; Ramachandran, S.; Kanagaraj, S.; Kothandan, G. Understanding the dual mechanism of bioactive peptides targeting the enzymes involved in Renin Angiotensin System (RAS): An *in-silico* approach. *J. Biomol. Struct. Dyn.* **2020**, *38*, 5044–5061. [CrossRef]
133. Pinheiro, S.V.; Simões, E.; Silva, A.C. Angiotensin converting enzyme 2, Angiotensin-(1-7), and receptor MAS axis in the kidney. *Int. J. Hypertens.* **2012**, *2012*, 414128. [CrossRef]
134. Santos, R.A.S.; Sampaio, W.O.; Alzamora, A.C.; Motta-Santos, D.; Alenina, N.; Bader, M.; Campagnole-Santos, M.J. The ACE2/Angiotensin-(1-7)/MAS Axis of the Renin-Angiotensin System: Focus on Angiotensin-(1-7). *Physiol. Rev.* **2018**, *98*, 505–553. [CrossRef] [PubMed]
135. Michel, M.C.; Brunner, H.R.; Foster, C.; Huo, Y. Angiotensin II type 1 receptor antagonists in animal models of vascular, cardiac, metabolic and renal disease. *Pharmacol. Ther.* **2016**, *164*, 1–81. [CrossRef]
136. Remuzzi, A.; Sangalli, F.; Macconi, D.; Tomasoni, S.; Cattaneo, I.; Rizzo, P.; Bonandrini, B.; Bresciani, E.; Longaretti, L.; Gagliardini, E.; et al. Regression of Renal Disease by Angiotensin II Antagonism Is Caused by Regeneration of Kidney Vasculature. *J. Am. Soc. Nephrol.* **2016**, *27*, 699–705. [CrossRef] [PubMed]
137. Simões, E.; Silva, A.C.; Lanza, K.; Palmeira, V.A.; Costa, L.B.; Flynn, J.T. 2020 update on the renin-angiotensin-aldosterone system in pediatric kidney disease and its interactions with coronavirus. *Pediatr. Nephrol.* **2021**, *36*, 1407–1426. [CrossRef] [PubMed]
138. Simões, E.; Silva, A.C.; Teixeira, M.M. ACE inhibition, ACE2 and angiotensin-(1-7) axis in kidney and cardiac inflammation and fibrosis. *Pharmacol. Res.* **2016**, *107*, 154–162. [CrossRef]
139. Ingraham, N.E.; Barakat, A.G.; Reilkoff, R.; Bezdicek, T.; Schacker, T.; Chipman, J.G.; Tignanelli, C.J.; Puskarich, M.A. Understanding the renin-angiotensin-aldosterone-SARS-CoV axis: A comprehensive review. *Eur. Respir. J.* **2020**, *56*, 2000912. [CrossRef]
140. Chen, Y.; Yang, D.; Shen, Z.; Wang, X.; Liu, X.; Xiong, M.; Chen, H.; Huang, K. Kidney injury molecule-1 is a potential receptor for SARS-CoV-2. *J. Mol. Cell Biol.* **2021**, *13*, 185–196. [CrossRef]
141. Slagelse, C.; Gammelager, H.; Iversen, L.H.; Liu, K.D.; Sørensen, H.T.; Christiansen, C.F. Renin-angiotensin system blockers and 1-year mortality in patients with post-operative acute kidney injury. *Acta Anaesthesiol. Scand.* **2020**, *64*, 1262–1269. [CrossRef]
142. South, A.M.; Brady, T.M.; Flynn, J.T. ACE2 (Angiotensin-Converting Enzyme 2), COVID-19, and ACE Inhibitor and Ang II (Angiotensin II) Receptor Blocker Use during the Pandemic: The Pediatric Perspective. *Hypertension* **2020**, *76*, 16–22. [CrossRef] [PubMed]
143. Wysocki, J.; Lores, E.; Ye, M.; Soler, M.J.; Batlle, D. Kidney and Lung ACE2 Expression after an ACE Inhibitor or an Ang II Receptor Blocker: Implications for COVID-19. *J. Am. Soc. Nephrol.* **2020**, *31*, 1941–1943. [CrossRef] [PubMed]
144. Salter, T.; Burton, H.; Douthwaite, S.; Newsholme, W.; Horsfield, C.; Hilton, R. Immune Complex Mediated Glomerulonephritis with Acute Thrombotic Microangiopathy following Newly Detected Hepatitis B Virus Infection in a Kidney Transplant Recipient. *Case Rep. Transplant.* **2016**, *2016*, 3152495. [CrossRef] [PubMed]

molecules MDPI

Communication

Inactivation Activities of Ozonated Water, Slightly Acidic Electrolyzed Water and Ethanol against SARS-CoV-2

Yohei Takeda [1,2], Dulamjav Jamsransuren [2], Yoshimasa Makita [3], Akihiro Kaneko [4], Sachiko Matsuda [2], Haruko Ogawa [2] and Hourei Oh [5,*]

1 Research Center for Global Agromedicine, Obihiro University of Agriculture and Veterinary Medicine, 2-11 Inada, Obihiro 080-8555, Japan; ytakeda@obihiro.ac.jp
2 Department of Veterinary Medicine, Obihiro University of Agriculture and Veterinary Medicine, 2-11 Inada, Obihiro 080-8555, Japan; duuya.dj@gmail.com (D.J.); chaka@obihiro.ac.jp (S.M.); hogawa@obihiro.ac.jp (H.O.)
3 Department of Chemistry, Osaka Dental University, 8-1 Kuzuha Hanazono Hirakata, Osaka 573-1121, Japan; makita@cc.osaka-dent.ac.jp
4 Department of Oral Surgery, Ikegami General Hospital, 6-1-19 Ikegami Ootaku, Tokyo 146-8531, Japan; kaneko@ikegamihosp.jp
5 Center of Innovation in Dental Education, Osaka Dental University, 8-1 Kuzuha Hanazono Hirakata, Osaka 573-1121, Japan
* Correspondence: ohoh@cc.osaka-dent.ac.jp

Abstract: This study aimed to compare the SARS-CoV-2-inactivation activity and virucidal mechanisms of ozonated water (OW) with those of slightly acidic electrolyzed water (SAEW) and 70% ethanol (EtOH). SARS-CoV-2-inactivation activity was evaluated in a virus solution containing 1%, 20% or 40% fetal bovine serum (FBS) with OW, SAEW or EtOH at a virus-to-test solution ratio of 1:9, 1:19 or 1:99 for a reaction time of 20 s. EtOH showed the strongest virucidal activity, followed by SAEW and OW. Even though EtOH potently inactivated the virus despite the 40% FBS concentration, virus inactivation by OW and SAEW decreased in proportion to the increase in FBS concentration. Nevertheless, OW and SAEW showed potent virucidal activity with 40% FBS at a virus-to-test solution ratio of 1:99. Real-time PCR targeting the viral genome revealed that cycle threshold values in the OW and SAEW groups were significantly higher than those in the control group, suggesting that OW and SAEW disrupted the viral genome. Western blotting analysis targeting the recombinant viral spike protein S1 subunit showed a change in the specific band into a ladder upon treatment with OW and SAEW. OW and SAEW may cause conformational changes in the S1 subunit of the SARS-CoV-2 spike protein.

Keywords: disinfectants; ethanol; hand hygiene; ozonated water; SARS-CoV-2; slightly acidic electrolyzed water; virucidal activity; virucidal mechanism

Citation: Takeda, Y.; Jamsransuren, D.; Makita, Y.; Kaneko, A.; Matsuda, S.; Ogawa, H.; Oh, H. Inactivation Activities of Ozonated Water, Slightly Acidic Electrolyzed Water and Ethanol against SARS-CoV-2. *Molecules* **2021**, *26*, 5465. https://doi.org/10.3390/molecules26185465

Academic Editors: Giovanni N. Roviello and Caterina Vicidomini

Received: 30 July 2021
Accepted: 7 September 2021
Published: 8 September 2021

1. Introduction

The global prevalence of severe acute respiratory syndrome coronavirus 2 (SARS-CoV-2) has not yet been fully controlled, despite the remarkable speed of development of therapeutic drugs and vaccines. To prevent further spread of infection, it is necessary to install individual infection prevention measures in daily life that can be implemented for society as a whole. SARS-CoV-2 may be transmitted through the mucous membranes of the mouth, nose and eyes through transfer onto hands, as well as via respiratory droplets. Therefore, in addition to countermeasures against the prevention of respiratory transmission, hand hygiene is also important for stricter infection control [1].

Hand hygiene involves hand washing with soap and running water. In addition, disinfection of viruses with rubbing alcohol has become mainstream worldwide [2,3]. Rubbing alcohol containing 70–80% ethanol (EtOH) can sufficiently inactivate viruses and ≥30% EtOH potently inactivated SARS-CoV-2 for 30 s when mixed with a virus solution at a virus-to-EtOH ratio of 2:8 [4]. Furthermore, a recent report showed that acidic electrolyzed water (AEW), a type of hypochlorous acid water, has an inactivating effect on SARS-CoV-2 [5,6]. Previous reports have detailed the SARS-CoV-2-inactivation activities of AEWs with different pH, namely, highly AEW (pH ~2.5) [5] and slightly AEW (SAEW; pH 5.0–5.6) [6]. SAEW with residual chlorine concentration (RCC) of ~30–50 ppm showed virucidal activity for 20 s when mixed with a virus solution at virus-to-SAEW ratios of 1:9–19 [6]. Furthermore, ozonated water (OW) with low concentrations (0.6 ppm) of ozone has been shown to reduce virus infectivity upon exposure to SARS-CoV-2 for 1 min when mixed with a virus solution at virus-to-OW ratio of 1:100 [7]. In addition, OW with a high ozone concentration (10 ppm) has been reported to inactivate SARS-CoV-2 within 5–20 s when mixed with a virus solution at virus-to-OW ratio of 1:99 [8]. Nevertheless, the SARS-CoV-2-inactivation activities of these different disinfectants have not been compared under the same experimental conditions. In this study, we evaluated the virucidal activity of OW with various ozone concentrations mixed with a SARS-CoV-2 solution at various liquid volume ratios, and this activity was compared to that of SAEW (pH 5.8) and EtOH. In addition, the impact of the organic substance content in the virus solution on these activities was evaluated. Moreover, the effects of these agents on the SARS-CoV-2 genome and S protein were analyzed to elucidate their mechanisms of action.

2. Materials and Methods

2.1. Test Solutions

OW (ozone concentration: 1–10 ppm) was generated from ultra-pure water (UPW) using a Handlex instrument (Nikka Micron Co., Ltd., Saitama, Japan). These ozone concentrations were determined based on the standard ozone concentrations of OW provided by an OW generator. The range of tested concentration of ozone (1–10 ppm) included low (1 ppm) and high (10 ppm) concentrations that are close to or consistent with previously reported virucidal concentrations [7,8]. The ozone concentration was measured using an OZ-20 ozone meter (DDK-TOA Co., Tokyo, Japan). SAEW (pH ~5.8, RCC ~34 ppm, oxidation-reduction potential ~950 mV) was generated by the electrolysis of hydrochloric acid using a PURESTER device (Morinaga Milk Industry Co., Ltd., Tokyo, Japan). The tested RCC was determined based on the recommended concentration for SARS-CoV-2 inactivation in Japan [9], which is empirically considered to have a low risk of skin irritation. The pH, RCC and oxidation-reduction potential were measured using a compact pH meter (Horiba Co., Ltd., Kyoto, Japan), AQUAB AQ-202 (Sibata Scientific Technology Ltd., Tokyo, Japan) and a waterproof oxidation-reduction potential meter (Custom Co., Tokyo, Japan), respectively. EtOH (70%) was prepared by diluting 100% EtOH (Fujifilm Wako Pure Chemical Co., Ltd., Osaka, Japan) with UPW.

2.2. Virus and Cells

SARS-CoV-2 (JPN/TY/WK-521 strain) was obtained from the National Institute of Infectious Diseases (Tokyo, Japan). VeroE6/TMPRSS2 cells [10] were obtained from the Japanese Collection of Research Bioresources (Osaka, Japan, Cell No. JCRB1819). SARS-CoV-2-infected VeroE6/TMPRSS2 cells were cultured in virus growth medium (VGM), the composition of which has been previously described [11].

2.3. Evaluation of the Virucidal Activity of Test Solutions

The VGM containing SARS-CoV-2 with 1–40% (v/v) fetal bovine serum (FBS), in which the viral titer was ~7.0 $\log_{10}$ 50% tissue culture infective dose ($TCID_{50}$)/mL, was mixed with the test solutions at virus-to-test solution ratios of 1:9, 1:19 and 1:99. As a control, the virus solution was mixed with the UPW. The mixtures were incubated for 20 s at 22 °C and then inoculated into the cells cultured in VGM containing 10 mM $Na_2S_2O_3$, which is an ozone and chlorine neutralizer. Ten-fold serial dilutions of the cell culture medium were performed. After incubation for 3 days, the viral titers ($TCID_{50}$/mL) were evaluated as previously described [11]. EtOH carried into the cell culture medium did not affect virus proliferation in infected cells and did not inactivate the viruses. The virus-to-test solution ratios and reaction times were set to broadly cover the experimental conditions used in related previous studies [4,6–8]. The tested FBS concentrations were determined based on our previous study, in which the SARS-CoV-2-inactivation activity of AEW was evaluated [5]. The detection limit of the viral titer in each test solution group was determined based on the cytotoxicity of each test solution. UPW, SAEW and OW did not show any cytotoxicity, and the detection limit of the viral titer in the groups treated with these solutions was set to 1.25 log_{10} $TCID_{50}$/mL according to our viral titer calculation. In contrast, 70% EtOH exhibited a moderate cytotoxicity. The detection limit in the 70% EtOH group was set to 2.25 log_{10} $TCID_{50}$/mL.

2.4. Real-Time Reverse Transcription-Polymerase Chain Reaction (RT-PCR) Analysis

The following two treatments were performed: In the first treatment (pre-addition of $Na_2S_2O_3$), $Na_2S_2O_3$ was added to the test solutions (34 ppm SAEW, 10 ppm OW and 70% EtOH), with a final $Na_2S_2O_3$ concentration of 10 mM. According to this treatment, the chlorine in the SAEW and ozone in the OW were neutralized. The VGM containing SARS-CoV-2 with 1% FBS was then mixed with the test solutions at a virus-to-test solution ratio of 1:29. The mixture was incubated for 20 s at 22 °C. In the second treatment (post-addition of $Na_2S_2O_3$), the VGM containing SARS-CoV-2 with 1% FBS was mixed with the test solutions at a virus-to-test solution ratio of 1:29. The mixture was incubated for 20 s at 22 °C. $Na_2S_2O_3$ was then added to the mixture.

RNA was extracted from these mixtures obtained by pre- and post-treatment with $Na_2S_2O_3$, and real-time RT-PCR targeting the SARS-CoV-2 nucleocapsid (*N*) gene was performed as previously described [11]. The viral titers of these mixtures were also evaluated.

2.5. Western Blotting (WB) Analysis

The following two treatments were performed using recombinant SARS-CoV-2 S protein S1 subunit (Catalog No. 40591-V08H; Sino Biological Inc., Beijing, China) and test solutions.

In the first treatment (pre-addition of $Na_2S_2O_3$), $Na_2S_2O_3$ was added to each test solution. The recombinant protein solution (250 μg/mL) was mixed with the test solutions at a protein-to-test solution ratio of 1:29. The mixture was incubated for 20 s at 22 °C. In the second treatment (post-addition of $Na_2S_2O_3$), the recombinant protein solution was mixed with the test solutions at a protein-to-test solution ratio of 1:29. The mixture was incubated for 20 s at 22 °C, after which $Na_2S_2O_3$ was added to the mixture.

The mixtures obtained by the two treatments were combined with sodium dodecyl sulfate (SDS) buffer containing 2-mercaptoethanol (FUJIFILM Wako Pure Chemical Co., Ltd.). These SDS samples were subjected to WB analysis targeting the S protein S1 subunit, as previously described [11].

2.6. Statistical Analysis

Student's *t*-test was performed to analyze the statistical significance of the differences in the viral titer between the UPW (control) group and each test solution group. Statistical significance was set at $p < 0.05$.

3. Results

3.1. Comparison of the Virucidal Activities of SAEW, OW and 70% EtOH

The SARS-CoV-2-inactivation activities of SAEW, OW and 70% EtOH were compared. When the virus solution (1% FBS concentration) and test solutions were mixed at a virus-to-test solution ratio of 1:9, the viral titer of the 70% EtOH group was below the detection limit ($\geq$3.53 log_{10} $TCID_{50}$/mL reduction of viral titer compared to that in the UPW group) after 20 s. In contrast, 34 ppm SAEW and 2–10 ppm OW showed weak and limited virucidal activity (Figure 1, left). At a ratio of 1:19, SAEW and 2–10 ppm OW showed more potent virucidal activities (SAEW: $\geq$3.73 log_{10} $TCID_{50}$/mL reduction; 2 ppm OW: 0.79 log_{10} $TCID_{50}$/mL reduction; 10 ppm OW: $\geq$2.85 log_{10} $TCID_{50}$/mL reduction) than those at a ratio of 1:9 after 20 s (Figure 1, middle). At a ratio of 1:99, the viral titers in the SAEW and 2–10 ppm OW groups were almost below the detection limit after 20 s ($\geq$3.25 $\log_{10}$ $TCID_{50}$/mL reduction) (Figure 1, right).

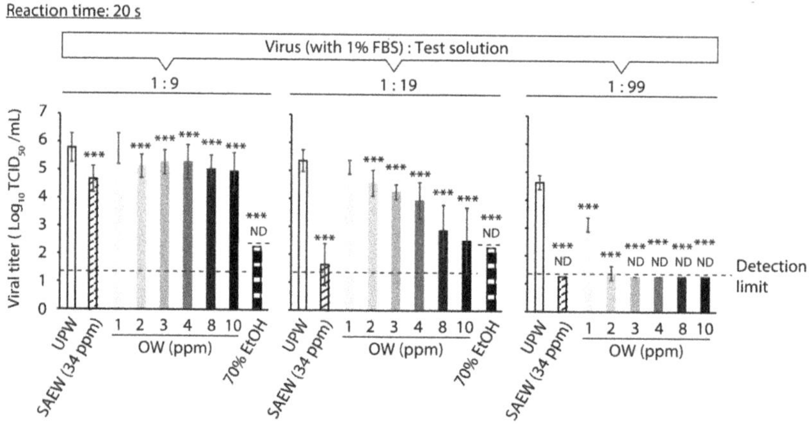

Figure 1. Comparison of virucidal activities of slightly acidic electrolyzed water (SAEW), ozonated water (OW) and 70% ethanol (EtOH). The viral titer of each reaction mixture was measured. Results are indicated as mean ± standard deviation (n = 3–30 per group). The detection limits of the viral titer are 1.25 $\log_{10}$ 50% tissue culture infective dose ($TCID_{50}$)/mL in the ultra-pure water (UPW), SAEW and OW groups, and 2.25 $\log_{10}$ $TCID_{50}$/mL in the 70% EtOH group. ND is the abbreviation for not detected, which means that the viral titer was below the detection limit. Student's t-test was performed to analyze the statistical significance between the UPW (control) group and each test solution group; *** $p < 0.001$. FBS, fetal bovine serum.

3.2. Evaluation of the Impact of FBS in Virus Solution on the Virucidal Activity of Test Solutions

The virucidal activities of the test solutions were evaluated against virus solutions with different FBS concentrations (1%, 20% and 40%). At a virus-to-test solution ratio of 1:19, the viral titer in the 70% EtOH group was below the detection limit after 20 s even though the FBS concentration in the virus solution was 40% ($\geq$3.13 $\log_{10}$ $TCID_{50}$/mL reduction compared to that in the UPW group). In contrast, the reduction in viral titer by 34 ppm SAEW and 10 ppm OW decreased in proportion to the increase in FBS concentration of the virus solution (Figure 2, left). At a ratio of 1:99, the viral titer in the SAEW group was below the detection limit after 20 s despite the 40% FBS setting ($\geq$3.63 $\log_{10}$ $TCID_{50}$/mL reduction). In contrast, the viral titer in the 10 ppm OW group was not below the detection limit in the 40% FBS setting. Nevertheless, the reduction in viral titer with OW treatment was $\geq$2.88 $\log_{10}$ $TCID_{50}$/mL (Figure 2, right).

3.3. Evaluation of the Impact of Test Solutions on the SARS-CoV-2 Genome

Real-time RT-PCR targeting the viral genome showed that the cycle threshold (Ct) values in all tested groups were comparable in the case of $Na_2S_2O_3$ pre-addition, whereas these values in the SAEW and OW groups were significantly higher than those in the UPW group in the case of $Na_2S_2O_3$ post-addition. However, the Ct value in the 70% EtOH group was comparable to that in the UPW group (Figure 3A). In this setting, the virucidal activities of SAEW and OW disappeared with pre-addition of $Na_2S_2O_3$, indicating that the chlorine in SAEW and the ozone in OW were neutralized by the addition of $Na_2S_2O_3$ (Figure 3B).

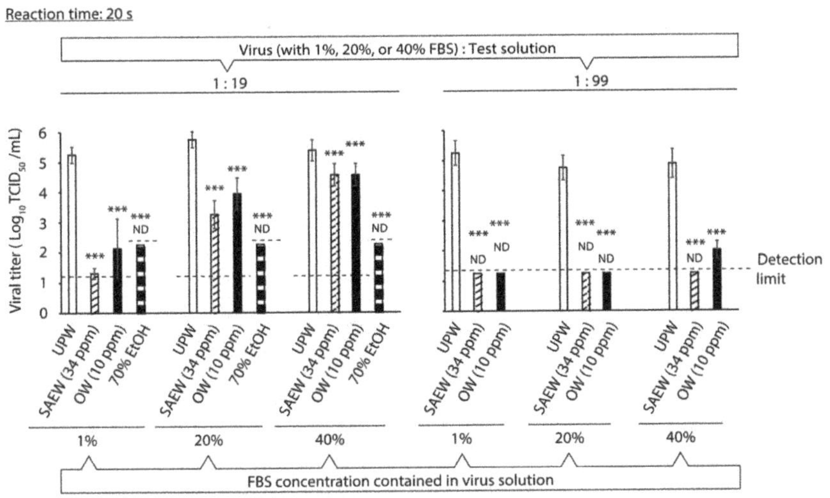

Figure 2. Evaluation of the impact of fetal bovine serum (FBS) concentration in the virus solution on the virucidal activities of test solutions. The viral titer of each reaction mixture was measured. Results are indicated as mean ± standard deviation (n = 4–8 per group). The detection limits of the viral titer are 1.25 $\log_{10}$ 50% tissue culture infective dose ($TCID_{50}$)/mL in the ultra-pure water (UPW), slightly acidic electrolyzed water (SAEW) and ozonated water (OW) groups, and 2.25 $\log_{10}$ $TCID_{50}$/mL in the 70% ethanol (EtOH) group. ND is the abbreviation for not detected, which means that the viral titer was below the detection limit. Student's t-test was performed to analyze the statistical significance between the UPW (control) group and each test solution group; *** $p < 0.001$.

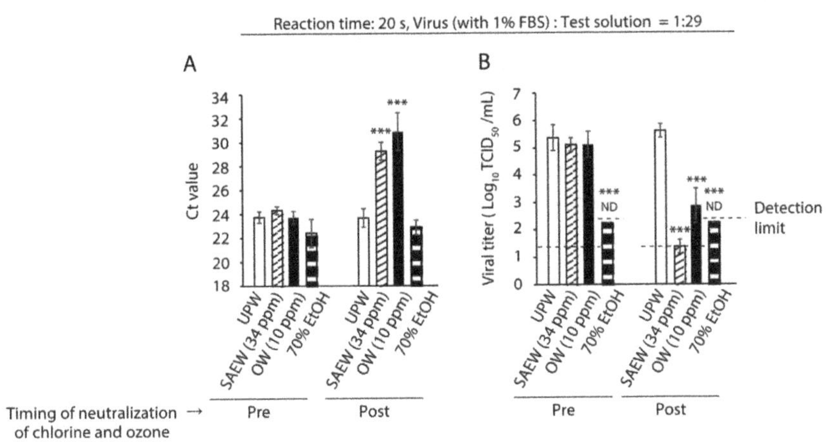

Figure 3. Evaluation of the impact of test solutions on the SARS-CoV-2 genome. (**A**) The cycle threshold (Ct) value of each reaction mixture was measured. (**B**) The viral titer of each reaction mixture was measured. The detection limits of the viral titer are 1.25 $\log_{10}$ 50% tissue culture infective dose ($TCID_{50}$)/mL in the ultra-pure water (UPW), slightly acidic electrolyzed water (SAEW) and ozonated water (OW) groups, and 2.25 $\log_{10}$ $TCID_{50}$/mL in the 70% ethanol (EtOH) group. ND is the abbreviation for not detected, which means that the viral titer was below the detection limit. (A, B) Results are indicated as mean ± standard deviation (n = 4 per group). Student's t-test was performed to analyze the statistical significance between the UPW (control) group and each test solution group; *** $p < 0.001$. Pre indicates that $Na_2S_2O_3$ was added to each test solution before mixing with the virus solution; Post indicates that $Na_2S_2O_3$ was added 20 s after mixing with the virus solution. FBS, fetal bovine serum.

3.4. *Impact of Test Solutions on the SARS-CoV-2 S Protein*

WB analysis showed no differences in the band patterns of the recombinant S protein S1 subunit among the proteins after treatment with all test solutions in the case of $Na_2S_2O_3$ pre-addition. The band pattern of the protein treated with 70% EtOH was comparable to that of the protein treated with UPW in the case of $Na_2S_2O_3$ post-addition. However, the band changed into a ladder for proteins following treatment with SAEW and OW in the case of $Na_2S_2O_3$ post-addition (Figure 4).

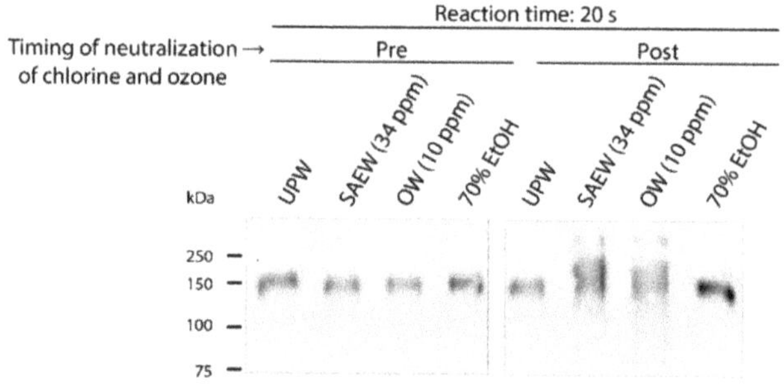

Figure 4. Impact of test solutions on the SARS-CoV-2 S protein. Images of the results of Western blotting analysis targeting the SARS-CoV-2 S protein S1 subunit are shown. Pre indicates that $Na_2S_2O_3$ was added to each test solution before mixing with the recombinant S protein S1 subunit; Post indicates that $Na_2S_2O_3$ was added 20 s after mixing with the S1 subunit. UPW, ultra-pure water; SAEW, slightly acidic electrolyzed water; OW, ozonated water; EtOH, ethanol.

4. Discussion

OW with a high concentration (10 ppm) of ozone has been reported to be effective in rapidly inactivating SARS-CoV-2 at a virus-to-test solution ratio of 1:99 [8]. However, the inactivation activity of OW against SARS-CoV-2 has not been fully evaluated. For example, the activity of disinfectants is greatly reduced by the influence of organic substances, and the effects of disinfectants largely differ depending on the test conditions [12]. Therefore, in this study, OW with various ozone concentrations and liquid volume ratios was mixed with virus solutions containing various concentrations of FBS, and its virucidal activity was compared with that of SAEW and 70% EtOH. The results showed that the virus inactivation activity of 70% EtOH was the strongest in comparison with that of 10 ppm OW and SAEW (RCC: ~34 ppm). In addition, the inactivation activity of SAEW (~34 ppm) was slightly stronger than that of OW (10 ppm). When 10 μL of the VGM containing ~5.0 $\log_{10}$ $TCID_{50}$ SARS-CoV-2 and 40% FBS was reacted with 990 μL of 10 ppm OW (virus-to-test solution ratio of 1:99), SARS-CoV-2 could be sufficiently inactivated (specifically, $\geq$2.88 $\log_{10}$ $TCID_{50}$/mL reduction of the viral titer; $\geq$99.87% inactivation of SARS-CoV-2) after 20 s, even in the presence of a high concentration of organic substances (Figure 2, right). The viral titer (~5.0 $\log_{10}$ $TCID_{50}$) of the tested virus solution was presumed to be higher than that of the virus on a COVID-19 patient's hand in the real world; Lin et al. [13] reported that the viral titer of SARS-CoV-2 detected on a COVID-19 patient's hand was up to ~3×10^3 plaque-forming unit/hand. Nevertheless, the current study has some limitations. First, our results cannot be fully translated to the efficacy of OW and SAEW in the real world because the compositions of substances contained in virus solutions are different. In addition, the influence of the substances contained in VGM on the virucidal activity of each test solution was not clarified in the current study. To assess the direct virucidal activity of each test solution, an additional analysis should be performed targeting purified SARS-CoV-2. Nevertheless, this in vitro study suggests that OW and SAEW may be useful SARS-CoV-2 disinfectants in situations such as hand disinfection under running water.

One of the disadvantages of OW is the short half-life of ozone after OW generation. The half-life of ozone in distilled water is reported to be 20 to 30 min at 20 °C; it is also reported to be 2–4 min in a pH 7.0 aqueous solution at 25 °C [14]. Owing to the short half-life of ozone, OW should be used for hand or object washing with running water. In contrast, our previous study showed that the RCC of SAEW does not dramatically decrease for 21 days in an enclosed container at 22 °C in the dark [6]. Nevertheless, AEW users need to consider that the available chlorine in AEW is decomposed by ultraviolet light, and this reduction speed is accelerated at high temperatures. Conversely, although EtOH is also lost by evaporation, it can be maintained for a long time under enclosed conditions. Even though EtOH is more stable than OW and SAEW, its skin irritation and the existence of alcohol-intolerant people are disadvantages. Previous reports suggest that the skin irritancy of OW is lower than that of EtOH [15,16]. In contrast, since SAEW is also considered to be safe and has low irritation to skin and mucous membranes, it has been proposed for use as a disinfectant for normal and wounded skin [17]. In addition to the low irritancy of OW and SAEW, they are less of a pollution risk for the environment, making them advantageous as disinfectants. Hence, OW and SAEW can be used as surrogate disinfectants of EtOH, especially for people who show adverse reactions to alcohol.

Previous studies have shown that ozone [18] and hypochlorous acid [19] damage viral proteins, genomes and lipid layers of biological membranes, such as the viral envelope. Alcohols show more substantial virucidal activity against enveloped viruses than non-enveloped viruses, suggesting that the viral lipid envelope is a potential target of EtOH [20]. However, there are no reports of comparative studies of these disinfectants for the inactivation of SARS-CoV-2 at the molecular level. In this study, real-time RT-PCR targeting the viral genome suggested that OW and SAEW may have destroyed the viral genome in a short time. In contrast, 70% EtOH had no effect on the viral genome at a reaction time of 20 s. This finding for EtOH is consistent with a report by Pfaender et al. [21], which showed that EtOH does not affect the integrity of the viral RNA of the hepatitis C virus. In general, when performing PCR screening of samples from patients suspected to be infected with SARS-CoV-2, it is important to prevent contamination of the viral genome between samples. Our real-time RT-PCR results suggest that OW and SAEW, which can rapidly inactivate SARS-CoV-2 with destruction of the viral genome, may be useful virucidal disinfectants in such situations.

The results of the WB analysis further demonstrated that OW and SAEW may have caused some conformational changes in the S1 subunit of the S protein within 20 s. Rowen et al. [22] stated that ozone can directly inactivate many viruses. They mentioned the possibility that ozone can oxidize the glycoprotein on viral particles, transforming it from the reduced form (R-S-H) to the oxidized form (R-S-S-R). We were not able to evaluate the impact of the test agents on the viral envelope in this study. Nevertheless, considering previous studies suggesting the impacts of ozone, hypochlorous acid and alcohol on biological membranes, the test agents might have affected the SARS-CoV-2 envelope, which also contributed to virus inactivation.

In conclusion, sufficient amounts of OW (10 ppm) and SAEW (34 ppm) could achieve effective SARS-CoV-2 inactivation rapidly, even in the presence of high concentrations of organic substances. Our results further revealed that OW and SAEW affected the genome and S protein of SARS-CoV-2; however, no such effect was observed with EtOH treatment under the same reaction conditions. Even though further clinical studies are needed in the future, we believe that both EtOH and OW or SAEW may be effective as SARS-CoV-2 disinfectants for actual hand hygiene, depending on the cleaning method, time and concentration.

Author Contributions: Y.T., H.O. (Haruko Ogawa) and H.O. (Hourei Oh) contributed to the conception and design of the study; Y.T., D.J., and S.M. contributed to the acquisition of data; Y.M., A.K., Y.T., H.O. (Haruko Ogawa) and H.O. (Hourei Oh) contributed to the analysis and interpretation of data; Y.T. and H.O. (Hourei Oh) contributed to drafting the manuscript; and all authors have read and agreed to the published version of the manuscript.

Funding: This work was supported by the Japanese Association for Dental Science and Sponsored Research [grant number 2020-A-4]. The funder played no role in the study design, collection, analysis and interpretation of data; in the writing of the report; or in the decision to submit the article for publication.

Institutional Review Board Statement: Not applicable.

Informed Consent Statement: Not applicable.

Data Availability Statement: Data is not available from the authors.

Conflicts of Interest: The authors declare no conflict of interest.

Sample Availability: Samples of the compounds are not available from the authors.

References

1. Ran, L.; Chen, X.; Wang, Y.; Wu, W.; Zhang, L.; Tan, X. Risk factors of healthcare workers with coronavirus disease 2019: A retrospective cohort study in a designated hospital of Wuhan in China. *Clin. Infect. Dis.* **2020**, *71*, 2218–2221. [CrossRef] [PubMed]
2. Boyce, J.M.; Pittet, D. Guideline for hand hygiene in health-care settings. Recommendations of the health care infections control practices advisory committee and the HICPAC/SHEA/APIC/IDSA Hand Hygiene Task Force. *Am. J. Infect. Control* **2002**, *30*, 77–81. [CrossRef] [PubMed]
3. World Health Organization. WHO Guidelines on Hand Hygiene in Health Care. 2009. Available online: https://apps.who.int/iris/bitstream/handle/10665/44102/9789241597906_eng.pdf?sequence=1&isAllowed=y (accessed on 5 June 2021).
4. Kratzel, A.; Todt, D.; V'kovski, P.; Steiner, S.; Gultom, M.; Thao, T.T.N.; Ebert, N.; Holwerda, M.; Steinmanm, J.; Niemeyer, D.; et al. Inactivation of severe acute respiratory syndrome coronavirus 2 by WHO-recommended hand rub formulations and alcohols. *Emerg. Infect. Dis.* **2020**, *26*, 1592–1595. [CrossRef] [PubMed]
5. Takeda, Y.; Uchiumi, H.; Matsuda, S.; Ogawa, H. Acidic electrolyzed water potently inactivates SARS-CoV-2 depending on the amount of free available chlorine contacting with the virus. *Biochem. Biophys. Res. Commun.* **2020**, *530*, 1–3. [CrossRef] [PubMed]
6. Takeda, Y.; Matsuda, S.; Jamsransuren, D.; Ogawa, H. Comparison of the SARS-CoV-2-inactivating activities of the differently manufactured hypochlorous acid water products with various pH. *J. Water Health* **2021**, *19*, 448–456. [CrossRef] [PubMed]
7. Martins, R.B.; Castro, I.A.; Pontelli, M.; Souza, J.P.; Lima, T.M.; Melo, S.R.; Siqueira, J.P.Z.; Caetano, H.; Arruda, E.; Almeida, M.T.G. SARS-CoV-2 inactivation by ozonated water: A preliminary alternative for environmental disinfection. *Ozone Sci. Eng.* **2020**, *42*, 378–385. [CrossRef]
8. Inagaki, H.; Saito, A.; Sudaryatma, P.E.; Sugiyama, H.; Okabayashi, T.; Fujimoto, S. Rapid inactivation of SARS-CoV-2 with ozonated water. *Ozone Sci. Eng.* **2021**, *43*, 208–212. [CrossRef]
9. The Committee on Efficacy Assessment of Disinfecting Substances Alternative to Alcohol for Use Against Severe Acute Respiratory Syndrome Coronavirus 2 (SARS-CoV-2). Final Report on Efficacy Assessment of Disinfecting Substances Alternative to Alcohol for Use Against Severe Acute Respiratory Syndrome Coronavirus 2 (SARS-CoV-2). Available online: https://www.nite.go.jp/data/000115863.pdf (accessed on 23 August 2021).
10. Nao, N.; Sato, K.; Yamagishi, J.; Tahara, M.; Nakatsu, Y.; Seki, F.; Katoh, H.; Ohnuma, A.; Shirogane, Y.; Hayashi, M.; et al. Consensus and variations in cell line specificity among human metapneumovirus strains. *PLoS ONE* **2019**, *14*, e0215822. [CrossRef] [PubMed]
11. Takeda, Y.; Jamsransuren, D.; Matsuda, S.; Crea, R.; Ogawa, H. The SARS-CoV-2-inactivating activity of hydroxytyrosol-rich aqueous olive pulp extract (HIDROX®) and its use as a virucidal cream for topical application. *Viruses* **2021**, *13*, 232. [CrossRef] [PubMed]
12. Centers for Disease Control and Prevention. Guideline for Disinfection and Sterilization in Healthcare Facilities. Available online: https://www.cdc.gov/infectioncontrol/pdf/guidelines/disinfection-guidelines-H.pdf (accessed on 10 March 2021).
13. Lin, Y.C.; Malott, R.J.; Ward, L.; Kiplagat, L.; Pabbaraju, K.; Gill, K.; Berenger, B.M.; Hu, J.; Fonseca, K.; Noyce, R.; et al. Detection and quantification of infectious severe acute respiratory coronavirus-2 in diverse clinical and environmental samples from infected patients: Evidence to support respiratory droplet, and direct and indirect contact as significant modes of transmission. *medRxiv* **2021**. [CrossRef]
14. Khadre, M.A.; Yuoseft, A.E.; Kim, J. Microbiological aspects of ozone applications in food: A review. *J. Food Sci.* **2001**, *6*, 1242–1252. [CrossRef]
15. Breidablik, H.J.; Lysebo, D.E.; Johannessen, L.; Skare, Å.; Andersen, J.R.; Kleiven, O.T. Ozonized water as an alternative to alcohol-based hand disinfection. *J. Hosp. Infect.* **2019**, *102*, 419–424. [CrossRef] [PubMed]
16. Kashiwazaki, J.; Nakamura, K.; Hara, Y.; Harada, R.; Wada, I.; Kanemitsu, K. Evaluation of the cytotoxicity of various hand disinfectants and ozonated water to human keratinocytes in a cultured epidermal model. *Adv. Skin Wound Care* **2020**, *33*, 313–318. [CrossRef] [PubMed]
17. Farah, R.I.; Ali, S.N.A. Electrolyzed water generated on-site as a promising disinfectant in the dental office during the COVID-19 pandemic. *Front. Public Health* **2021**, *9*, 629142. [CrossRef] [PubMed]

18. Bayarri, B.; Cruz-Alcalde, A.; López-Vinent, N.; Micó, M.M.; Sans, C. Can ozone inactivate SARS-CoV-2? A review of mechanisms and performance on viruses. *J. Hazard. Mater.* **2021**, *415*, 125658. [CrossRef] [PubMed]
19. Block, M.S.; Rowan, B.G. Hypochlorous acid: A review. *J. Oral Maxillofac. Surg.* **2020**, *78*, 1461–1466. [CrossRef] [PubMed]
20. van Engelenburg, F.A.C.; Terpstra, F.G.; Schuitemaker, H.; Moorer, W.R. The virucidal spectrum of a high concentration alcohol mixture. *J. Hosp. Infect.* **2002**, *51*, 121–125. [CrossRef] [PubMed]
21. Pfaender, S.; Brinkmann, J.; Todt, D.; Riebesehl, N.; Steinman, J.; Steinman, J.; Pietschmann, T.; Steinman, E. Mechanisms of methods for hepatitis C virus inactivation. *Appl. Environ. Microbiol.* **2015**, *81*, 1616–1621. [CrossRef] [PubMed]
22. Rowen, R.J.; Robins, H.A. A plausible penny costing effective treatment for coronavirus-ozone therapy. *J. Infect. Dis. Epidemiol.* **2020**, *6*, 113. [CrossRef]

molecules

MDPI

Review

An Update on COVID-19 Vaccine Induced Thrombotic Thrombocytopenia Syndrome and Some Management Recommendations

Amin Islam [1,2,*], Mohammed Sheraz Bashir [1], Kevin Joyce [1], Harunor Rashid [3,4], Ismail Laher [5] and Shereen Elshazly [1,6]

1 Department of Haematology, Mid & South Essex University Hospital NHS Foundation Trust, Prittlewell Chase, Westcliff-on-Sea SS0 0RY, UK; mohammed.bashir3@nhs.net (M.S.B.); kevin.joyce@nhs.net (K.J.); shereen.elshazly@nhs.net (S.E.)
2 Department of Haematology, Queen Mary University of London, Mile End Road, London E1 3NS, UK
3 National Centre for Immunisation Research and Surveillance(NCIRS) Kids Research, The Children's Hospital at Westmead, Westmead, NSW 2145, Australia; harunor.rashid@health.nsw.gov.au
4 The Discipline of Child and Adolescent Health, Faculty of Medicine and Health, The University of Sydney, Camperdown, NSW 2145, Australia
5 Department of Anesthesiology, Pharmacology & Therapeutics, Faculty of Medicine, The University of British Columbia, Vancouver, BC V6T 1Z3, Canada; ismail.laher@ubc.ca
6 Adult Haemato-Oncology Unit, Faculty of Medicne, Ainshams University, Cairo 11566, Egypt
* Correspondence: amin.islam@qmul.ac.uk

Citation: Islam, A.; Bashir, M.S.; Joyce, K.; Rashid, H.; Laher, I.; Elshazly, S. An Update on COVID-19 Vaccine Induced Thrombotic Thrombocytopenia Syndrome and Some Management Recommendations. *Molecules* **2021**, *26*, 5004. https://doi.org/10.3390/molecules26165004

Academic Editors: Caterina Vicidomini and Giovanni N. Roviello

Received: 1 June 2021
Accepted: 13 August 2021
Published: 18 August 2021

Publisher's Note: MDPI stays neutral with regard to jurisdictional claims in published maps and institutional affiliations.

Abstract: The thrombotic thrombocytopenia syndrome (TTS), a complication of COVID-19 vaccines, involves thrombosis (often cerebral venous sinus thrombosis) and thrombocytopenia with occasional pulmonary embolism and arterial ischemia. TTS appears to mostly affect females aged between 20 and 50 years old, with no predisposing risk factors conclusively identified so far. Cases are characterized by thrombocytopenia, higher levels of D-dimers than commonly observed in venous thromboembolic events, inexplicably low fibrinogen levels and worsening thrombosis. Hyper fibrinolysis associated with bleeding can also occur. Antibodies that bind platelet factor 4, similar to those associated with heparin-induced thrombocytopenia, have also been identified but in the absence of patient exposure to heparin treatment. A number of countries have now suspended the use of adenovirus-vectored vaccines for younger individuals. The prevailing opinion of most experts is that the risk of developing COVID-19 disease, including thrombosis, far exceeds the extremely low risk of TTS associated with highly efficacious vaccines. Mass vaccination should continue but with caution. Vaccines that are more likely to cause TTS (e.g., Vaxzevria manufactured by AstraZeneca) should be avoided in younger patients for whom an alternative vaccine is available.

Keywords: COVID-19; COVID-19 vaccine; disseminated intravascular coagulation; heparin induced thrombocytopenia; platelet factor 4; thrombosis; thrombotic thrombo-cytopenic purpura; vaccine induced thrombocytopenia and thrombosis

1. Background

A new deadly virus of the coronavirus family was first identified in December 2019 and named SARS-2-CoV-2; this virus caused severe acute respiratory syndrome and is now known as COVID-19. Patients presented with variable symptoms, ranging from asymptomatic carriers to life-threatening/changing consequences. Several vaccines have been developed and are currently being used to reduce disease incidence and mortality in many countries. Lately, rare but life-threatening events such as thrombosis with thrombocytopenia syndrome (TTS) (also called VITT—vaccine induced thrombocytopenia and thrombosis) have been reported with some COVID-19 vaccines. Recent reviews of TTS following COVID-19 vaccinations have not included clinical management guidelines [1,2]. To this end, this review summarizes the available data on the pathophysiology of COVID-19 and thrombosis, the different types of vaccines used to prevent COVID-19, the proposed mechanisms of TTS and some clinical management recommendations.

2. COVID-19 and Thrombosis

Approximately 40–45% of SARS-2-CoV-2 infected individuals are asymptomatic [3], while 14% develop severe illness, and 5% are critically unwell [4]. The clinical course of COVID-19 can be severe and sometimes associated with complications such as venous thromboembolism events (VTEs), severe inflammatory response syndrome, acute respiratory distress syndrome and multi-organ dysfunction syndromes, particularly in the elderly or those with co-morbidities such as diabetes mellitus, renal disease and cardiovascular conditions [5]. More than one-fifth of all patients with COVID-19 develop VTEs [6].

The pathophysiology of COVID-19 associated thrombosis is not completely understood. The associated hyper-inflammatory response produced by COVID-19 is inter-woven with other pathways, involving pro-inflammatory mediators, endothelial damage and direct invasion of cells such as type 2 pneumocytes, and coagulopathies such as disseminated intravascular coagulation (DIC) [7].

Platelets contain chemokines, chemotactic factors, various adhesion molecules, co-stimulatory molecules in their membranes and granules to support their role in haemostasis and immunomodulation. Platelets trigger blood coagulation and inflammation and initiate innate immune responses through the expression of Toll-like receptors (TLRs) to release inflammatory cytokines, trigger adaptive immune responses and activate T cells through the expression of key costimulatory molecules and major histocompatibility complex (MHC) molecules. Platelets release large amounts of extracellular vesicles which can interact with multiple immune mediators. Therefore, the function of platelets extends beyond aggregation, and the interaction with haemostasis, inflammation and the immune response results in the amplification of the body's defence mechanisms [8].

An abnormal viral coagulopathy typically results in a pro-thrombotic phenotype that can cause micro-thrombi and macro-thrombi, resulting in both venous and arterial occlusions [9]. There have also been rare isolated cases of thrombotic thrombocytopenic purpura (TTP) [10]. Thrombotic complications were reported early in our understanding of viral induced clinical syndromes of COVID-19 infection, and they are now a leading cause of death during the COVID-19 pandemic.

Early research identified various laboratory parameters to help stratify thrombotic risk, resulting in many hospitals in the United Kingdom (UK) adopting a COVID-19 test panel consisting of D-dimer tests, clotting screen, fibrinogen, serum ferritin, lactate dehydrogenase and troponin levels. This enabled VTE prophylaxis to be adapted to individual patient risk, with low molecular weight heparin used as an anticoagulant for inpatients and continued post discharge for periods of up to 12 weeks. Oral anticoagulants are also useful in some patients post discharge [11].

Thrombocytopenia occurred in almost one in three COVID-19 (31.6%) infected inpatients and at a greater rate in individuals with severe COVID-19 (57.7%) [12]. The majority of patients presented with mild thrombocytopenia, with some cases of severe thrombocytopenia requiring a careful balance between bleeding and thrombotic risk. Several pathophysiological mechanisms have been postulated for thrombocytopenia in such cases. These include bone marrow suppression that is somewhat similar to that observed in sepsis; there is likely direct infection of the marrow by SARS-2-CoV-2, affecting megakaryocytic and haematopoietic precursor cell function and synthesis. Other mechanisms include reduced thrombopoietin synthesis in the liver, generation of micro-thrombi with subsequent platelet consumption and, finally, immune thrombocytopenia (ITP) resulting in peripheral platelet destruction [13]. Infection with COVID-19 and injury to the type II alveolar cells results in increased expression of p53, suppression of uPA and uPAR and increases in PAI-1. In addition, the type II alveolar cells are also the source of surfactant; viral infection diminishes the release of surfactant. There is a temporal relationship between COVID-19 infection and fibrinolysis. In the acute phase of the infection, the inflow of fibrinogen and coagulation factors result in fibrin deposition and hyaline membrane formation. Acute inflammatory cytokines consisting of IL-1, IL-6 and IL-17A upregulates plasminogen activator inhibitor-1 (PAI-1) and suppresses the expression of urokinase-type plasminogen activator (uPA) and urokinase-type plasminogen activator receptor (uPAR). Reduced surfactant levels activate the p53 pathway, resulting in increases in PAI-1 and decreases in uPA and uPAR. The fibrinolytic balance is then shifted to a hypofibrinolytic state, which stimulates fibrin deposition, hyaline membrane formation and microvascular thrombosis [14–16].

3. Vaccinations

The World Health Organization (WHO) estimates that over 200 million people have so far been diagnosed with COVID-19, with over 4 million deaths worldwide [17]. The identification of this virus stimulated a global race to develop a vaccine. The UK was the first country to start a mass vaccination campaign with Comirnaty (the Pfizer-BioNTech vaccine, Mainz, Germany) for COVID-19 in December 2020; this was then quickly followed by approval of Vaxzevria (AstraZeneca, Oxford, UK). The Moderna COVID-19 vaccine was approved in January 2021 [18]; other vaccines include Sputnik V (Gamaleya Institute, Moscow, Russia), BBBP-CorV (SinoPharm, Shanghai, China), Janssen COVID-19 vaccine (Johnson & Johnson, New Brunswick, NJ, USA) and Covovax (Novavax, Gaithersburg, MD, USA).

There are currently at least nine different technology platforms used to create vaccines against COVID-19. Most vaccine candidates undergoing clinical trials focus on the coronavirus spike protein and its variants as the primary antigen of COVID-19 infection. These methods involve nucleic acid technologies (nucleoside-modified messenger RNA and DNA), non-replicating viral vectors, peptides, recombinant proteins, live attenuated viruses and inactivated viruses [19].

The vaccine types are listed as follows:

- **RNA vaccines**: An RNA vaccine contains RNA which, when introduced into a tissue, acts as messenger RNA (mRNA) in order to cause cells to synthesize spike proteins and stimulate an adaptive immune response. RNA vaccines often, but not always, use the nucleoside-modified messenger RNA. The delivery of mRNA is achieved by using lipid nanoparticles to protect the RNA strands and to facilitate their absorption into cells. The best-known examples are the Comirnaty and the Moderna COVID-19 vaccines [20].
- **Adenovirus vector vaccines:** These vaccines are examples of non-replicating viral vector vaccines that use an adenovirus shell containing DNA encoding a SARS-CoV-2 protein. These vaccines are non-replicating and produce only the antigens that elicit a systemic immune response [21]. Examples of this group include Vaxzevria, Sputnik V [22], Convidecia (CanSino Biologics, Tianjin, China) and Janssen COVID-19 vaccines [23].
- **Inactivated viral vaccines:** These vaccines include viral particles grown in culture and subsequently inactivated toward non-pathogenic particles with immunogenic properties. Vaccines of this type include CoronaVac (Sinovac Biotech, Beijing, China); BBIBP-CorV and WIBP-CorV (Sinopharm, Beijing, China); Covaxin (Bharat Biotech, Hyderabad, India); and CoviVac (Chumakov Centre, Moscow, Russia) [24].

Despite recent surges of COVID-19 in some countries (e.g., Israel, UK) that have been fueled by more transmissible variants, mass vaccination programs have nonetheless been generally successful with fewer deaths reported after implementation of the vaccination program. A single inoculation with Comirnaty or Vaxzevria in the UK resulted in an 85% reduction in hospitalization from COVID-19 infections [13]. It is estimated that programs in the UK (Comirnaty and Vaxzevria inoculations) have prevented at least 10,400 deaths as of March 2021 [25]. The number of deaths in the UK within 28 days of a positive COVID-19 test has been steadily falling since mid-January 2021.

Data from the official UK government dashboard suggests that deaths from coronavirus in those who are 80 years or older in England fell by 62% between 24 January and 12 February 2021 [26]. This compares to a drop of 47% in those aged between 20 and 64 years old and of 51% in those between aged 65 and 79 years old. Further evidence comes from Scotland which has seen deaths from COVID-19 falling in all locations, with the fastest decreases occurring in long-term care homes where deaths fell by 62% in the three weeks leading to 14 February 2021, approaching a level last observed near the end of October 2020. Older residents in care homes were prioritized when the vaccination program began. The report from the National Records of Scotland shows that the number of deaths in those aged 85 and over has fallen by 45%, which is steeper than in younger age groups [27].

4. Vaccines and Serious Adverse Events

Adverse events have been reported with most of the COVID-19 vaccines; these are typically mild (pain at the injection site, myalgia, headaches, fatigue and tiredness) and usually resolved within a few days. The most frequent adverse reactions in adolescents 12 to 15 years of age were injection site pain (>90%), fatigue and headache (>70%), myalgia and chills (>40%) and arthralgia and pyrexia (>20%). These are the data published and regularly updated on UK government central adverse drugs reactions reporting systems [28].

However, some patients experienced rare side-effects that sometimes were fatal. The early adoption of Comirnaty as frequently associated with reports of anaphylaxis type reactions and some safety concerns were raised, resulting in some recommendations for increased monitoring of patients after receiving the injection and even avoiding Comirnaty in those with a history of allergic reactions. These measures substantially reduced the rates of serious adverse events.

Attention has now centred on VTEs that occur 5 to 24 days after the first inoculation with Vaxzevria. It is important to note that VTE is not an uncommon event and can occur naturally at any age. Even as new waves of COVID-19 emerge in many countries, apprehension remains relative to the safety and efficacy of Vaxzevria, with several countries in the European Union either entirely suspending inoculations or restricting its use to selected age groups [20,21]. For example, Austria suspended the use of a single batch (ABV5300) of vaccines pending further investigation [29], while the UK has advised that an alternative should be sought for those aged below 30 years, and Germany has suspended its use in those aged 60 years and under [30]. At the same time, these concerns led France and Germany to consider using Sputnik V [31]. The marketing authorization holder report concluded in its observed-to-expected analysis that the number of deep vein thrombosis (DVT) or pulmonary embolisms (PE) cases observed was in fact significantly lower than expected, suggesting no causal association between VTEs and Vaxzevria [32]. However, the pharmacovigilance risk assessment committee stressed caution in this interpretation due to concerns related to quality, sensitivity and appropriate stratification in the marketing authorization holder report. The recently approved Janssen COVID-19 vaccine (which also uses an adenovirus vector) includes VTEs in its risk management plan [33].

5. Thrombosis with Thrombocytopenia Syndrome

TTS is an extremely rare but increasingly recognized serious adverse event related to unusual sites of thromboembolism, such as cerebral venous sinus thrombosis (CVST) or abdominal thromboses (splanchnic, mesenteric or portal vein), all of which are associated with thrombocytopenia. 'CVST with thrombocytopenia' is a rare subtype of cerebrovascular accident, with an incidence of 5.0 per million in those receiving Vaxzevria and 4.1 per million in those receiving mRNA based vaccines, and the prevalence is three times greater in younger to middle aged women (mean age 35) [34–37]. An international study on cerebral venous and dural sinus thrombosis, as one of the largest prospective cohort studies on CVST, confirmed this gender bias for CVST. The presentation of CVST was initially described in case-controlled studies, but later confirmed in a meta-analysis (pooled odds ratio: 5.59) that proposed gender-specific risk factors such as oral contraceptives [38], hormone replacement therapies and pregnancy (including the post-partum period) [39]. This is further supported by epidemiological studies demonstrating a correlation between the use of oral contraceptive and CVST in younger to middle aged females [40]. The epidemiology of CVST has not been carefully studied, making it difficult to comprehend and compare the data in the context of the COVID-19 era.

At least 12 cases of CVST with thrombocytopenia have been reported in the USA from 2 March to 21 April 2021 [41], and case reports were also published from European countries [42,43]; starting in the first week of April 2021, with a total of 169 cases of CVST with thrombocytopenia reported among 34 million people vaccinated Vaxzevria in the European Economic Area (EEA) and the UK [44].

Recent estimates suggest a 100-fold increased risk of developing a CVST after being infected with COVID-19, with a third of cases occurring in those aged less than 30 years old [37]. A non-peer reviewed report identified a 30-fold increased risk of splanchnic vein thrombosis in recipients of mRNA based vaccines such as the Comirnaty or Moderna COVID-19 vaccines (1 per 1.6 million for Vaxzevria recipients vs. 1 per 44.9 million for mRNA vaccines) [37]. It is important to note that the report focusses on cases of CVST and splanchnic thrombosis without associated thrombocytopenia.

A Danish population study of mostly female healthcare workers who received Vaxzevria identified no clear causal association between the vaccine and blood clots [45]. Records on the time course of VTEs in relation to the vaccine administration are unknown. It is also unclear if the patients who developed VTE were infected with COVID-19 (asymptomatic or otherwise) prior to or immediately before immunity developed [45]. It is difficult to apply data from this Danish study to other regions or to gauge the prevalence of different COVID-19 strains or to determine the viral pathogenicity and associated secondary complications. The underlying co-morbidities or pre-existing risk factors such as a history of previous VTEs or thrombophilia were also not considered. The rapid spread of the COVID-19 pandemic limited the Danish study to only pre-COVID-19 era data for the comparisons of VTEs. Importantly, the study reinforced early concerns about the unusual sites of thrombosis with thrombocytopenia shortly after receiving Vaxzevria.

The overall interim recommendation and shared agreement between the WHO, European Medicines Agency (EMA) and MHRA is that the intended benefits of receiving a vaccine for COVID-19 still far outweighs its rare side-effects. More recently, TTS was reported in six patients receiving the Janssen vaccine [46]. Vaxzevria and the Janssen vaccine both consist of recombinant adenoviral vectors based on a chimpanzee adenovirus (Vaxzevria) or a human adenovirus (Janssen vaccine) that encodes the SARS-CoV-2 spike protein immunogen. While a few individuals receiving the Moderna COVID-19 vaccine lipid nanoparticle encapsulated mRNA vaccine were diagnosed with CVST, the FDA reports that there are currently no reports of patients who developed TTS after being vaccinated with Comirnaty or the Moderna COVID-19 vaccine. The striking clinical similarities of TTS and heparin-induced thrombocytopenia (HIT) and the uniformly positive platelet factor 4 (PF4)-heparin enzyme-linked immunosorbent assays (ELISAs) in these cases may be due to circulating PF4-reactive antibodies that can directly activate platelets in the absence of heparin. It is important to note that non-ELISA-based commercial assays (e.g., HemosIL AcuStar HIT-IgG, HemosIL HIT-Ab, Diamed PaGIA gel and STic Expert assays) are weakly sensitive for anti-PF4 antibodies in samples from patients with suspected TTS [47]. Intravenous immune globulin and a monoclonal antibody to the Fc receptor can block platelet activation by these antibodies, at least in vitro. These clinical and laboratory features are similar to rare cases of an HIT-like syndrome previously described after some medications or infections in patients not receiving heparin [48].

6. Possible Pathophysiology of TTS

Antibody-mediated thrombotic thrombocytopenia during COVID-19 is presumed to be an autoimmune reaction induced by SARS-CoV-2. The high incidence of thrombotic thromboembolic events during severe COVID-19 results in the frequent administration of heparin in affected patients [49]. HITT is a possible cause when thrombocytopenia is associated with thrombosis in this setting [50]. Several studies report the presence of anti-PF4/heparin antibodies in COVID-19 patients, these antibodies can also be found without any history of heparin administration [51]. Furthermore, these antibodies do not always activate platelets in the presence of heparin/PF4 complexes [52], although they can do so in presence of PF4 alone [53], suggesting that their production is likely unrelated to HIT [54]. Related to this notion is that IgG antibodies in the serum of severe cases of COVID-19 infections induce platelet apoptosis and procoagulant activity via FcγRIIA (CD32) receptor-dependent mechanisms [55]. The antigenic specificity of these antibodies is unclear, although it is likely that at least some of them are directed against PF4.

The model we support is based on the hyperactivation of platelets during COVID-19, which results in the release of PF4 into the circulation [56]. Circulating PF4 forms complexes with endogenous polyanionic proteoglycans released by damaged endothelial cells. Syndecan-1 and endocan are potential proteoglycans candidates as their serum levels are increased in severely ill COVID-19 patients in association with other markers of endothelial injury [57]. Complexes formed between PF4 and endothelial cell-derived polyanionic proteoglycans can then stimulate extra follicular B cells that produce anti-PF4 antibodies, as suggested by previous reports that autoimmune responses elicited by extra follicular B cells may be involved in the pathophysiology of severe COVID-19 [58].

A recent study by Kowarz and colleagues suggested a different slicing mechanism for spike open reading frame in adeno vector vaccines, which results in soluble spike variants that can initiate severe side effects when binding to ACE2-expressing vascular endothelial cells. They compared this phenomenon to thromboembolic events caused by spike protein encoded by the SARS-CoV-2 virus and termed this possible mechanism as the "Vaccine-Induced COVID-19 Mimicry" syndrome (VIC19M syndrome) [59].

Anti-phospholipid antibodies could also additionally contribute to platelet activation, as suggested by increases in anti-SARS-CoV-2 antibodies in other viral diseases [60]. The rare prothrombotic thrombocytopenic events following vaccination with Vaxzevria (~1 in 100 000 recipients) has a clinical presentation similar to HIT, suggesting that a vaccine-induced autoimmune response to PF4 may be plausible. Supporting this hypothesis is a recent study identifying platelet-activating anti-PF4 antibodies in the sera of patients suffering from unusual thrombotic events associated with thrombocytopenia within 4 to 16 days after receiving Vaxzevria [61]. The progression of this possible vaccine-induced anti-PF4 autoimmune response could be related to mechanisms similar to those for prothrombotic thrombocytopenia induced by the SARS-CoV-2 virus itself. Other possible mechanisms include adenoviral vector entry in megakaryocytes and the subsequent expression of spike protein on platelet surfaces and also direct platelet activation by the vector [62].

7. Heparin Induced Thrombocytopenia with Thrombosis

HIT with thrombosis is a severe prothrombotic condition that occurs in less than five percent of patients receiving intravenous unfractionated heparin, less commonly with low molecular weight heparin and usually between 4 and 10 days after initiation of treatment [63]. Thrombocytopenia is a hallmark of HIT with thrombosis, with platelet counts decreased by more than half in most patients. After exclusion of other causes of thrombocytopenia and HIT, a clinical diagnosis of HIT with thrombosis is established by immune enzymatic detection of circulating antibodies to PF4/heparin complexes, followed by a functional assay demonstrating platelet activation by the patient's serum in the presence of heparin [63]. Risk factors for HIT with thrombosis include the following: (1) the duration of heparin therapy (>5 days); (2) the type (unfractionated heparin > low molecular weight heparin > fondaparinux) and dosage of heparin; (3) the indication for treatment (surgical and trauma patients at highest risk); and (4) gender (female > male) [64]. Thrombotic complications can develop in unusual locations such as cerebral venous sinuses [65].

The specificity of PF4 autoantibodies causing HIT with thrombosis was confirmed in studies demonstrating that the epitope recognized on PF4 tetramers was exposed after interaction with heparin or other long polyanions [54]. The injection of heparin releases PF4 [66], resulting in the assembly of PF4/heparin complexes which activate complement and bind circulating B lymphocytes in a complement-dependent manner [67]. B cells responsible for the synthesis of PF4 autoantibodies rapidly mount an IgG immune response following their first exposure to heparin [68]. B cells that produce anti-PF4 antibodies are present in healthy individuals in an anergic state in which there is an absence of an immune response to an antigen such as PF4. This B cell tolerance could be disturbed after exposure to heparin and also in some inflammatory conditions [69], where anti-PF4 IgG antibodies elicit thrombus formation and thrombocytopenia via multiple mechanisms (Figure 1).

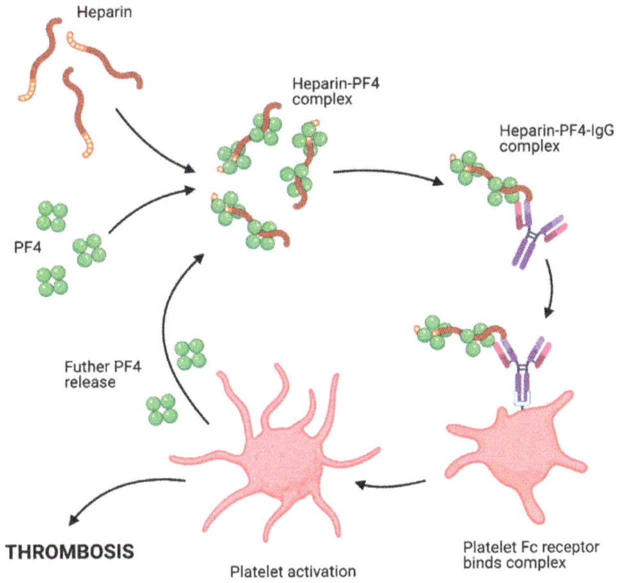

Figure 1. Diagrammatic representation of the mechanism by which HIT can cause thrombosis. (HIT: heparin induced thrombocytopenia.)

Immune complexes assembled with PF4 bound to heparin induce platelet activation and aggregation by crosslinking with FcγRIIa receptors [63]. The pathogenesis of HIT is shown in the schematic Figure 1. Anti-PF4 antibodies also activate the pro-coagulant activity of monocytes by cross-linking their FcγR1 receptors and increase the thrombotic activity of endothelial cells via recognition of PF4 firmly attached to surface proteoglycans [70]. Thrombocytopenia results from enhanced apoptosis and clearance of antibody-decorated platelets [71]. The similarities and differences between HIT with thrombosis and TTS are highlighted in Table 1.

Table 1. Comparison between HIT with thrombosis and TTS.

	HIT with Thrombosis	TTS
Responsible Agent	Heparin, more likely with unfractionated rather than low molecular weight heparin. Acidic glycosaminoglycan 10–15 kilodaltons.	Vaxzevria. Adenovirus vector (approximately 150 megadaltons) or other vaccine constituent.
Onset	5–14 days following administration of heparin [63].	4–28 days following Vaxzevria vaccination [72], although majority of reported cases occurred 5–16 days following administration [61].
Presentation	Thrombosis and thrombocytopenia. Usually venous thrombosis, which can extend to unusual sites (e.g., cerebral venous sinus thrombosis, splanchnic vein thrombosis) [63].	Thrombosis and thrombocytopenia. Preponderance of unusual sites of venous thrombosis in the currently reported cases [61].
Pathophysiology	Conformational change of PF4 upon binding heparin known to be crucial to the generation of HIT antibodies and the subsequent activation of platelets [63].	Unknown.
Investigations	HIT antibodies detected with ELISA. Functional platelet assays used to confirm diagnosis [63].	HIT antibodies detected with ELISA [61].

HIT: heparin induced thrombocytopenia; TTS: thrombosis with thrombocytopenia syndrome; ELISA: enzyme-linked immunosorbent assay; PF4: platelet factor 4.

8. Prevalence of Platelet Factor 4 Antibody in the General Population

An assessment of immunoassay results (11 studies; 860 subjects) on the prevalence of PF4/heparin antibody (IgG/M/A) in healthy subjects concluded that commercial immunoassays are able to detect PF4/heparin antibodies in 1.0–4.3% of healthy subjects [73]. Another prospective study measured PF4/heparin antibody levels in approximately 4000 blood donors by using a commercial enzyme-linked immunosorbent assay for initial findings and then repeated for confirmatory testing. Antibody levels were initially detected in 249 of 3795 donors (6.6%; 95% confidence interval [CI], 5.8–7.4%) and then confirmed in 163 of 3789 evaluable donors (4.3%; 95% CI, 3.7–5.0%). Of the 163 repeated positive samples, 116 (71.2%) were low positives, and 124 (76.1%) exhibited heparin-dependent binding. Predominant isotypes of intermediate to high seropositive samples (OD > 0.6) were IgG (20/39 (51%)), IgM (9/39 (23%)) and indeterminate (10/39 (26%)). The high background seroprevalence of PF4/heparin antibody (4.3–6.6%) with the preponderance of low (and frequently nonreproducible) positives in blood donors suggests the need for greater assay calibration, categorization of antibody levels and evaluation of the clinical relevance of "naturally occurring" PF4/heparin antibodies [74].

However, although anti-PF4–polyanion antibodies are common—for example, they are detected in 25 to 50% of patients after cardiovascular surgery—HIT is uncommon. Cerebral venous sinus thrombosis or thrombi in abdominal vessels rarely develop in patients with HIT. The prevalence of thrombocytopenia and anti-PF4 antibodies was studied in 492 Norwegian health care workers 11 to 35 days after Vaxzevria vaccination; anti-PF4 antibodies with optical density values over a cutoff of $\geq$0.4 were detected in six (1.2%) vaccines [75]. This suggests that our understanding of the pathogenesis of TTS is incomplete, and the usefulness of measuring pathogenic anti-PF4-related antibodies in all vaccine recipients is unclear. Data are needed to confirm that the anti-PF4 antibodies described here can cause thrombosis and thrombocytopenia in vivo [76].

9. Other Differential Diagnoses to Consider

The unusual clinical presentation of simultaneous thrombosis with thrombocytopenia, while appearing to be contradictory, is not a medical curiosity. There are a number of serious medical conditions where this can occur, as described below.

9.1. *Microangiopathic Haemolytic Anaemia*

Microangiopathic haemolytic anaemia describes non-immune (negative Coomb's Test) haemolysis (schistocytes) and red blood cell fragmentation on a peripheral blood film and can occur in many conditions such as prosthetic heart valves and primary thrombotic microangiopathies (TMAs), including thrombotic thrombocytopenic purpura (TTP), haemolytic uraemic syndrome, complement mediated-TMA and also drug-induced-TMA [77].

Although both TTP and hemolytic uremic syndrome have many similarities, they have different aetiologies. TTP is more common in adults and is caused by severe ADAMTS13 (a Von Willebrand factor cleaving protease) deficiency, thereby invoking potent systemic thrombotic activity [78]. Only 40% of all TTP cases present with the classical 'pentad' of thrombosis, thrombocytopenia, fever, renal impairment and neurological symptoms [79]. By contrast, haemolytic uremic syndrome is caused by the Shiga-like toxin and mostly affects children. Renal function is invariably always impaired with less than one-third of all patients developing a fever or having neurological symptoms [79]. Primary TMA is a medical emergency with high mortality and morbidity and, as such, requires astute recognition and prompt treatment. The PLASMIC score is a validated tool using easily testable blood markers to predict the likelihood of TTP [80]. In the absence of other explainable causes whilst awaiting ADAMTS13 results, a PLASMIC score >5 warrants urgent empirical treatment such as using plasma exchange and corticosteroids. Other treatments include rituximab and anti-von Willebrand factor (caplacizumab) [81,82], usually at a specialized tertiary medical center. Other systemic disorders can also present as microangiopathic hemolytic anemia; the examples include preeclampsia and HELLP syndrome (a complication of pregnancy accompanied by hemolysis, increased liver enzymes, low platelet counts and severe hypertension), infection, solid or stem cell transplant recipients, systemic rheumatic diseases and catastrophic antiphospholipid syndrome (a rare hypercoagulative state caused by a catastrophic complement activation resulting in widespread production of microthrombi). Management of these cases requires the treatment of the underlying disorders and do not usually require TMA-specific interventions [83].

9.2. *Disseminated Intravascular Coagulopathy*

There have been unusual case reports of COVID-19 patients with 'DIC like' coagulopathies [84]. The mechanisms are not clearly understood but current hypotheses suggest endothelial injury related to SARS-CoV-2 invasion. DIC is a consumptive coagulopathy resulting from unregulated and abnormally activated coagulation and fibrinolysis and can range from acute to subclinical presentations. DIC always occurs in secondary underlying disorders such as infections, malignancies and severe intravascular hemolysis, as observed in an acute hemolytic transfusion reaction [85]. The generation of thrombin and the consumption of coagulation factors including platelets results in variable phenotypic expression of bleeding and/or clotting. Thus, while acute management is challenging, it is usually guided by the dominating phenotype. Unlike in TMA where the clotting profile is almost universally normal, patients with DIC present with prolonged prothrombin time and activated partial thromboplastin time with low fibrinogen levels. The severity of microangiopathic hemolysis is generally lower than TMA's [86].

10. Thrombocytopenia following Vaccine Administration in Children

Thrombocytopenia is an adverse event associated with vaccine administration and can limit vaccine use due to several factors such as uncertainly about which vaccines are likely causative triggers, its incidence and severity, the risk of chronic disease and the possibility of recurrences after additional doses of the same vaccine. Vaccine-related thrombocytopenia is considered to be of immune origin because antibodies can be detected on platelets in about 79% of cases, making it a part of secondary ITPs in the subgroup of drug-induced ITPs. Thrombocytopenia following vaccine administration depends on the development of autoantibodies that cross-react with naturally present antigenic targets on platelets [87]. A comprehensive review on vaccine administrations and very rare development of ITP in children concluded that it can occur after the administration of vaccines. The only vaccine that is currently known to cause ITP is the mumps, measles and rubella (MMR) vaccine, but again the incidence of ITP is significantly lower than caused by mumps, measles and rubella, which are the diseases for which the vaccine provides 99% protection. Thus, ITP, regardless of its association with vaccination, should not limit the use of MMR vaccines, and a careful risk-benefit analysis performed particularly in children with persistent or chronic ITP should be performed. It is possible that newer technologies such as reverse vaccinology could prepare protein vaccines with a lower risk of causing ITP [88].

The role of adenoviral vectors in the development of thrombocytopenia has been described early in the pandemic. Adenoviral vectors remain ideal candidates as vaccine carriers and in cancer gene therapy due to their ability to effectively activate CD8+ T cells [89]. Early innate immune responses related to adenoviral vectors are associated with the activation of vascular endothelial cells, resulting in the release of ultra-large-molecular-weight multimers of the von Willebrand factor, a blood protein that is critical for platelet adhesion. This also activates platelets and induces the exposure of the adhesion molecule P-selectin and formation of platelet-leukocyte aggregates, ultimately causing thrombocytopenia and, thus, a risk for bleeding [90].

11. Post COVID-19 Syndrome and Risk of Thrombosis

Scientific and clinical evidence is evolving on the subacute and long-term effects of COVID-19, which can affect multiple organ systems [91]. Early reports suggest residual effects of SARS-CoV-2 infection, such as fatigue, dyspnea, chest pain, cognitive disturbances, arthralgia and decline in quality of life [92]. Cellular damage, a robust innate immune response with inflammatory cytokine production, and a pro-coagulant state induced by SARS-CoV-2 infection may contribute to these sequelae [93]. Long-term outcomes of patients with COVID-19 and VTE are unknown. A recent prospective study evaluated long-term bleeding, recurrence and death of COVID-19-associated VTE reported high rates of mortality (24%) and major bleeding (11%) in the first 30 days. ICU admission, thrombocytopenia and cancer indicated a poor prognosis [94].

12. Current Management Recommendations

The most recent updated recommendations from the Expert Hematology Panel (UK) and the American Society of Hematology suggest careful assessment of patients who present with symptoms of thrombosis 4–30 days after receiving Vaxzevria or Janssen COVID-19 vaccine [48,72]. The four diagnostic criteria below must be met:

1. Receipt of a COVID vaccine (Janssen/Vaxzevria) 4 to 30 days previously;
2. Thrombosis (often cerebral or abdominal);
3. Thrombocytopenia;
4. Positive PF4-HIT test using ELISA.

While the incidence of this thrombotic complication remains very rare, the risk of death and serious effects including thrombosis due to contracting COVID-19 nonetheless remains high. Current recommendations from both the UK and American regulators are for urgent medical evaluation for TTS if any of the following symptoms develop 4 to 30 days after vaccination: severe headache, visual changes, abdominal pain, nausea and/or vomiting, backache, shortness of breath, leg pain or swelling, petechiae or easy bruising.

If TTS is suspected, urgent diagnostic workup should be arranged, including a complete blood count with a platelet count and peripheral blood smear, imaging for thrombosis based on signs/symptoms, PF4-ELISA (HIT assay) using blood drawn prior to any therapies and fibrinogen level.

The Expert Hematology Panel (UK) classifies clinical presentations as follows:

Possible case: Any patient presenting with acute thrombosis and new onset thrombocytopenia within 28 days of receiving COVID-19 vaccination;

Unlikely case: Patients with either a reduced platelet count without thrombosis or with a D-dimer count at or near normal levels (<2000 μg/L) but with and normal fibrinogen (2–4 g/L) levels;
Probable case: Elevated D-dimers (>4000 μg/L > 2000 with a strong clinical suspicion);
Definite case: Cases usually present 5–28 days after vaccination and are characterized by thrombocytopenia, raised D-dimers and thrombosis, which often rapidly deteriorate.

There is a high preponderance of cerebral venous sinus thrombosis. Portal vein and splanchnic vein thrombosis, pulmonary embolism and arterial ischaemia are also common, as are adrenal infarction and hemorrhage. Intracranial hemorrhage can be significant and unexpected. Typical laboratory features include a low platelet count ($<150 \times 10^9$/L) and greatly increased D-dimer levels (above those usually expected for VTE) and many develop low fibrinogen levels contrary to hyper-fibrinogenemia observed in the acute stages of COVID-19 [95]. Antibodies associated with PF4 have been identified as in HIT but without exposure to heparin treatment. Antibodies associated with PF4 are detected by ELISA HIT assays but rarely by other HIT assay methods. The platelet count at which it is safe to initiate anticoagulation therapy is made on a case-by-case basis. The Expert Hematology Panel (UK) and American Society of Hematology revise their guidance on a regular basis as evidence emerges, with the following general recommendations: (1) Low fibrinogen or bleeding associated with TTS should not preclude anticoagulation, particularly if the platelet count is >20,000/μL or increasing following intravenous immunoglobulin initiation; (2) concurrent replacement of fibrinogen in patients with bleeding and/or very low values should be considered; (3) avoid platelet transfusions due to the similarities with HIT where platelets transfusion is relatively contraindicated unless bleeding and associated with paradoxical thrombosis. However, risk/benefit assessment in individual patients with serious bleeding and/or the need for surgical intervention may favour platelet transfusion following the initiation of intravenous immunoglobulin (IvIG), non-heparin anti-coagulation and fibrinogen replacement if level < 1.5 g/L. Platelet transfusion is an option to support therapeutic anticoagulation. However there is insufficient evidence to state that this is superior to critical care argatroban (low dose) without platelet transfusion. If urgent neurosurgery is required, then transfuse the platelets to $>100 \times 10^9$/L and cryoprecipitate to maintain fibrinogen >1.5 g/L [48,72].

Management involves all of the following, including probable cases while awaiting confirmatory tests:

- **Intravenous immunoglobulin**: Initiate urgently as this could most likely influence the disease process, using 1 g/kg (over two days if needed) irrespective of the degree of thrombocytopenia, and continue to review the clinical course. Steroids can also be helpful although it is unclear if its benefits outweigh potential harm.
- **Anticoagulants**: Use non-heparin-based therapies such as direct acting oral anticoagulants, fondaparinux, danaparoid or argatroban depending on the clinical presentation. Bleeding and thrombotic risk needs to be carefully assessed; low dose treatment with fondaparinux or critical illness dose argatroban may be appropriate when platelet count is $<30 \times 10^9$/L.
- **Plasma exchange**: May be considered in cases of severe or resistant disease. This may be required daily for up to 5 days if recovery is delayed.
- **Transfer patients with cerebral venous thrombosis**: Transfer to a neurosurgical unit and consider early recourse to neuroradiology and/or neurosurgery in cases of further deterioration/ bleeding. If urgent neurosurgery is required then transfuse platelets (to $>100 \times 10^9$/L) and cryoprecipitate to maintain fibrinogen levels at >1.5g/L.
- While it is unclear if platelet transfusion will exacerbate cerebral venous thrombosis, the risks/benefits are unknown in patients with platelets $<50 \times 10^9$/L on anticoagulation treatment and who have secondary cerebral bleeding not requiring procedures; therefore, clear advice cannot be given at present. Consider platelet transfusion in life threatening bleeding situations.

- It is unknown whether heparin exacerbates the condition but until further evidence is available, heparin is best avoided (including line flushes) as the syndrome mimics HIT with thrombosis.
- Replace fibrinogen to ensure levels do not drop below 1.5 g/L (using fibrinogen concentrate or cryoprecipitate where fibrinogen concentrates are not readily available).
- For patients who are refractory relative to repeated doses of intravenous immune globulin treatment and plasma exchange, treatment with rituximab can be considered although there is currently no evidence of its efficacy in TTS.

Further management, including intravenous immune globulin and subsequent immunomodulation, hinges on the diagnosis of thrombotic complications in the presence of thrombocytopenia driven by a positive PF4 test. TTS should also be considered in the absence of signs, symptoms or imaging documenting thrombosis with a low platelet count and greatly increased or progressively increasing D-dimer counts or a positive PF4-ELISA test. If the PF4-ELISA produces a negative result and if there is no thrombocytopenia, then TTS is ruled out. The advice is to treat for standard VTEs in such cases. Thrombocytopenia with negative PF4-ELISA in the absence of thrombosis should be managed as possible ITP.

Patients presenting with thrombosis and a normal platelet count post-vaccination might be in an early stage of TTS. These patients should be continuously assessed for tbe development of thrombocytopenia. The use of non-heparin anticoagulant may be indicated 4 to 30 days after receiving either the Vaxzevria or Janssen COVID-19 vaccine [48,72].

13. Evolving Clinical and Laboratories Studies

As this continues to be an evolving condition, additional data collection is underway, including serial PF4 testing, serum sample for COVID-19 antibody testing and whole genome sequencing used in tandem with standard diagnostic criteria. The addition of antiplatelet therapy to standard anticoagulation treatment may add some clinical benefits for young patients presenting with premature coronary artery disease or other arterial thrombotic conditions. As more clinical cases are being managed in the UK, argatroban monitoring has proved to be logistically cumbersome as activated partial thromboplastin time correlates poorly with benefits of argatroban (a small molecule direct thrombin inhibitor) due to high levels of Factor VIII [72]. Using argatroban can also provide false low fibrinogen levels when using some Clauss fibrinogen assays. Thus, it is recommended to use fondaparinux or another direct oral anticoagulant once bleeding risks are reduced [72].

14. Some Unanswered Questions about Vaccine Induced Thrombocytopenia and Thrombosis

- It appears that these clots are primarily caused by autoantibodies against PF4 (currently, it is unclear why these form). These are probably IgG and can cross the blood–brain barrier, causing platelets to aggregate and collect in specific locations such as the cerebral sinus vein. 'HIT with thrombosis' is another rare disorder with the same constellation of findings (low platelet counts and thrombosis) that is caused by treatment with heparin but at a much higher frequency (1–5%) than TTS. Whereas HIT with thrombosis has clear risk factors, no clear risk factors have so far been identified for TTS. Heparin complexes with PF4 during HIT with thrombosis, which then attaches to platelets to trigger Fc receptor mediated platelet activation. Much about TTS still remains unclear. It appears that the mechanisms of TTS and HIT with thrombosis are quite similar in being mediated by PF4 antibodies, but clots in the brain are uncommon in HIT with thrombosis for reasons that are not clear. One possibility is that there may be differences in blood flow patterns and the size of the platelet aggregates in the two conditions.
- It is unclear if there is a relationship between the new variants of COVID-19 and TTS when using adenovirus vector derived recombinant vaccines.
- Any genetic predisposition to develop TTS has not been reported yet. Next generation sequencing may be useful in addressing this important issue.

- It is not clear why the cerebral circulation may be a preferred site for the formation of blood clots with some vaccines for COVID-19.

15. Conclusions

TTS is a dynamic target requiring urgent medical attention. Autoantibodies appear to play a critical role in the prothrombotic thrombocytopenic events that can occur during the treatment of COVID-19, particularly when using Vaxzevria and Janssen COVID-19 vaccines. National and international regulatory bodies strongly advocate for mass vaccination programs but with careful selection of patients and the avoidance of the use of Vaxzevria in patients less than 30 years of age, which may be a reasonable approach provided that alternative vaccines are available. It must be emphasized that mortality and morbidity from COVID-19 are much higher than the rare risk of developing TTS or other immune mediated complications related to the vaccines. A rapid assessment of patients presenting with the symptoms of TTS is advised, and management should be initiated without delay. It is important to report confirmed and suspected cases to regulatory bodies so a clearer picture of risk factors and vulnerable populations will emerge.

Author Contributions: Conceptualization, A.I. and M.S.B.; resources, A.I., S.E., K.J., I.L. and M.S.B.; writing—A.I., M.S.B., K.J. and S.E.; review—writing and editing, H.R. and I.L.; supervision, A.I.; Figure 1 original crafted by K.J. All authors have read and agreed to the published version of the manuscript.

Funding: This work received no external funding.

Institutional Review Board Statement: Not applicable.

Informed Consent Statement: Not applicable.

Data Availability Statement: No new data was created or analyzed in this study. Data sharing is not applicable to this article.

Acknowledgments: Figure 1 created with BioRender.com.

Conflicts of Interest: The authors declare no conflict of interest.

Sample Availability: Not applicable.

References

1. Douxfils, J.; Favresse, J.; Dogné, J.M.; Lecompte, T.; Susen, S.; Cordonnier, C.; Lebreton, A.; Gosselin, R.; Sié, P.; Pernod, G.; et al. Hypotheses behind the Very Rare Cases of Thrombosis with Thrombocytopenia Syndrome after SARS-CoV-2 Vaccination. *Thromb. Res.* **2021**, *203*, 163–171. [CrossRef] [PubMed]
2. Rzymski, P.; Perek, B.; Flisiak, R. Thrombotic Thrombocytopenia after COVID-19 Vaccination: In Search of the Underlying Mechanism. *Vaccines* **2021**, *9*, 559. [CrossRef] [PubMed]
3. Lavezzo, E.; Franchin, E.; Ciavarella, C.; Cuomo-Dannenburg, G.; Barzon, L.; Del Vecchio, C.; Rossi, L.; Manganelli, R.; Loregian, A.; Navarin, N.; et al. Suppression of a SARS-CoV-2 Outbreak in the Italian Municipality of Vo'. *Nature* **2020**, *584*, 425–429. [CrossRef]
4. Wu, Z.; McGoogan, J.M. Characteristics of and Important Lessons from the Coronavirus Disease 2019 (COVID-19) Outbreak in China: Summary of a Report of 72314 Cases from the Chinese Center for Disease Control and Prevention. *JAMA* **2020**, *323*, 1239–1242. [CrossRef] [PubMed]
5. Wu, F.; Zhao, S.; Yu, B.; Chen, Y.M.; Wang, W.; Song, Z.G.; Hu, Y.; Tao, Z.W.; Tian, J.H.; Pei, Y.Y.; et al. A New Coronavirus Associated with Human Respiratory Disease in China. *Nature* **2020**, *579*, 265–269. [CrossRef] [PubMed]
6. Al-Ani, F.; Chehade, S.; Lazo-Langner, A. Thrombosis Risk Associated with COVID-19 Infection. A Scoping Review. *Thromb. Res.* **2020**, *192*, 152–160. [CrossRef]
7. Karim, S.; Islam, A.; Rafiq, S.; Laher, I. The COVID-19 Pandemic: Disproportionate Thrombotic Tendency and Management Recommendations. *Trop. Med. Infect. Dis.* **2021**, *6*, 26. [CrossRef] [PubMed]
8. Rolla, R.; Puricelli, C.; Bertoni, A.; Boggio, E.; Gigliotti, C.L.; Chiocchetti, A.; Cappellano, G.; Dianzani, U. Platelets: "multiple Choice" Effectors in the Immune Response and Their Implication in COVID-19 Thromboinflammatory Process. *Int. J. Lab. Hematol.* **2021**, 1–12. [CrossRef]
9. Tang, N.; Bai, H.; Chen, X.; Gong, J.; Li, D.; Sun, Z. Anticoagulant Treatment Is Associated with Decreased Mortality in Severe Coronavirus Disease 2019 Patients with Coagulopathy. *J. Thromb. Haemost.* **2020**, *18*, 1094–1099. [CrossRef]

10. Bhattacharjee, S.; Banerjee, M. Immune Thrombocytopenia Secondary to COVID-19: A Systematic Review. *SN Compr. Clin. Med.* **2020**, *2*, 2048–2058. [CrossRef] [PubMed]
11. Moores, L.K.; Tritschler, T.; Brosnahan, S.; Carrier, M.; Collen, J.F.; Doerschug, K.; Holley, A.B.; Jimenez, D.; Le Gal, G.; Rali, P.; et al. Prevention, Diagnosis, and Treatment of VTE in Patients With Coronavirus Disease 2019: CHEST Guideline and Expert Panel Report. *Chest* **2020**, *158*, 1143–1163. [CrossRef]
12. Bao, C.; Tao, X.; Cui, W.; Yi, B.; Pan, T.; Young, K.H.; Qian, W. SARS-CoV-2 Induced Thrombocytopenia as an Important Biomarker Significantly Correlated with Abnormal Coagulation Function, Increased Intravascular Blood Clot Risk and Mortality in COVID-19 Patients. *Exp. Hematol. Oncol.* **2020**, *9*, 16. [CrossRef]
13. European Medicines Agency Vaxzevria (Previously COVID-19 Vaccine AstraZeneca). Available online: https://www.ema.europa.eu/en/medicines/human/EPAR/vaxzevria-previously-covid-19-vaccine-astrazeneca (accessed on 26 April 2021).
14. Bhandary, Y.P.; Shetty, S.K.; Marudamuthu, A.S.; Ji, H.L.; Neuenschwander, P.F.; Boggaram, V.; Morris, G.F.; Fu, J.; Idell, S.; Shetty, S. Regulation of Lung Injury and Fibrosis by P53-Mediated Changes in Urokinase and Plasminogen Activator Inhibitor-1. *Am. J. Pathol.* **2013**, *183*, 131–143. [CrossRef]
15. Puthusseri, B.; Marudamuthu, A.; Tiwari, N.; Fu, J.; Idell, S.; Shetty, S. Regulation of P53-Mediated Changes in the UPA-Fibrinolytic System and in Lung Injury by Loss of Surfactant Protein C Expression in Alveolar Epithelial Cells. *Am. J. Physiol.—Lung Cell. Mol. Physiol.* **2017**, *312*, L783–L796. [CrossRef] [PubMed]
16. Kwaan, H.C.; Lindholm, P.F. The Central Role of Fibrinolytic Response in COVID-19—A Hematologist's Perspective. *Int. J. Mol. Sci.* **2021**, *22*, 1283. [CrossRef]
17. World Health Organisation. Coronavirus Disease (COVID-19). Available online: https://www.who.int/emergencies/diseases/novel-coronavirus-2019 (accessed on 8 August 2021).
18. Covid-19: England Gets Third Jab as Moderna Rollout Begins. Available online: https://www.bbc.co.uk/news/uk-56727510 (accessed on 28 April 2021).
19. Thanh Le, T.; Andreadakis, Z.; Kumar, A.; Gómez Román, R.; Tollefsen, S.; Saville, M.; Mayhew, S. The COVID-19 Vaccine Development Landscape. *Nat. Rev. Drug Discov.* **2020**, *19*, 305–306. [CrossRef] [PubMed]
20. A Multi-Site, Phase I/II, 2-Part, Dose Escalation Trial Investigating the Safety and Immunogenicity of Four Prophylactic SARS-CoV-2 RNA Vaccines against COVID-19 Using Different Dosing Regimens in Healthy and Immunocompromised Adults. Available online: https://www.clinicaltrialsregister.eu/ctr-search/trial/2020-001038-36/DE (accessed on 17 May 2021).
21. A Phase 2/3 Study to Determine the Efficacy, Safety and Immunogenicity of the Candidate Coronavirus Disease (COVID-19) Vaccine ChAdOx1 NCoV-19. Available online: https://www.clinicaltrialsregister.eu/ctr-search/trial/2020-001228-32/GB (accessed on 17 May 2021).
22. Baraniuk, C. Covid-19: What Do We Know about Sputnik v and Other Russian Vaccines? *BMJ* **2021**, *372*, n743. [CrossRef] [PubMed]
23. A Study of Ad26.COV2.S in Adults (COVID-19). Available online: https://clinicaltrials.gov/ct2/show/NCT04436276 (accessed on 17 May 2021).
24. Ivanova, P. Russia Approves Its Third COVID-19 Vaccine, CoviVac. Available online: https://www.reuters.com/article/us-health-coronavirus-russia-vaccine-idUSKBN2AK07H (accessed on 17 May 2021).
25. Public Health England—Impact of COVID-19 Vaccines on Mortality in England. Available online: https://assets.publishing.service.gov.uk/government/uploads/system/uploads/attachment_data/file/977249/PHE_COVID-19_vaccine_impact_on_mortality_March.pdf (accessed on 1 June 2021).
26. Wise, J. Covid-19: Is Vaccination Roll out Reducing Cases and Deaths in the UK? *BMJ* **2021**, *372*, n506. [CrossRef] [PubMed]
27. National Records of Scotland. Deaths Involving COVID-19 Week 6: 8–14 February. Available online: https://www.nrscotland.gov.uk/news/2021/deaths-involving-covid-19-week-6-8-feb-14-feb (accessed on 17 May 2021).
28. Medicines & Healthcare Products Regulatory Agency Information for Healthcare Professionals on COVID-19 Vaccine Pfizer/BioNTech (Regulation 174). Available online: https://www.gov.uk/government/publications/regulatory-approval-of-pfizer-biontech-vaccine-for-covid-19/information-for-healthcare-professionals-on-pfizerbiontech-covid-19-vaccine (accessed on 4 August 2021).
29. European Medicines Agency COVID-19 Vaccine AstraZeneca: PRAC Preliminary View Suggests No Specific Issue with Batch Used in Austria. Available online: https://www.ema.europa.eu/en/news/covid-19-vaccine-astrazeneca-prac-preliminary-view-suggests-no-specific-issue-batch-used-austria (accessed on 17 May 2021).
30. Covid: Germany Limits Use of AstraZeneca Covid Jab for under-60s. Available online: https://www.bbc.co.uk/news/world-europe-56580728 (accessed on 28 April 2021).
31. Dyer, O. Covid-19: EMA Defends AstraZeneca Vaccine as Germany and Canada Halt Rollouts. *BMJ* **2021**, *373*, n883. [CrossRef] [PubMed]
32. European Medicines Agency COVID-19 Vaccine AstraZeneca: PRAC Investigating Cases of Thromboembolic Events—Vaccine's Benefits Currently Still Outweigh Risks. Available online: https://www.ema.europa.eu/en/news/covid-19-vaccine-astrazeneca-prac-investigating-cases-thromboembolic-events-vaccines-benefits (accessed on 26 April 2021).
33. COVID-19 Vaccine (Ad26.COV2-S [Recombinant]) Risk Managmenet Plan. Available online: https://www.ema.europa.eu/en/documents/rmp-summary/covid-19-vaccine-janssen-epar-risk-management-plan_en.pdf (accessed on 17 May 2021).
34. Stam, J. Thrombosis of the Cerebral Veins and Sinuses. *N. Engl. J. Med.* **2005**, *352*, 1791–1798. [CrossRef]

35. Capecchi, M.; Abbattista, M.; Martinelli, I. Cerebral Venous Sinus Thrombosis. *J. Thromb. Haemost.* **2018**, *16*, 1918–1931. [CrossRef]
36. Coutinho, J.M.; Zuurbier, S.M.; Aramideh, M.; Stam, J. The Incidence of Cerebral Venous Thrombosis: A Cross-Sectional Study. *Stroke* **2012**, *43*, 3375–3377. [CrossRef]
37. Taquet, M.; Husain, M.; Geddes, J.R.; Luciano, S.; Harrison, P.J. Cerebral Venous Thrombosis and Portal Vein Thrombosis: A Retrospective Cohort Study of 537,913 COVID-19 Cases. *medRxiv* **2021**. [CrossRef]
38. Bousser, M.G.; Ferro, J.M. Cerebral Venous Thrombosis: An Update. *Lancet Neurol.* **2007**, *6*, 162–170. [CrossRef]
39. Alvis-Miranda, H.; Castellar-Leones, S.; Alcala-Cerra, G.; Moscote-Salazar, L. Cerebral Sinus Venous Thrombosis. *J. Neurosci. Rural Pract.* **2013**, *4*, 427–438. [CrossRef] [PubMed]
40. Krayenbühl, H.A. Cerebral Venous and Sinus Thrombosis. *Clin. Neurosurg.* **1966**, *14*, 1–24. [CrossRef] [PubMed]
41. See, I.; Su, J.R.; Lale, A.; Woo, E.J.; Guh, A.Y.; Shimabukuro, T.T.; Streiff, M.B.; Rao, A.K.; Wheeler, A.P.; Beavers, S.F.; et al. US Case Reports of Cerebral Venous Sinus Thrombosis With Thrombocytopenia After Ad26.COV2.S Vaccination, March 2 to April 21, 2021. *JAMA* **2021**, *325*, 2448–2456. [CrossRef] [PubMed]
42. Mehta, P.R.; Apap Mangion, S.; Benger, M.; Stanton, B.R.; Czuprynska, J.; Arya, R.; Sztriha, L.K. Cerebral Venous Sinus Thrombosis and Thrombocytopenia after COVID-19 Vaccination – A Report of Two UK Cases. *Brain. Behav. Immun.* **2021**, *95*, 514–517. [CrossRef]
43. Castelli, G.P.; Pognani, C.; Sozzi, C.; Franchini, M.; Vivona, L. Cerebral Venous Sinus Thrombosis Associated with Thrombocytopenia Post-Vaccination for COVID-19. *Crit. Care* **2021**, *25*, 137. [CrossRef] [PubMed]
44. European Medicines Agency AstraZeneca's COVID-19 Vaccine: EMA Finds Possible Link to Very Rare Cases of Unusual Blood Clots with Low Blood Platelets. Available online: https://www.ema.europa.eu/en/news/astrazenecas-covid-19-vaccine-ema-finds-possible-link-very-rare-cases-unusual-blood-clots-low-blood (accessed on 4 August 2021).
45. Østergaard, S.D.; Schmidt, M.; Horváth-Puhó, E.; Thomsen, R.W.; Sørensen, H.T. Thromboembolism and the Oxford–AstraZeneca COVID-19 Vaccine: Side-Effect or Coincidence? *Lancet* **2021**, *397*, 1441–1443. [CrossRef]
46. Schuchat, A.; Marks, P. Joint CDC and FDA Statement on Johnson & Johnson COVID-19 Vaccine. Available online: https://www.cdc.gov/media/releases/2021/s0413-JJ-vaccine.html (accessed on 17 May 2021).
47. Platton, S.; Bartlett, A.; MacCallum, P.; Makris, M.; McDonald, V.; Singh, D.; Scully, M.; Pavord, S. Evaluation of Laboratory Assays for Anti-Platelet Factor 4 Antibodies after ChAdOx1 NCOV-19 Vaccination. *J. Thromb. Haemost.* **2021**, *19*, 2007–2013. [CrossRef]
48. Bussel, J.B.; Connors, J.M.; Cines, D.B.; Dunbar, C.; Michaelis, L.; Kreuziger, L.B.; Lee, A.Y.Y.; Pabinger-Fasching, I. Thrombosis with Thrombocytopenia Syndrome (Also Termed Vaccine-Induced Thrombotic Thrombocytopenia). Available online: https://www.hematology.org/covid-19/vaccine-induced-immune-thrombotic-thrombocytopenia (accessed on 17 May 2021).
49. Hippensteel, J.A.; LaRiviere, W.B.; Colbert, J.F.; Langou t-Astri, C.J.; Schmidt, E.P. Heparin as a Therapy for COVID-19: Current Evidence and Future Possibilities. *Am. J. Physiol.-Lung Cell. Mol. Physiol.* **2020**, *319*, L211–L217. [CrossRef] [PubMed]
50. Liu, X.; Zhang, X.; Xiao, Y.; Gao, T.; Wang, G.; Wang, Z.; Zhang, Z.; Hu, Y.; Dong, Q.; Zhao, S.; et al. Heparin-Induced Thrombocytopenia Is Associated with a High Risk of Mortality in Critical COVID-19 Patients Receiving Heparin-Involved Treatment. *medRxiv* **2020**. [CrossRef]
51. Lingamaneni, P.; Gonakoti, S.; Moturi, K.; Vohra, I.; Zia, M. Heparin-Induced Thrombocytopenia in COVID-19. *J. Investig. Med. High Impact Case Rep.* **2020**, *8*, 232470962094409. [CrossRef] [PubMed]
52. Brodard, J.; Kremer Hovinga, J.A.; Fontana, P.; Studt, J.D.; Gruel, Y.; Greinacher, A. COVID-19 Patients Often Show High-Titer Non-Platelet-Activating Anti-PF4/Heparin IgG Antibodies. *J. Thromb. Haemost.* **2021**, *19*, 1294–1298. [CrossRef]
53. Nazy, I.; Jevtic, S.D.; Moore, J.C.; Huynh, A.; Smith, J.W.; Kelton, J.G.; Arnold, D.M. Platelet-Activating Immune Complexes Identified in Critically Ill COVID-19 Patients Suspected of Heparin-Induced Thrombocytopenia. *J. Thromb. Haemost.* **2021**, *19*, 1342–1347. [CrossRef] [PubMed]
54. Nguyen, T.H.; Medvedev, N.; Delcea, M.; Greinacher, A. Anti-Platelet Factor 4/Polyanion Antibodies Mediate a New Mechanism of Autoimmunity. *Nat. Commun.* **2017**, *8*, 1–12. [CrossRef]
55. Althaus, K.; Marini, I.; Zlamal, J.; Pelzl, L.; Singh, A.; Häberle, H.; Mehrländer, M.; Hammer, S.; Schulze, H.; Bitzer, M.; et al. Antibody-Induced Procoagulant Platelets in Severe COVID-19 Infection. *Blood* **2021**, *137*, 1061–1071. [CrossRef]
56. Comer, S.P.; Cullivan, S.; Szklanna, P.B.; Weiss, L.; Cullen, S.; Kelliher, S.; Smolenski, A.; Murphy, C.; Altaie, H.; Curran, J.; et al. COVID-19 Induces a Hyperactive Phenotype in Circulating Platelets. *PLoS Biol.* **2021**, *19*, e3001109. [CrossRef]
57. Suzuki, K.; Okada, H.; Tomita, H.; Sumi, K.; Kakino, Y.; Yasuda, R.; Kitagawa, Y.; Fukuta, T.; Miyake, T.; Yoshida, S.; et al. Involvement of Syndecan-1 in the State of COVID-19 Related to Endothelial Injury. *Thromb. J.* **2021**, *19*, 1–5. [CrossRef]
58. Woodruff, M.C.; Ramonell, R.P.; Nguyen, D.C.; Cashman, K.S.; Saini, A.S.; Haddad, N.S.; Ley, A.M.; Kyu, S.; Howell, J.C.; Ozturk, T.; et al. Extrafollicular B Cell Responses Correlate with Neutralizing Antibodies and Morbidity in COVID-19. *Nat. Immunol.* **2020**, *21*, 1506–1516. [CrossRef] [PubMed]
59. Kowarz, E.; Krutzke, L.; Reis, J.; Bracharz, S.; Kochanek, S.; Marschalek, R. "Vaccine-Induced Covid-19 Mimicry" Syndrome: Splice Reactions within the SARS-CoV-2 Spike Open Reading Frame Result in Spike Protein Variants That May Cause Thromboembolic Events in Patients Immunized with Vector-Based Vaccines. *Res. Sq.* **2021**. [CrossRef]
60. Boilard, E.; Paré, G.; Rousseau, M.; Cloutier, N.; Dubuc, I.; Lévesque, T.; Borgeat, P.; Flamand, L. Influenza Virus H1N1 Activates Platelets through FcγRIIA Signaling and Thrombin Generation. *Blood* **2014**, *123*, 2854–2863. [CrossRef] [PubMed]

61. Greinacher, A.; Greifswald, U.; Thiele, T.; Warkentin, T.E.; Weisser, K.; Eichinger, S. A Prothrombotic Thrombocytopenic Disorder Resembling Heparin-Induced Thrombocytopenia Following Coronavirus-19 Vaccination. *Res. Sq.* **2021**. [CrossRef]
62. Tsilingiris, D.; Vallianou, N.G.; Karampela, I.; Dalamaga, M. Vaccine Induced Thrombotic Thrombocytopenia: The Shady Chapter of a Success Story. *Metab. Open* **2021**, *11*, 100101. [CrossRef]
63. Greinacher, A. Heparin-Induced Thrombocytopenia. *N. Engl. J. Med.* **2015**, *373*, 252–261. [CrossRef]
64. Salter, B.S.; Weiner, M.M.; Trinh, M.A.; Heller, J.; Evans, A.S.; Adams, D.H.; Fischer, G.W. Heparin-Induced Thrombocytopenia: A Comprehensive Clinical Review. *J. Am. Coll. Cardiol.* **2016**, *67*, 2519–2532. [CrossRef]
65. Fesler, M.J.; Creer, M.H.; Richart, J.M.; Edgell, R.; Havlioglu, N.; Norfleet, G.; Cruz-Flores, S. Heparin-Induced Thrombocytopenia and Cerebral Venous Sinus Thrombosis: Case Report and Literature Review. *Neurocrit. Care* **2011**, *15*, 161–165. [CrossRef]
66. Dawes, J.; Smith, R.C.; Pepper, D.S. The Release, Distribution, and Clearance of Human β-Thromboglobulin and Platelet Factor 4. *Thromb. Res.* **1978**, *12*, 851–861. [CrossRef]
67. Khandelwal, S.; Lee, G.M.; Hester, C.G.; Poncz, M.; McKenzie, S.E.; Sachais, B.S.; Rauova, L.; Kelsoe, G.; Cines, D.B.; Frank, M.; et al. The Antigenic Complex in HIT Binds to B Cells via Complement and Complement Receptor 2 (CD21). *Blood* **2016**, *128*, 1789–1799. [CrossRef]
68. Staibano, P.; Arnold, D.M.; Bowdish, D.M.E.; Nazy, I. The Unique Immunological Features of Heparin-Induced Thrombocytopenia. *Br. J. Haematol.* **2017**, *177*, 198–207. [CrossRef] [PubMed]
69. Zheng, Y.; Wang, A.W.; Yu, M.; Padmanabhan, A.; Tourdot, B.E.; Newman, D.K.; White, G.C.; Aster, R.H.; Wen, R.; Wang, D. B-Cell Tolerance Regulates Production of Antibodies Causing Heparin-Induced Thrombocytopenia. *Blood* **2014**, *123*, 931–934. [CrossRef] [PubMed]
70. Madeeva, D.; Cines, D.B.; Poncz, M.; Rauova, L. Role of Monocytes and Endothelial Cells in Heparin-Induced Thrombocytopenia. *Thromb. Haemost.* **2016**, *116*, 806–812. [CrossRef] [PubMed]
71. Rollin, J.; Pouplard, C.; Gruel, Y. Risk Factors for Heparin-Induced Thrombocytopenia: Focus on Fcγ Receptors. *Thromb. Haemost.* **2016**, *116*, 799–805. [CrossRef] [PubMed]
72. Pavord, D.S.; Lester, W.; Makris, M.; Scully, M.; Hunt, B. Guidance from the am Panel (EHP) on Covid-19 Vaccine-Induced Immune Thrombocytopenia and Thrombosis (VITT). Available online: https://b-s-h.org.uk/media/19590/guidance-version-17-on-mngmt-of-vitt-20210420.pdf (accessed on 17 May 2021).
73. Arepally, G.M.; Hursting, M.J. Platelet Factor 4/Heparin Antibody (IgG/M/A) in Healthy Subjects: A Literature Analysis of Commercial Immunoassay Results. *J. Thromb. Thrombolysis* **2008**, *26*, 55–61. [CrossRef]
74. Hursting, M.J.; Pai, P.J.; McCracken, J.E.; Hwang, F.; Suvarna, S.; Lokhnygina, Y.; Bandarenko, N.; Arepally, G.M. Platelet Factor 4/Heparin Antibodies in Blood Bank Donors. *Am. J. Clin. Pathol.* **2010**, *134*, 774–780. [CrossRef]
75. Sørvoll, I.H.; Horvei, K.D.; Ernstsen, S.L.; Lægreid, I.J.; Lund, S.; Grønli, R.H.; Olsen, M.K.; Jacobsen, H.K.; Eriksson, A.; Halstensen, A.M.; et al. An Observational Study to Identify the Prevalence of Thrombocytopenia and Anti-PF4/Polyanion Antibodies in Norwegian Health Care Workers after COVID-19 Vaccination. *J. Thromb. Haemost.* **2021**, *19*, 1813–1818. [CrossRef]
76. Cines, D.B.; Bussel, J.B. SARS-CoV-2 Vaccine-Induced Immune Thrombotic Thrombocytopenia. *N. Engl. J. Med.* **2021**, *384*, 2254–2256. [CrossRef]
77. George, J.N.; Nester, C.M. Syndromes of Thrombotic Microangiopathy. *N. Engl. J. Med.* **2014**, *371*, 654–666. [CrossRef]
78. Furlan, M.; Robles, R.; Galbusera, M.; Remuzzi, G.; Kyrle, P.A.; Brenner, B.; Krause, M.; Scharrer, I.; Aumann, V.; Mittler, U.; et al. Von Willebrand Factor–Cleaving Protease in Thrombotic Thrombocytopenic Purpura and the Hemolytic–Uremic Syndrome. *N. Engl. J. Med.* **1998**, *339*, 1578–1584. [CrossRef]
79. McCrae, K.R. Chapter 44—Thrombocytopenia in Pregnancy. In *Platelets*, 3rd ed.; Michelson, A.D., Ed.; Academic Press: Cambridge, MA, USA, 2013; pp. 909–928. ISBN 978-0-12-387837-3.
80. Paydary, K.; Banwell, E.; Tong, J.; Chen, Y.; Cuker, A. Diagnostic Accuracy of the PLASMIC Score in Patients with Suspected Thrombotic Thrombocytopenic Purpura: A Systematic Review and Meta-Analysis. *Transfusion* **2020**, *60*, 2047–2057. [CrossRef]
81. Lim, W.; Vesely, S.K.; George, J.N. The Role of Rituximab in the Management of Patients with Acquired Thrombotic Thrombocytopenic Purpura. *Blood* **2015**, *125*, 1526–1531. [CrossRef] [PubMed]
82. Chander, D.P.; Loch, M.M.; Cataland, S.R.; George, J.N. Caplacizumab Therapy without Plasma Exchange for Acquired Thrombotic Thrombocytopenic Purpura. *N. Engl. J. Med.* **2019**, *381*, 92–94. [CrossRef] [PubMed]
83. George, J.N.; Cuker, A. Acquired TTP: Clinical Manifestations and Diagnosis—UpToDate. Available online: https://www.uptodate.com/contents/acquired-ttp-clinical-manifestations-and-diagnosis (accessed on 1 June 2021).
84. The Lancet Haematology. COVID-19 Coagulopathy: An Evolving Story. *Lancet Haematol.* **2020**, *7*, e425. [CrossRef]
85. Martinod, K.; Wagner, D.D. Thrombosis: Tangled up in NETs. *Blood* **2014**, *123*, 2768–2776. [CrossRef]
86. Levi, M. Current Understanding of Disseminated Intravascular Coagulation. *Br. J. Haematol.* **2004**, *124*, 567–576. [CrossRef]
87. Fujita, H. Idiopathic thrombocytopenic purpura following viral infection. *Nihon Rinsho.* **2003**, *61*, 650–654.
88. Cecinati, V.; Principi, N.; Brescia, L.; Giordano, P.; Esposito, S. Vaccine Administration and the Development of Immune Thrombocytopenic Purpura in Children. *Hum. Vaccines Immunother.* **2013**, *9*, 1158–1162. [CrossRef]
89. Shirley, J.L.; de Jong, Y.P.; Terhorst, C.; Herzog, R.W. Immune Responses to Viral Gene Therapy Vectors. *Mol. Ther.* **2020**, *28*, 709–722. [CrossRef]
90. Othman, M.; Labelle, A.; Mazzetti, I.; Elbatarny, H.S.; Lillicrap, D. Adenovirus-Induced Thrombocytopenia: The Role of von Willebrand Factor and P-Selectin in Mediating Accelerated Platelet Clearance. *Blood* **2007**, *109*, 2832–2839. [CrossRef]

91. Gupta, A.; Madhavan, M.V.; Sehgal, K.; Nair, N.; Mahajan, S.; Sehrawat, T.S.; Bikdeli, B.; Ahluwalia, N.; Ausiello, J.C.; Wan, E.Y.; et al. Extrapulmonary Manifestations of COVID-19. *Nat. Med.* **2020**, *26*, 1017–1032. [CrossRef] [PubMed]
92. Carfì, A.; Bernabei, R.; Landi, F. Persistent Symptoms in Patients after Acute COVID-19. *JAMA* **2020**, *324*, 603–605. [CrossRef] [PubMed]
93. McElvaney, O.J.; McEvoy, N.L.; McElvaney, O.F.; Carroll, T.P.; Murphy, M.P.; Dunlea, D.M.; Choileáin, O.N.; Clarke, J.; O'Connor, E.; Hogan, G.; et al. Characterization of the Inflammatory Response to Severe COVID-19 Illness. *Am. J. Respir. Crit. Care Med.* **2020**, *202*, 812–821. [CrossRef]
94. Demelo-Rodríguez, P.; Ordieres-Ortega, L.; Ji, Z.; del Toro-Cervera, J.; de Miguel-Díez, J.; Álvarez-Sala-Walther, L.A.; Galeano-Valle, F. Long-Term Follow-up of Patients with Venous Thromboembolism and COVID-19: Analysis of Risk Factors for Death and Major Bleeding. *Eur. J. Haematol.* **2021**, *106*, 716–723. [CrossRef] [PubMed]
95. Thachil, J. Lessons Learnt from COVID-19 Coagulopathy. *eJHaem* **2021**, *2021*, 1–8. [CrossRef]

molecules MDPI

Article

Pharmacophore-Based Virtual Screening, Quantum Mechanics Calculations, and Molecular Dynamics Simulation Approaches Identified Potential Natural Antiviral Drug Candidates against MERS-CoV S1-NTD

Thamer A. Bouback [1,†], Sushil Pokhrel [2,†], Abdulaziz Albeshri [1], Amal Mohammed Aljohani [1], Abdus Samad [3,4], Rahat Alam [3,4], Md Saddam Hossen [3,4,5], Khalid Al-Ghamdi [1], Md. Enamul Kabir Talukder [3,4], Foysal Ahammad [1,3,4,*], Ishtiaq Qadri [1,*] and Jesus Simal-Gandara [6,*]

1 Department of Biological Science, Faculty of Science, King Abdul-Aziz University, Jeddah 21589, Saudi Arabia; tbouback@kau.edu.sa (T.A.B.); aalbishri0061@stu.kau.edu.sa (A.A.); amal.m.aljohany@gmail.com (A.M.A.); kalghamdy@kau.edu.sa (K.A.-G.)
2 Department of Biomedical Engineering, State University of New York (SUNY), Binghamton, NY 13902, USA; sushilpokhrel@binghamton.edu
3 Department of Genetic Engineering and Biotechnology, Faculty of Biological Science and Technology, Jashore University of Science and Technology, Jashore 7408, Bangladesh; kazisamad50@gmail.com (A.S.); rahatalam1643@gmail.com (R.A.); hossensaddam621@gmail.com (M.S.H.); talukder.mek@gmail.com (M.E.K.T.)
4 Laboratory of Computational Biology, Biological Solution Centre (BioSol Centre), Jashore 7408, Bangladesh
5 Department of Microbiology, Faculty of Life Sciences and Medicine, Zhejiang Sci-Tech University, Hangzhou 310018, China
6 Nutrition and Bromatology Group, Department of Analytical Chemistry and Food Science, Faculty of Food Science and Technology, University of Vigo—Ourense Campus, E-32004 Ourense, Spain
* Correspondence: foysalgebt@gmail.com (F.A.); ishtiaq80262@yahoo.com (I.Q.); jsimal@uvigo.es (J.S.-G.)
† These authors contributed equally to this work.

Citation: Bouback, T.A.; Pokhrel, S.; Albeshri, A.; Aljohani, A.M.; Samad, A.; Alam, R.; Hossen, M.S.; Al-Ghamdi, K.; Talukder, M.E.K.; Ahammad, F.; et al. Pharmacophore-Based Virtual Screening, Quantum Mechanics Calculations, and Molecular Dynamics Simulation Approaches Identified Potential Natural Antiviral Drug Candidates against MERS-CoV S1-NTD. *Molecules* **2021**, *26*, 4961. https://doi.org/10.3390/molecules26164961

Academic Editors: Giovanni N. Roviello and Caterina Vicidomini

Received: 24 April 2021
Accepted: 7 May 2021
Published: 17 August 2021

Abstract: Middle East respiratory syndrome coronavirus (MERS-CoV) is a highly infectious zoonotic virus first reported into the human population in September 2012 on the Arabian Peninsula. The virus causes severe and often lethal respiratory illness in humans with an unusually high fatality rate. The N-terminal domain (NTD) of receptor-binding S1 subunit of coronavirus spike (S) proteins can recognize a variety of host protein and mediates entry into human host cells. Blocking the entry by targeting the S1-NTD of the virus can facilitate the development of effective antiviral drug candidates against the pathogen. Therefore, the study has been designed to identify effective antiviral drug candidates against the MERS-CoV by targeting S1-NTD. Initially, a structure-based pharmacophore model (SBPM) to the active site (AS) cavity of the S1-NTD has been generated, followed by pharmacophore-based virtual screening of 11,295 natural compounds. Hits generated through the pharmacophore-based virtual screening have re-ranked by molecular docking and further evaluated through the ADMET properties. The compounds with the best ADME and toxicity properties have been retrieved, and a quantum mechanical (QM) based density-functional theory (DFT) has been performed to optimize the geometry of the selected compounds. Three optimized natural compounds, namely Taiwanhomoflavone B (Amb23604132), 2,3-Dihydrohinokiflavone (Amb23604659), and Sophoricoside (Amb1153724), have exhibited substantial docking energy >−9.00 kcal/mol, where analysis of frontier molecular orbital (FMO) theory found the low chemical reactivity correspondence to the bioactivity of the compounds. Molecular dynamics (MD) simulation confirmed the stability of the selected natural compound to the binding site of the protein. Additionally, molecular mechanics generalized born surface area (MM/GBSA) predicted the good value of binding free energies (ΔG bind) of the compounds to the desired protein. Convincingly, all the results support the potentiality of the selected compounds as natural antiviral candidates against the MERS-CoV S1-NTD.

Keywords: MERS-CoV; S1-NTD; pharmacophore modeling; ADME; quantum mechanical calculation; DFT; FMO; MM/GBSA; molecular docking; Molecular Dynamics Simulation; HOMO; LUMO

1. Introduction

Coronavirus (CoVs) is a member of the family Coronaviridae, containing a single-stranded positive-sense RNA genome [+ssRNA] that has a length between ~27 kb to 32 kb. The virus causes illness ranging from the common cold to more severe diseases in humans and animals [1]. Genetically diverse coronaviruses cycles in nature among its three principal hosts, which are the natural host (bats, mice), intermediate host (camels, masked palm civets, swine, dogs, and cats), and humans [2]. Four common human coronaviruses (HCoVs) NL63 and 229E (α-CoVs); OC43 and HKU1 (β-CoVs) circulate widely in the human population, each capable of causing severe disease ranging from common colds to self-limiting upper respiratory infections in immunocompetent people.

Severe acute respiratory syndrome coronavirus (SARS-CoV), SARS-CoV-2, and MERS-CoV are other known human beta CoVs (β-CoVs) capable of causing epidemics [3,4]. They are zoonotic in characters and infections with the virus resulting in various clinical severity featuring respiratory and extra-respiratory manifestations [5]. The number of global people infected with coronavirus has risen rapidly, which began with the pandemic of SARS-CoV in 2003, followed by the MERS-CoV in 2012 and, most recently, the SARS-CoV-2 outbreaks with a fatality rate of ~10%, ~35%, and 0.1% to over 25%, respectively [6,7]. Among the animal coronaviruses, MERS-CoV has the highest fatality rate in humans and animals, but effective antiviral candidates are not available to treat the infection caused by the pathogen.

Receptor recognition is an initial and key step of a virus to infections into the host cells. MERS-CoV class I membrane fusion trimeric S glycoprotein of the virion can recognize host receptor that mediates entry into the cells. The S trimer of highly pathogenic MERS-CoV can recognize the host cellular receptor dipeptidyl peptidase 4 (DPP4), resulting in membrane fusion and viral entry [8]. The trimeric ectodomain segment of MERS-CoV-S protein can be divided into two subunits, the first one is the receptor binding S1 subunit and another one is the membrane-fusion S2 subunit. The receptor binding S1 subunit of the virus can also be divided into two independent domains, namely the N-terminal domain (S1-NTD) and C-terminal domain (S1-CTD), which can function as a receptor-binding domain (RBD) for the S protein [9]. The S1-NTD of MERS-CoV can identify specific sugar moieties upon primary attachment and help in the prefusion-to-post fusion transition, which is critical in determining tissue tropism, host ranges, and cross-species infection [10]. Therefore, targeting the S1-NTD of the MERS-CoV S protein can inhibit the primary attachment to the host and block the prefusion-to-post fusion transition and will be an effective prophylactic against the virus [11].

The S1-NTD targeting natural compounds with potent inhibitory activity can be a focus on the improvement of therapeutic interventions of the virus [12]. Many studies reported different neutralizing antibodies and chemically synthesized compounds as drug candidates previously, for example, folic acid showed activity against NTD from the mammalian expression medium [10,13]. Sometimes, this type of antibody can induce resistance against the virus and chemically synthesized compounds can causes adverse side effect of the host [14]. Natural compounds having low toxicity and side effect can be developed as antiviral candidates by targeting S1-NTD that will be novel therapeutics for MERS-CoV [15]. We thus sought to identify potential natural antiviral drug candidates against the MERS-CoV by targeting S1-NTD.

Nowadays, computer-aided drug design (CADD) has become an effective and powerful technique in different therapeutic development. The technique has helped to overcome the long-term and expensive process that costs billions of dollars previously during drug design and development [16]. The importance of the in-silico drug design technique is greater than ever before in the modern drug design process [17]. Therefore, the study utilized different in-silico technique includes pharmacophore modeling, virtual screening, molecular docking, ADMET, QM calculation, MD simulation, and MM/GBSA to identify effective and potential natural drug candidates against MERS-CoV.

2. Results

2.1. Results of Pharmacophore Modeling

Pharmacophore can be defined as an ensemble of common steric and electronic chemical features that indicates a compound-specific mode of action to the active site of a targeted biological macromolecule. The pharmacophore features can be observed during ligand–protein interaction and helps in screening a large chemical database for retrieving novel scaffolds as a lead compound [18].

To identify novel scaffolds as a lead compound against MERS-CoV S1-NTD, two different pharmacophore models were generated based on the protein PDB ID: 5VYH and 6PXH in complex with folic acid (FOL409) and dihydro-folic acid (DHF428), respectively. The ligand (FOL409) in complex with the protein 5VYH generated a total of 14 pharmacophore features includes two aromatic ring (AR), eight hydrogen bond acceptor (HBA), and four hydrogen bond donor (HBD) features, where complex of DHF428 and 6PXH produced 10 pharmacophore features includes one AR, one hydrophobic (H), four HBA, and four HBD features shown in Figure 1A,B. This two-pharmacophore model was aligned and merged to interpolate overlapping features, which generated a total of 20 pharmacophore features including three AR, one H, two negative ionizable area (NI), 10 HBA, and four HBD features shown in Figure 1C. The overlapped and duplicate pharmacophore features from the aligned pharmacophore models have been removed to optimize and relaxed the geometrical confirmation of the model. After removing the duplicate pharmacophore features a total of 11 features includes three HBD, one H, three HBA, three AR, and one NI feature were selected for further study shown in Figure 1D. Exclusion volume coat generated during the pharmacophore modeling process has not been considered in the study.

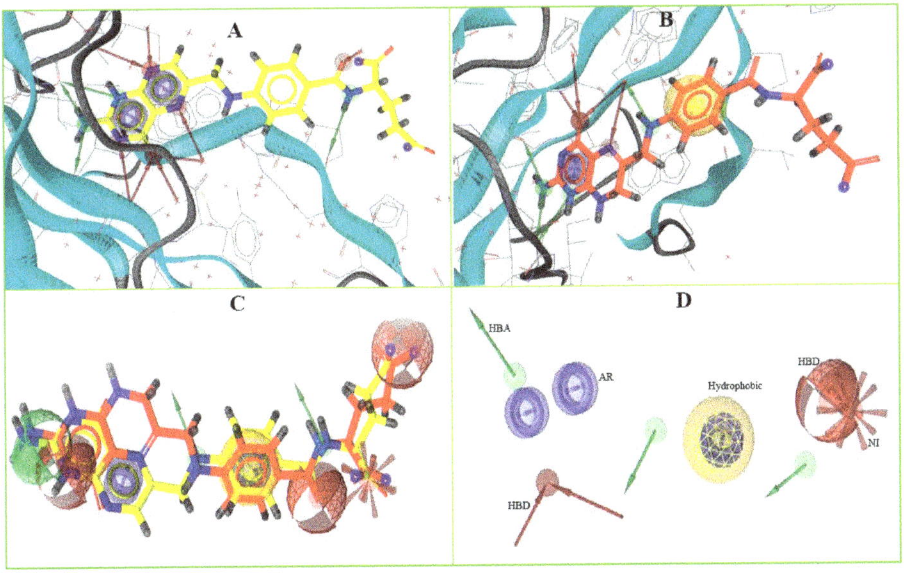

Figure 1. The overall pharmacophore model generated during the study. Herein, (**A**) pharmacophore features generated from 5VYH (PDB)-Folic Acid (yellow) complex interaction, (**B**) 6PXH (PDB)-Dihydrofolic Acid (red) complex interaction, (**C**) merge pharmacophore features, and (**D**) final pharmacophore features utilized for virtual screening. The hydrogen bond donor (HBD) features have shown in green, hydrogen bond acceptor (HBA) in red, negative ionizable area (NI) in red astricts, aromatic ring (AR) in blue, and hydrophobic (H) features in yellow color.

2.2. Molecule Library Preparation

Virtual screening can be defined as a cheminformatics technology that utilizes different computational techniques to screen a large number of molecules and identify the structures of interest for biological assays [19]. The accuracy of a cheminformatics model depends on the data mining process that is related to database preparation. Therefore, to accurately mine the database, a total of 11,295 natural compounds have been retrieved from the Ambinter, and a library has prepared for virtual screening. The geometry of all the molecular structures has been optimized by conforming MMFF94 force field available at the LigandScout tool and a molecular library has been prepared [20]. The library prepared through the software has further utilized for the virtual screening process.

2.3. Active Compounds Identification and Decoy Set Generation

Validation of a pharmacophore model is essential before a large database screening process can provide reliable outcomes on a real-life project. The SBPM can be validated through known active compounds together with inactive compounds called "decoys". Ideally, active compounds for model validation should be selected based on experimental data [19]. Therefore, 12 experimentally active compounds against MERS-CoV S protein have been identified and retrieved from the ChEMBL database. The active compounds have been selected based on their half maximal inhibitory concentration IC_{50} (nM) value shown in Figure 2. The active compounds then submitted into the DUDE-E decoy database and a total of 1326 decoys correspondence to active compounds has been retrieved. The geometry of the compounds has also been optimized by using the MMFF94 force field and converted into the LDB file format through LiganScout software.

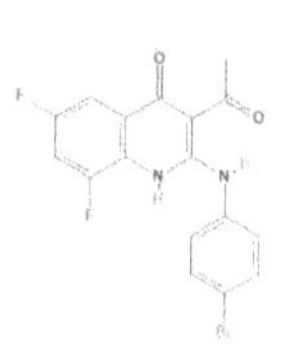
CHEMBL4444279 (IC_{50}=1130.0 nM)

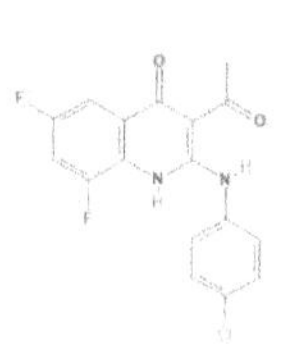
CHEMBL4454132 (IC_{50}=1440.0 nM)

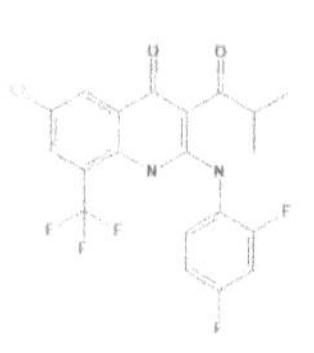
CHEMBL4545243 (IC_{50}=130.0 nM)

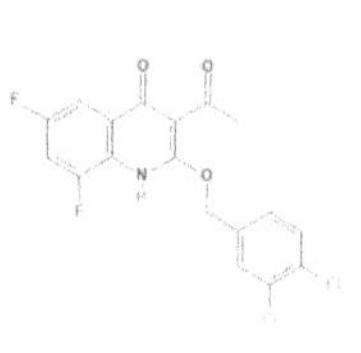
CHEMBL4453755 (IC_{50}=7800.0 nM)

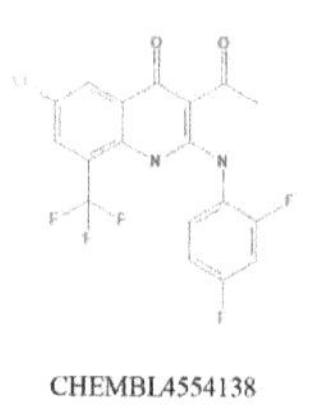
CHEMBL4554138 (IC_{50}=150.00 nM)

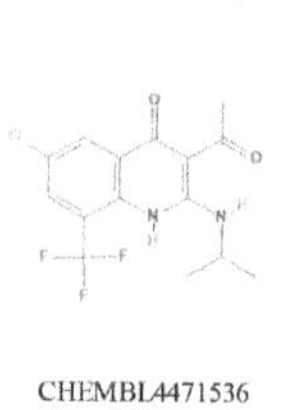
CHEMBL4471536 (IC_{50}=770.00 nM)

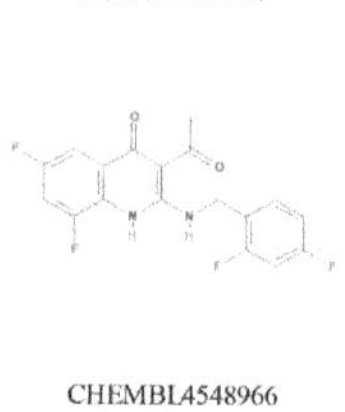
CHEMBL4548966 (IC_{50}=5900.0 nM)

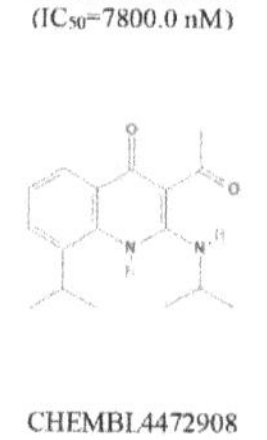
CHEMBL4472908 (IC_{50}=25000.0 nM)

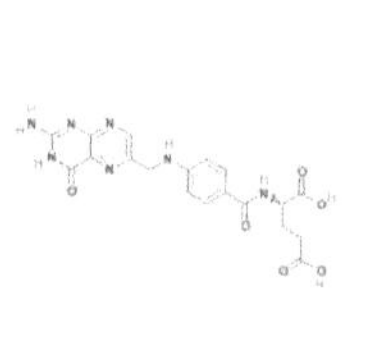
CHEMBL1622 (IC_{50}=5420.00 nM)

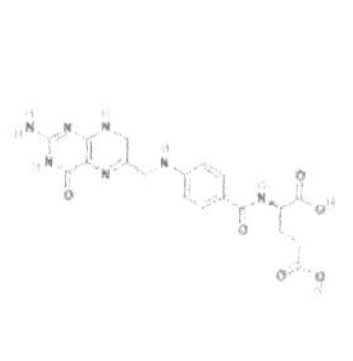
CHEMBL46294 (IC_{50}=3760.00 nM)

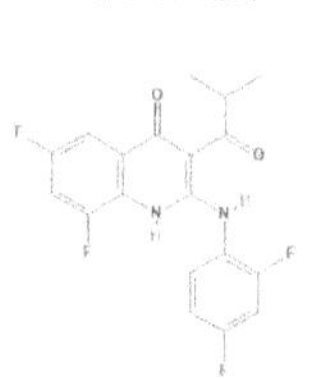
CHEMBL4463570 (IC_{50}=790.00 nM)

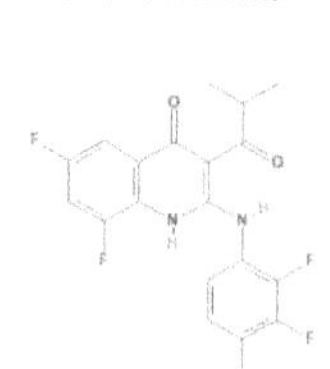
CHEMBL4526734 (IC_{50}=86.00 nM)

Figure 2. List of active compounds identified against MERS coronavirus S1-NTD protein. The IC_{50} value and correspondence ChEMBL identity for each compound has also been provided.

2.4. Pharmacophore-Based Virtual Screening

In-silico virtual screening is a type of computational approach by which molecules with desired properties can retrieve structures with similar properties from large molecule libraries. During the drug design and development process, this technique helps to identify small molecules as hits and further optimization as lead candidates [21]. Furthermore, this process can help to reduce the assay-to-lead attrition rate that has excluded time and expensive experiments require during the drug design and development process. A specific 3D pharmacophoric pattern searching approach to screen large molecule libraries is now being considered as the next step in the drug design process. Therefore, the study utilized a 3D pharmacophore models-based virtual screening process to identify hit compounds against the targeted protein. The structure-based virtual screening process retrieved 32 active compounds as hits with a geometric fit score of 65.46 to 67.75, where the number of conformations generated during the screening was a minimum of eight and a maximum of 25 shown in Table S1.

2.5. Pharmacophore Model Performance Analysis

To determine the performance of the pharmacophore model, the ROC curve generated during the virtual screening process has been analyzed. Receiver-operating characteristic (ROC) is a simple and useful graphical tool for evaluating the accuracy of a statistical model. The ROC curve in the virtual screening process provides information regarding the discrimination ability of the model from active to inactive (decoy) set [19]. The overall summary of the model accuracy can be calculated from the Area Under the Curve (AUC) that represents the degree of discrimination ability. The AUC value ranges between 0.0 to 1.0, where a value between 0 to 0.5 indicates random chance of discrimination, 0.51 to 0.7 indicates acceptable, 0.71 to 0.8 indicates good, and 0.81 to 1.0 indicates the excellence of the model [19]. The enrichment factor in the pharmacophore model provides an idea about the number of active compounds found from a specific model compared to hypothetically active compounds found from a randomly screened model. The EF factor can range from 1 to >100, where 1 indicates the number of randomly sorted molecules and >100 indicates the least number of compounds need to screen in vitro to find a large number of active compounds. The AUC and EF values found in the study were 0.74 and 1.1, respectively, indicating good discrimination ability and robustness of the pharmacophore model shown in Figure 3.

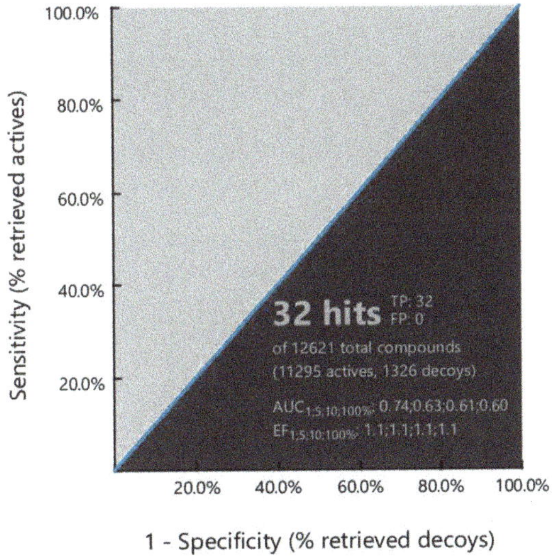

Figure 3. Showing the ROC curve generated during pharmacophore model-based virtual screening. The curve presents the relationship between sensitivity (true positive fraction to the *Y*-axis) and specificity (false positive fraction to the *x*-axis) for every possible cut-off.

2.6. Binding Site Identification and Receptor Grid Generation

A binding site can be defined as a specific amino acid (AA) residue in a protein to which ligands can binds and is fundamentally important for guiding drug design. Identification of the location of protein binding sites is essential during molecular docking simulation, which helps to generate enough contact points with the protein and significantly increases the docking efficiency [22]. Binding site is evolved to be optimized to bind a particular substrate, therefore the binding site of the protein has been identified in this study. Analysis of previously identified complex protein–ligands (PDB: 5VYH) interaction found eight binding site residues in the protein. The eight-binding site residues was resided at TRP44, PRO45, ALA123, GLY128, THR129, ILE140, TRP310, and ALA312 in the S1-NTD protein has been represented in a ball shape and shown in Figure 4.

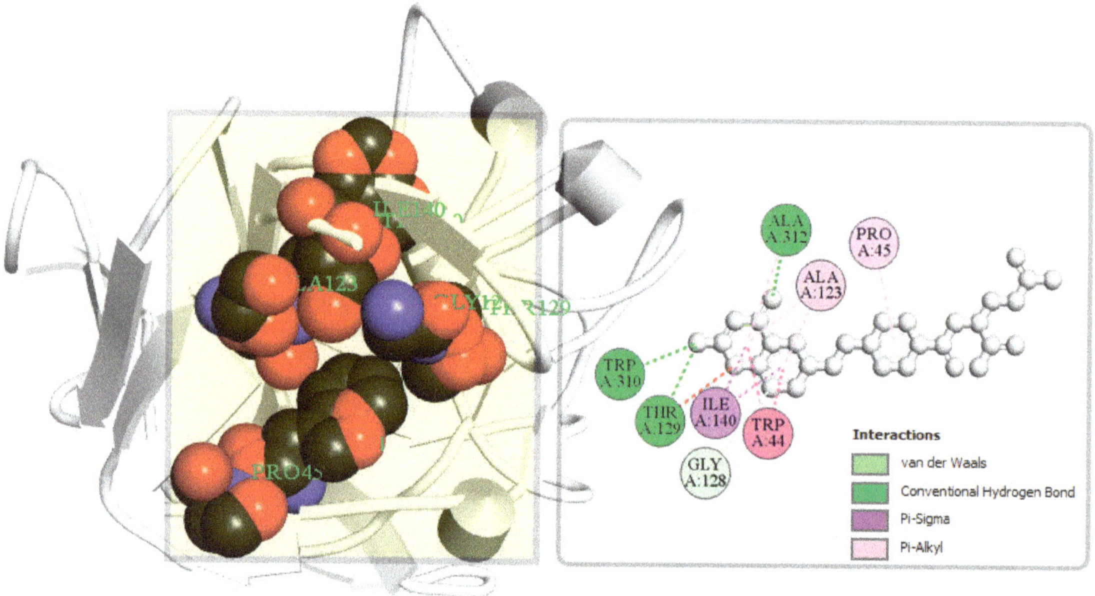

Figure 4. The binding site position of MERS-CoV S1-NTD identified from the protein–ligand complex (PDB ID: 5VYH) structure. Ball shape 3D representation of the binding site with the grid box shown on the left side in the figure, where 2D binding site position has also been represented on the right side of the figure.

PyRx is a grid-based docking program that requires the definition of receptor grid box size before initiating the molecular docking process. Grid box fixation before the molecular docking process helps to generate more reliable scoring to the ligand poses. Therefore, to identify more reliable ligand poses towards the protein, a receptor grid box with a dimension X = 30.69 (Å), Y = 33.36 (Å), and Z = 43.41(Å) has been generated based on previously identified binding residues position of the protein.

2.7. Molecular Docking Simulation

Molecular docking in CADD is an important technique that helps to determine the bound geometry and interaction between a small molecule and a protein at the atomic level. The technique has become an increasingly important tool for drug discovery due to the ability to screen large compound libraries [23]. The technique also helps to determine the behavior and predict how a protein (enzyme) interacts with small molecules (ligands) to the binding site of target proteins. To elucidate the ligand–receptor binding mechanism, a molecular docking simulation has been performed in this study. The 32 hits identified previously through the structure-based virtual screening process have been docked to the binding site of the MERS-CoV S1-NTD protein. The docking score found for the 32 hits has a range between −6.4 and −9.2 kcal/mol provided in Table S1. Based on the binding affinity top (10%), four compounds Amb6600135 (−9.2 kcal/mol), Amb23604132 (−9.1 kcal/mol), Amb23604659 (−8.6 kcal/mol), and Amb1153724 (−8.1 kcal/mol) with zero upper and lower RMSD value have been chosen for further evaluation listed in Table 1.

Table 1. List of the top four compounds and their chemical name, molecular formula, binding affinity (kcal/mol), and pharmacophore fit score.

Ambinter ID	Molecule Name	Formula	Structure	Binding Affinity (kcal/mol)	Pharmacophore Fit Score
Amb6600135	Nicotinamide adenine dinucleotide	$C_{21}H_{28}N_7O_{14}P_{2+}$	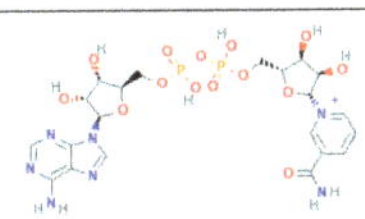	−9.2	66.6
Amb23604132	Taiwanhomoflavone B	$C_{32}H_{24}O_{10}$	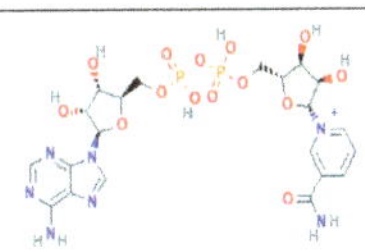	−9.1	65.48
Amb23604659	2,3-Dihydrohinokiflavone	$C_{30}H_{20}O_{10}$	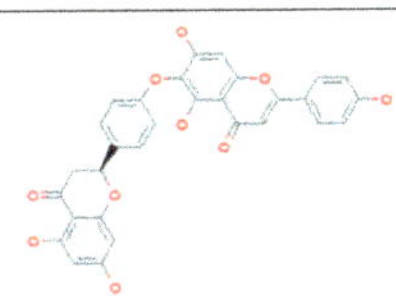	−8.6	65.55
Amb1153724	Sophoricoside	$C_{21}H_{20}O_{10}$	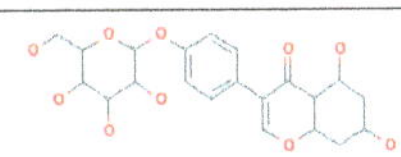	−8.1	66.64

2.8. ADME Analysis

ADME properties of chemical compounds play an important role in the likelihood of success of a drug. Optimization of the ADME properties can reduce the pharmacokinetics-related failure in the clinical phases, which is difficult and challenging in the drug development and discovery process [24]. It has been found that early-stage evaluation of ADME can reduce the attrition rates during the clinical drug development phase. Therefore, the study utilized the SwissADME web tool for the early-stage evaluation of ADME properties for selected four compounds. The server evaluated the ADME properties of selected four (Amb6600135, Amb23604132, Amb23604659, and Amb1153724) compounds based on lipophilicity, solubility, pharmacokinetics, medicinal chemistry, and drug-likeness properties.

All the compounds have maintained an optimum pharmacokinetics property except the compounds Amb6600135, which has negative Log $P_{o/w}$ value, active P-glycoprotein (P-GP) substrate (Figure S1) and violated the maximum Lipinski's rule of five (RO5) listed in Table 2. On the other hand, the synthesis accessibility of the compound (Amb6600135) was higher (difficult to synthesize) than the other three compounds. Therefore, the compound has not been considered for further stages of evaluation.

Table 2. List of pharmacokinetics (ADME) properties includes physicochemical properties, lipophilicity, water-solubility, drug-likeness, and medicinal chemistry of selected four compounds.

Properties		Amb6600135	Amb23604132	Amb23604659	Amb1153724
Physico-chemical Properties	MW (g/mol)	664.43	568.53	540.47	432.38
	Heavy atoms	44	42	40	31
	Aro. atoms	15	28	28	16
	Rotatable bonds	11	5	4	4
	H-bond acceptors	17	10	10	10
	H-bond donors	8	4	5	6
	TPSA (Å^2)	337.88	155.89	166.89	170.05
Lipophilicity	Log $P_{o/w}$ (Cons)	-5.39	4.37	3.70	0.45
Water Solubility	Log S (ESOL)	High	Soluble	Soluble	Moderate
Pharmacokinetics	GI absorption	Low	Moderate	Moderate	Low
	BBB permeant	No	No	No	No
	P-GP substrate	Yes	No	No	No
Drug likeness	Lipinski violations	3	1	1	1
Medi. Chemistry	Synth. accessibility	Medium	Easy	Easy	Medium

2.9. *Toxicity Test*

Analysis of toxicity is an important and one of the main steps in drug design that helps to identify the harmful effects of chemical substances on humans, animals, plants, or the environment. Traditional assessment of compounds toxicity requires in vivo animal model, which is time-consuming, expensive, and subject to be ethical considerations [25]. Therefore, computer-aided in silico toxicity measurement of chemical substances can be considered useful in the drug design process. The study utilized the ProTox-II web server to identify the toxicity of the compound computationally, as it is not time-consuming, non-expensive, and requires no ethical considerations. The three compounds (Amb23604132, Amb23604659, and Amb1153724) selected previously through different screening process have been submitted in the ProTox-II web server that determines the acute toxicity, hepatotoxicity, cytotoxicity, carcinogenicity, and mutagenicity of the compounds listed in Table 3. All three compounds have shown no oral toxicity or organ toxicity effect.

Table 3. List of compounds toxicity endpoints includes acute toxicity, hepatotoxicity, cytotoxicity, carcinogenicity, and mutagenicity of selected three compounds.

Classification	Target	Amb23604132	Amb23604659	Amb1153724
Oral toxicity	LD_{50} (mg/kg)	5000	5000	5000
	Toxicity Class	5	5	5
Organ toxicity	Hepatotoxicity	Inactive	Inactive	Inactive
Toxicity endpoints	Carcinogenicity	Inactive	Inactive	Inactive
	Mutagenicity	Inactive	Inactive	Inactive
	Cytotoxicity	Inactive	Inactive	Inactive

2.10. *Theoretical Calculation*

Geometry Optimization

Geometry optimization is a quantum chemical technique used by most computational biologist, chemists, academics, and researchers to find the configuration of minimum energy with the most stable form of a chemical structure. It is a method of taking rough geometric approximations and making them as exact as possible [26]. The geometry with the lowest energy is the most stable because molecules with lowest energy state spontaneously decrease its energy by emitting. Therefore, the best optimized molecular geometry with the lowest energy value has been determined by using the default basis set 6-31G(d,p) in Jaguar. The 2D structures and 3D optimized geometries of the compounds Amb23604659, Amb23604132, and Amb1153724 have been plotted in Figure 5. Additionally, the bond angles, bond lengths (bohr, angstroms), and torsional angles optimized during the process have been provided in Supplementary text file format (renamed as Geometry). The optimized structure has been retrieved for further evaluation through molecular docking simulation.

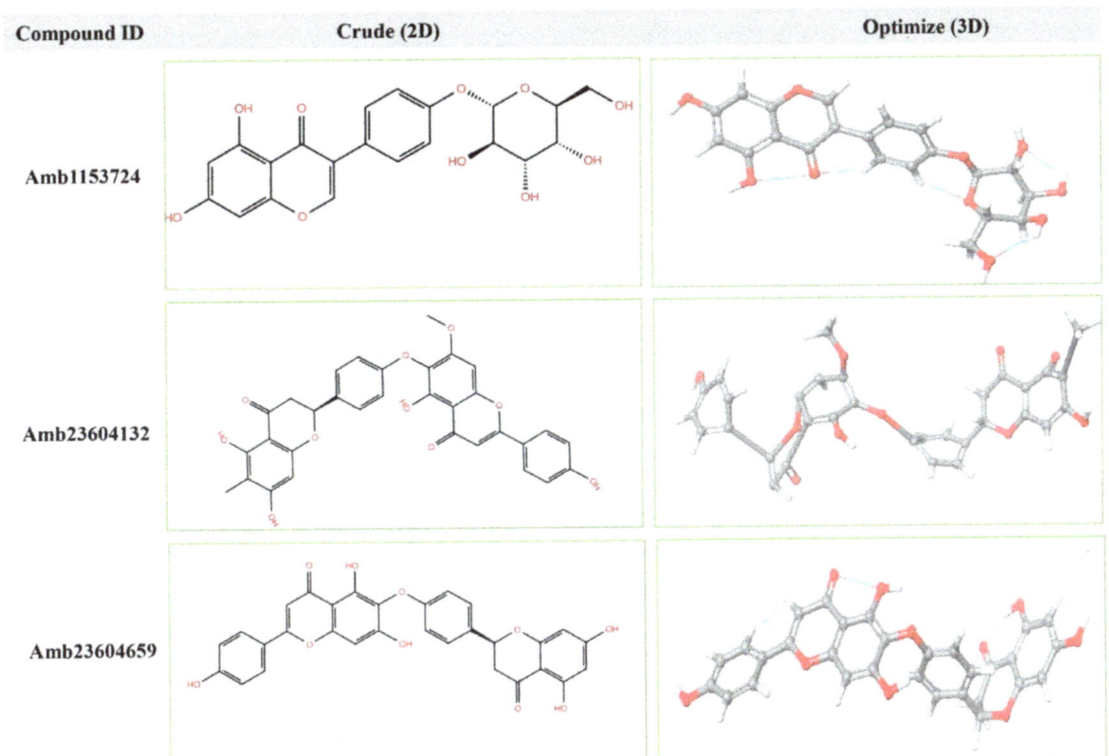

Figure 5. The 2D structures and 3D optimized molecular geometries of selected three compounds, Amb1153724, Amb23604132, and Amb23604659 calculated using B3LYP/6-31G(d,p) level of DFT calculations.

2.11. Frontier Molecular Orbital HOMO/LUMO Calculation

The FMO is now significantly used in organic chemistry to explain the structure and reactivity of molecules. The theory can describe the electronic and optical properties of molecules by utilizing HOMO-LUMO bandgap energy [27]. The energy gap between the two orbitals HOMO and LUMO also helps to determine the sensitivity of atoms toward electrophilic and nucleophilic attacks, chemical kinetic stability, chemical hardness, and softness of a molecule [26]. The electrons localized from the HOMO orbital is most free to participate in the nucleophilic reaction, where the LUMO participates in the electrophilic reaction. A molecule with low HOMO-LUMO gap energy should have a high chemical reactivity and low kinetic stability that can be considered as a soft molecule. In this process, a molecule with a high frontier (HOMO-LUMO) orbital gap should have low chemical reactivity or bioactive and high kinetic stability due to the low probability of adding an electron to the high-energy LUMO. The molecules with a high FMO energy gap are energetically stable related to low chemical reactivity and high kinetic stability compared to a molecule having a low FMO energy gap [27]. Therefore, to evaluate the chemical reactivity and kinetic stability of the selected three compounds the HOMO, LUMO, and HOMO-LUMO, gap energy was calculated from Equation (3) and shown in Figure 6, the hardness and softness of the molecules have also been calculated and listed in Table S2. The calculated FMO energy band gap values found for the compounds Amb1153724, Amb23604132, and Amb23604659 was 4.48 eV, 3.60 eV, and 4.35 eV, respectively, which was considerably higher, indicating kinetic stability and low chemical reactivity of the molecules.

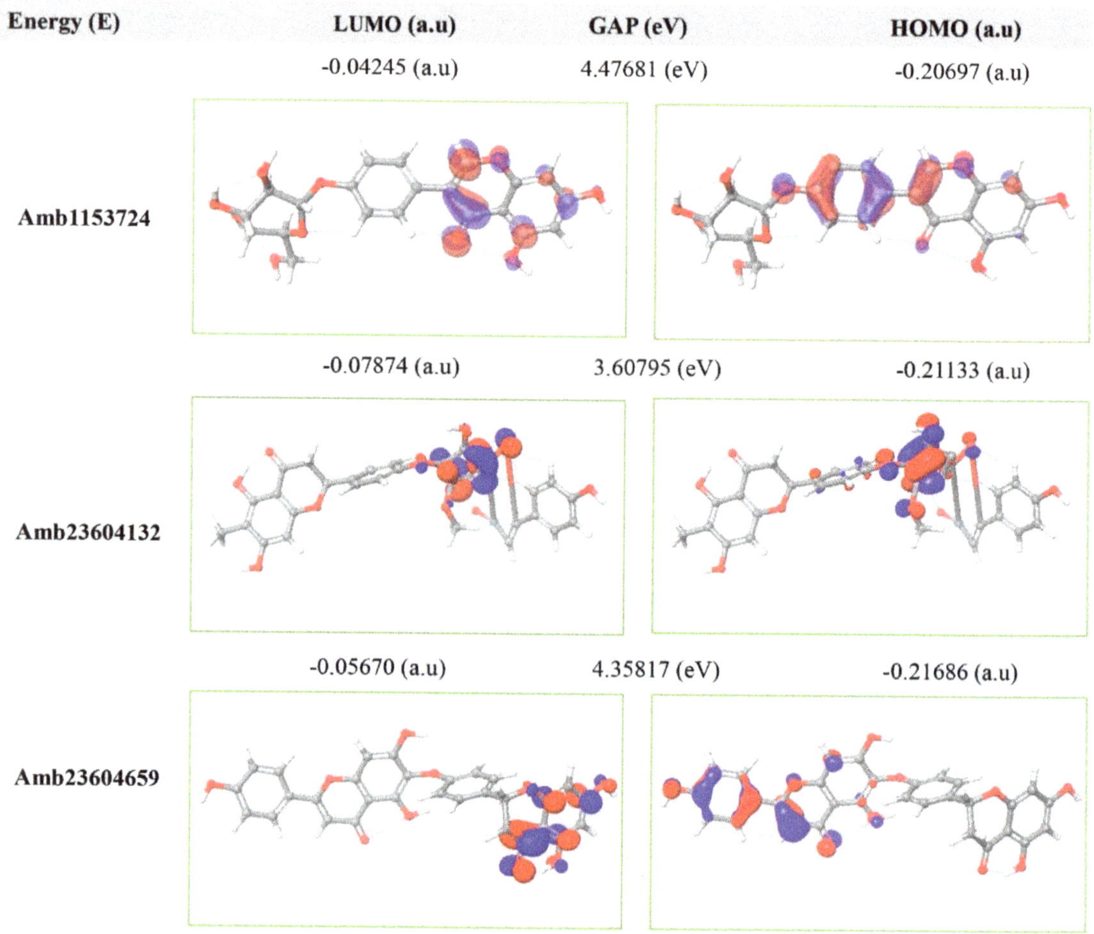

Figure 6. Representing the asymmetric HOMO, LUMO, and HOMO-LUMO gap energy for selected three natural compounds, where red is representing negative and blue represents positive phases of the molecular frontier orbital wave function.

2.12. Re-Docking, Interaction, and Pharmacophore Analysis

2.12.1. Redocking Score

The re-docking process has been performed to identify the possible docking poses in a restricted area by using the previously obtained binding sites of the S1 protein. The geometry optimized structure has been docked and the score found for the selected three compounds Amb23604132, Amb23604659, and Amb1153724 were −10.2 kcal/mol, −9.5 kcal/mol, and −9.2 kcal/mol, which was better than the previously obtained binding score (Table 1). Therefore, it can be considered that the QM-based optimization of the compounds was effective for the selected three compounds.

2.12.2. Protein–Ligands Interaction Interpretation

Understanding the potential interactions between a protein–ligand complex is an important part of the field of drug discovery, which helps to identify hits to leads as a potential drug candidate. Interaction analysis also helps to navigate the position of small molecules in a protein and determine the behavior on biological networks. Accurate identification of protein–ligand interactions play a key role in drug development and disease treatment. Therefore, the interaction between the selected three compounds and desire S1-NTD protein has been analyzed by using the BIOVIA Discovery Studio Visualizer tools. Analysis of the complex structure identified different bonding interaction includes hydrogen bond (Conventional H-B, Carbon H-B, and Pi-Donor H-B), electrostatic (Pi-Anion), hydrophobic (Alkyl, Pi-Alkyl, Pi-Pi T-shaped, and Pi-Sigma) between the protein and ligand listed in Table 4 and depicted in Figure 7.

Complex structure analysis of Amb1153724 found a total of eleven bonding interaction including seven hydrogen bonds (four conventional H-B, two carbon H-B, and one Pi-Donor H-B), one electrostatic (Pi-Anion), and three hydrophobic (two Pi-Alkyl, one Pi-Pi T-shaped) bonds to the different binding site residues of the protein, which have bonds distance range between minimum 1.99 Å to maximum 5.61 Å shown in Figure 7A and listed in Table 4.

The compounds Amb23604659 have been found to form a total of ten bands including five hydrogen bonds (four conventional H-B and one carbon H-B), one electrostatic (Pi-Anion), and four hydrophobic bonds (one Alkyl, two Pi-Alkyl, one Pi-Pi T-shaped) with the protein in different residual position having a distance between 2.0 Å to 5.10 Å shown in Figure 7B.

For the compound Amb23604659, a total of seven bonds have been found to form, which have a bond distance range between a minimum of 2.0 Å to a maximum of 5.10 Å. The compounds formed four conventional hydrogen bonds, an electrostatic (Pi-Anion) bond, and one Pi-Sigma hydrophobic bond with the S1 protein shown in Figure 7C.

Table 4. List of the interaction between the selected three compounds and MERS-CoV S1-NTD protein found during the complex structure analysis and generated through the docking simulation.

Compound	Residues	Bond Distance (Å)	Category	Bond Types
Amb1153724	GLN37	2.93083	Hydrogen Bond	Conventional H-B
	TRP44	1.96777	Hydrogen Bond	Conventional H-B
	HIS81	2.39932	Hydrogen Bond	Conventional H-B
	LYS42	1.99029	Hydrogen Bond	Conventional H-B
	GLN37	3.53458	Hydrogen Bond	Carbon H-B
	ASN104	3.56047	Hydrogen Bond	Carbon H-B
	ASP41	4.46727	Electrostatic	Pi-Anion
	MET84	2.6909	Hydrogen Bond	Pi-Donor H-B
	TYR314	5.61379	Hydrophobic	Pi-Pi T-shaped
	MET84	4.9701	Hydrophobic	Pi-Alkyl
	MET84	4.95866	Hydrophobic	Pi-Alkyl
Amb23604132	GLN37	2.54483	Hydrogen Bond	Conventional H-B
	LYS42	2.04393	Hydrogen Bond	Conventional H-B
	MET84	2.12397	Hydrogen Bond	Conventional H-B
	PHE40	2.49011	Hydrogen Bond	Conventional H-B
	ASP108	3.4188	Hydrogen Bond	Carbon H-B
	ASP108	4.6773	Electrostatic	Pi-Anion
	TYR314	5.74845	Hydrophobic	Pi-Pi T-shaped
	MET161	4.84458	Hydrophobic	Alkyl
	MET84	4.15196	Hydrophobic	Pi-Alkyl
	MET84	5.10244	Hydrophobic	Pi-Alkyl
Amb23604659	GLN37	2.99559	Hydrogen Bond	Conventional H-B
	ARG46	2.60436	Hydrogen Bond	Conventional H-B
	MET84	2.18527	Hydrogen Bond	Conventional H-B
	ASP41	2.59464	Hydrogen Bond	Conventional H-B
	ASP108	4.05492	Electrostatic	Pi-Anion
	MET161	3.87544	Hydrophobic	Pi-Sigma
	MET84	4.80802	Hydrophobic	Pi-Alkyl
	MET84	5.05152	Hydrophobic	Pi-Alkyl

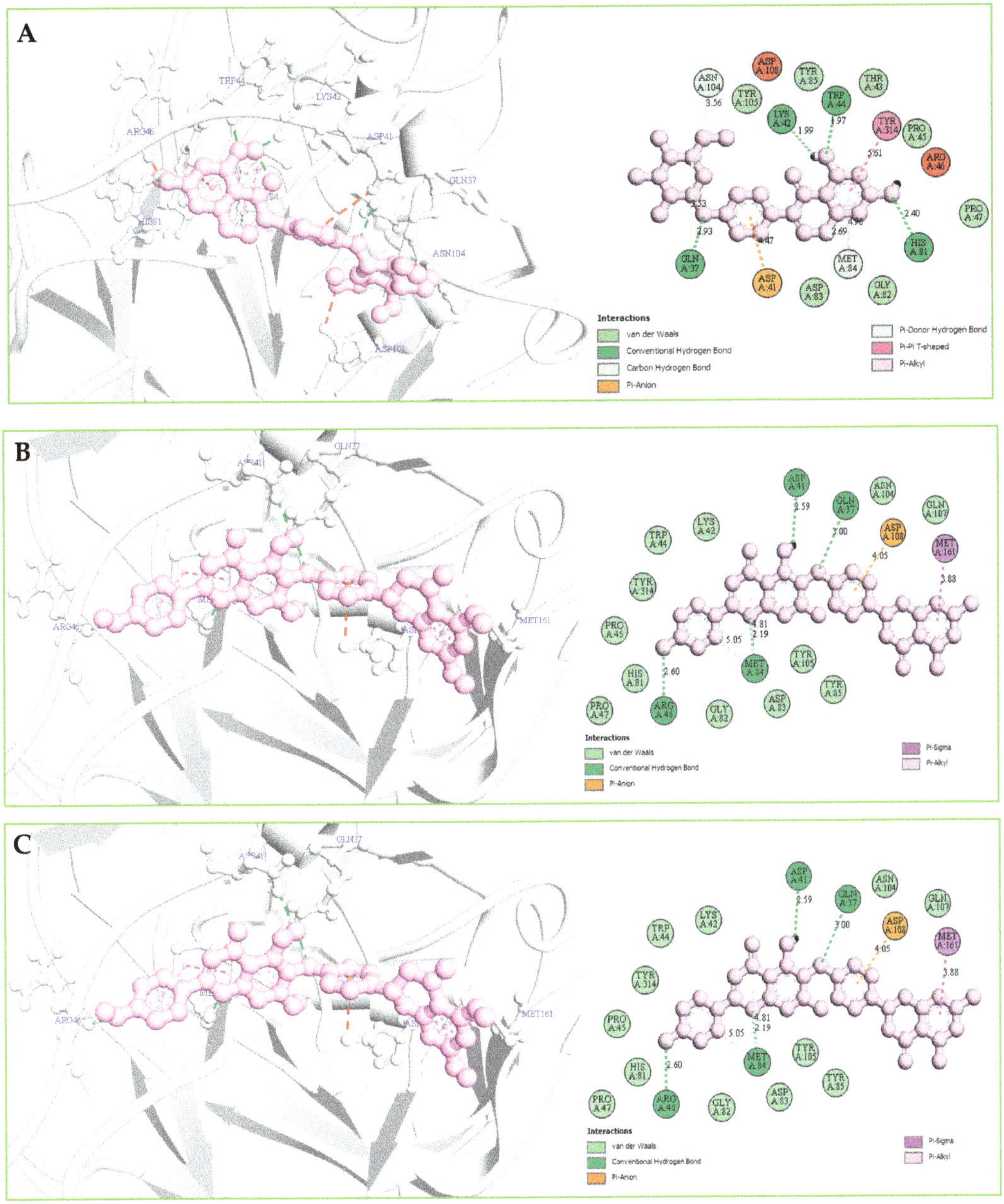

Figure 7. (**A**): The interaction between the MERS-CoV S1-NTD and Amb1153724 compounds. The 3D interaction has represented left side of the figure, where 2D interaction has depicted in right side of the figure accordingly. (**B**): The interaction between the MERS-CoV S1-NTD and Amb23604659 compounds. The 3D interaction has represented left side of the figure, where 2D interaction has depicted in right side of the figure accordingly. (**C**): The interaction between the MERS-CoV S1-NTD and Amb23604659 compounds. The 3D interaction has represented left side of the figure, where 2D interaction has been depicted in the right side of the figure accordingly.

2.12.3. Pharmacophore Features Analysis

Pharmacophore features of a compound play an important role during the molecular recognition process of targeted biological macromolecules. The pharmacophore of a compound can be described based on the H, AR, HBA or HBD, PI, NI features. These features can derive from the ligand or projected points believe to reside in the protein that helps to identify and design a new drug for the treatment of a selected disease.

These features retain the necessary geometric arrangement of atoms requires to producing a specific biological response. Therefore, the pharmacophore features of the selected three compounds include Amb23604659, Amb23604132, and Amb1153724 have been analyzed and compared with the query pharmacophore features shown in Figure 8. The 14 pharmacophore features used to screen the compounds generated 32 hits, which has further screened through the different screening process and identified three compounds as potential drug candidates. Each of the compounds generated 25 confirmations that have better pharmacophore properties than the query pharmacophore features. Therefore, the selected compounds should be effective to our target protein.

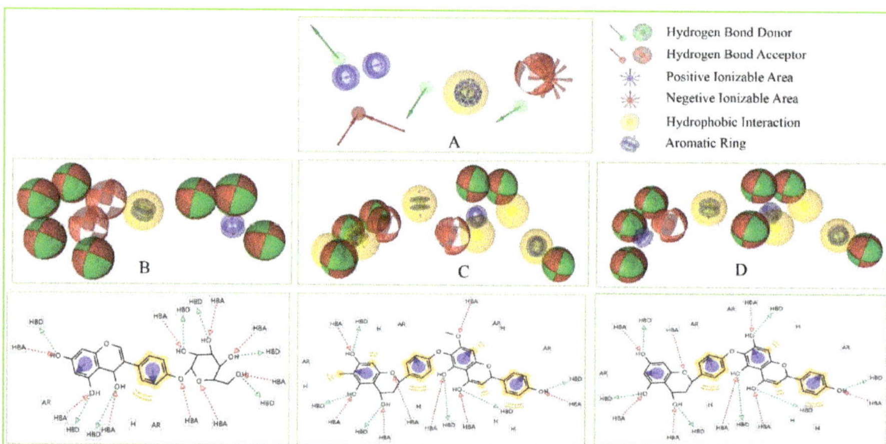

Figure 8. Showing the pharmacophore features of selected three compounds in 3D and 2D format along with the query features that have previously been used to screen the compounds. Herein, the figure representing (**A**) Query pharmacophore features and features generated from the compounds (**B**) Amb1153724, (**C**) Amb23604132, and (**D**) Amb23604659 during the screening process.

2.13. MD Simulations Analysis

MD simulations help to determine the physical movements of atoms and molecules by simulating the system at an atomistic scale. The invaluable technique for observing biomolecular structure and dynamics has expanded dramatically in recent years. The MD simulation offers a great and distinct approach to investigate the stability of a ligand to a targeted macromolecule. Therefore, to identify the stability of the selected three compounds with the desired protein, a 200 ns MD simulation has been performed for each complex structure and described based on the RMSD, RMSF, and protein–ligand contact mapping.

2.13.1. RMSD Analysis

RMSD of a protein–ligand complex system helps to determine the average distance generated through the dislocation of an elected atom during a specific time compared to a mentioned time [4]. RMSD of the selected three compounds has been observed to identify the changes in protein structure as compared to the starting point. It also helps to determine the equilibration state of the protein determined from the flattening of the RMSD curve. Initially, the protein frames and the reference frame backbone were aligned during the MD simulation and then the RMSD of the system has been calculated based on the atom selection. The complex system with a time frame x should have the RMSD that can be calculated from the following Equation (1).

$$RMSD_x = \sqrt{\frac{1}{N}\sum_{i=1}^{N}(r'_i(t_x)) - r_i\left(t_{ref}\right))^2} \quad (1)$$

Here, the $RMSD_x$ is the calculation of RMSD for the specific number of frames, N is the number of selected atoms; t_{ref} is the reference or mentioned time, and r' is the selected atom in the frame x after superimposing on the reference frame, t_x is the recording intervals.

The RMSD of the selected three compounds and the protein has been analyzed to determine the system has equilibrated or not. The RMSD of selected three compounds Amb23604659, Amb23604132, and Amb1153724 complex structure has been compared with the native S1 protein structure to observe the changes of the order shown in Figure 9. The RMSD for all the compounds was in a range between 1.0 Å to 2.5 Å that was perfectly acceptable compared to the structure of the native protein. The highest fluctuations (<3.0 Å) found for the compounds Amb1153724 between 185 ns to 200 ns simulation time and gradually stabilize, however, indicate that the compound has undergone a small conformational change during the simulation. It has been found that the simulation was converged between 20 ns to 160 ns for all the compounds and the RMSD values have been stabilized around a fixed value within the time. The fluctuations for all the selected compounds towards the end of the simulation were around some thermal average structure. Therefore, the selected compounds can be considered as stable to the targeted protein. Additionally, the RMSD for all the selected three ligands was observed to show how stable the ligand was concerning the desired protein and its binding site (Figure S3). The values observed for the ligand were closer than the RMSD of the S1-NTD protein, then it has been considered that the ligand will not be diffused away from its initial binding site.

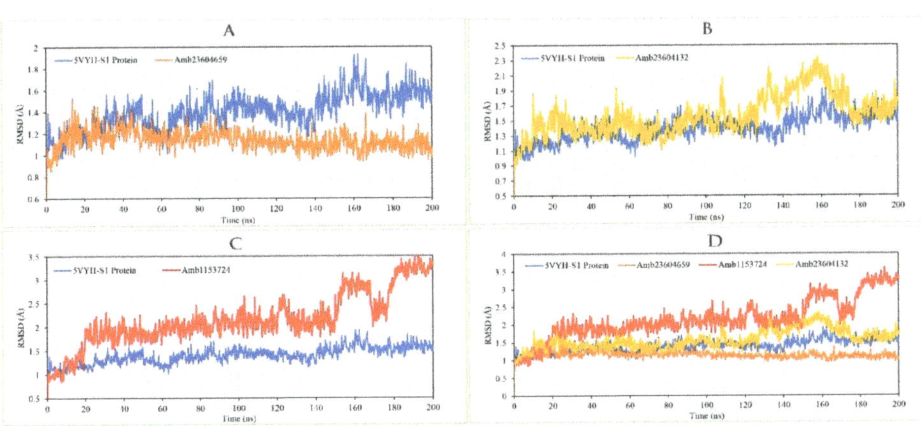

Figure 9. Depicted the RMSD values extracted from the Cα atoms of the selected three compounds in complex with the MERS-CoV S1-NTD protein. Herein, showing the RMSD of S1-NTD (Blue) in complex with the compounds (**A**) Amb23604659 (orange), (**B**) Amb23604132 (Yellow), and (**C**) Amb1153724 (Red), where (**D**) representing all the compounds and protein RMSD together.

2.13.2. RMSF Analysis

The RMSF is important to observe the local changes of a protein that helps to measure the displacement of a specific atom compared to the reference structure by calculating the average change observe over the number of atoms [19,22]. This is a numerical calculation like RMSD useful for characterizing a protein, which can determine the residue flexibility and fluctuation during the simulation. The RMSF for residue *i* has been calculated from the following Equation (2).

$$RMSF_i = \sqrt{\frac{1}{T}\sum_{t=1}^{T} < (r'_i(t)) - r_i\left(t_{ref}\right))^2 >} \quad (2)$$

where T is the overall trajectory time, r_i is the residue location, t_{ref} is the reference time, r' is the location of atoms in residue *i* after aligned on the reference, and the angle brackets (< >) are the average of the square distance.

The RMSF of the selected three complex structures has been analyzed to measure the displacement of a particular atom during the simulation. The RMSF of the selected three complex structures has been compared with the native S1-NTD protein structure to observe the atomic changes of the order shown in Figure 10. On this figure, the peaks indicate the protein fluctuation of the Amb23604659, Amb23604132, and Amb1153724 complex structure, which found minimal between 30 to 340 AA residue of the most rigid secondary structure elements includes alpha-helices and beta-strands. The highest fluctuation founds for all the three compounds before 30 AA and after 340 AA residue due to the location of the N- and C-terminal domain. Therefore, it can be considered that the displacement of a particular atom or a group of atoms will be lower in a real-life environment for all the selected three compounds.

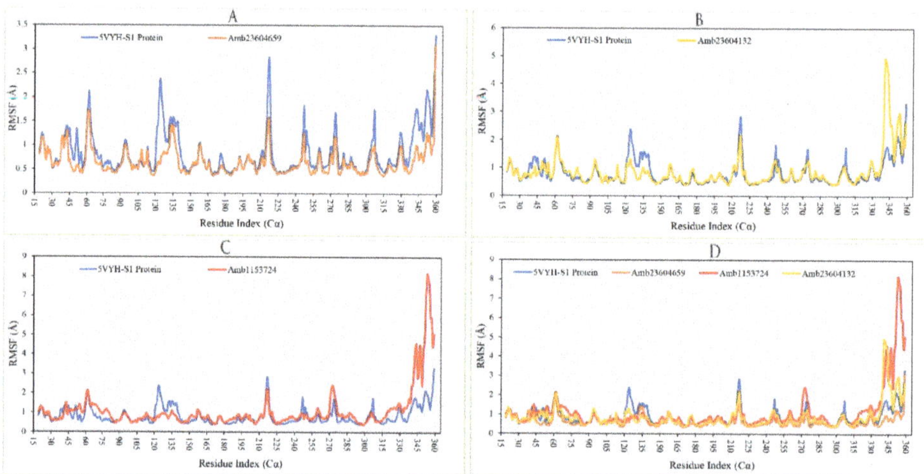

Figure 10. Showing the RMSF values extracted from the Cα atoms of the selected three complex structures. Herein, showing the RMSF of S1-NTD protein (Blue) in complex with the compounds (**A**) Amb23604659 (orange), (**B**) Amb23604132 (Yellow), and (**C**) Amb1153724 (Red), where (**D**) representing all the compounds and protein RMSF together.

2.13.3. Protein–Ligands Contact Analysis

Protein interactions with the selected three ligands Amb23604659, Amb23604132, and Amb1153724 have been monitored throughout the SID. The hydrogen bonds, hydrophobic, ionic, and water bridge interactions found during the MD simulation have been observed and shown in the stacked bar charts (Figure 11). The different types of bonding interaction play a significant role in ligand binding to the targeted protein, where hydrogen-bonding properties in drug design play an important role to influence drug specificity, metabolization, and adsorption. The hydrogen bonding interaction found for all three compounds was observable until the last AA residue during the simulation. For all the complex structures, it has also been found to form multiple interactions (hydrogen bonds, hydrophobic, ionic, and water bridges) at the same residue position of the protein with the ligand represented by a darker shade of orange, according to the scale to the right of the plot (Figure S2). The compound Amb23604659 generated multiple (more than two) interactions at ASP41, HIS81, MET84, TYR85, ASP108, VAL109, and LYS110 residues with an interaction fraction (IF) value 0.50, 0.45, 0.75, 0.25, 0.40, and 0.98, respectively indicating that 50%, 45%, 75%, 25%, 40%, and 98% of the simulation time the specific interaction is maintained by the multiple contacts of the same subtype with the ligand accordingly. The compound Amb23604132 formed multiple interaction at ASP41 (0.1), LYS42 (0.09), ARG62 (0.5), THR63 (0.68), HIS81 (0.1), LYS99 (0.38), GLN261 (0.2), TYR270 (0.05), GLN 280 (0.3) residues maintained by 10%, 9%, 50%, 68%, 10%, 38%, 20%, 5%, and 30% simulation time accordingly. In the case of the compound Amb1153724, it has found to form multiple interactions at the position of ASP41 (0.85), HIS81 (0.99), MET84 (1.3), and GLN107 (0.58) suggests that 85%, 99%, 130%, and 58% of the simulation time the specific interaction is maintained and helped to make a stable binding with the desired protein.

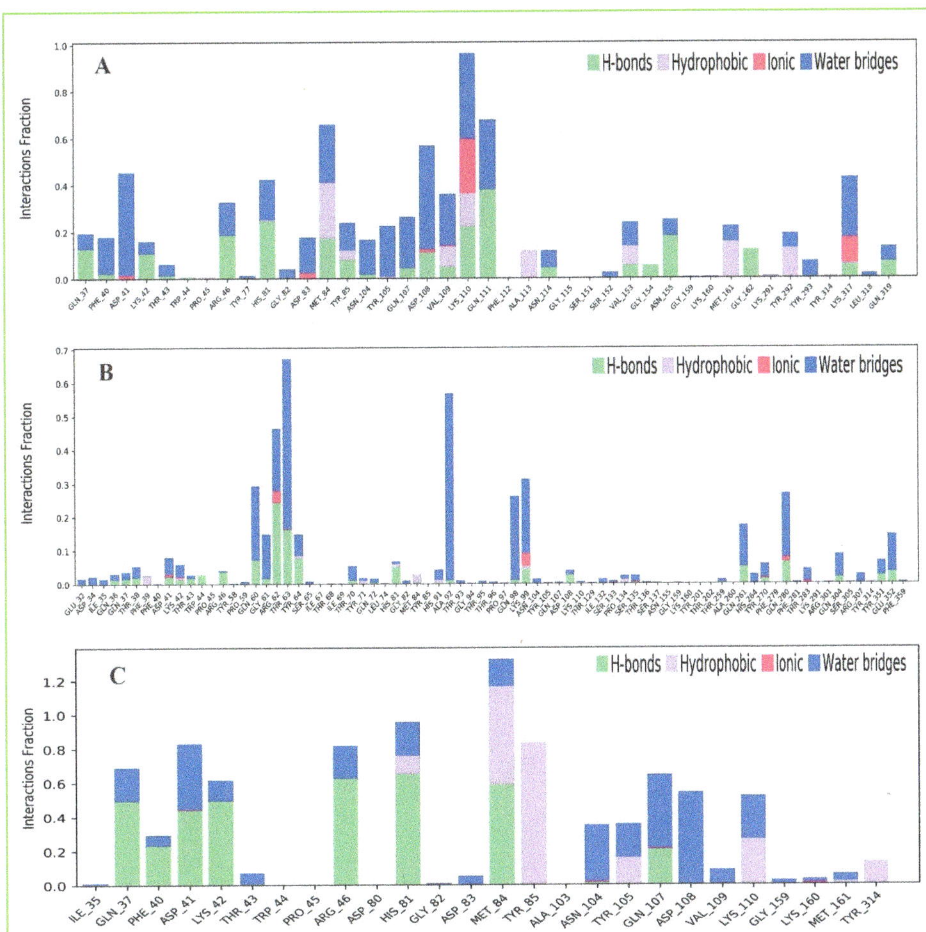

Figure 11. The stacked bar charts showing the protein–ligands interactions found during the 200 ns simulation run. Herein, showing the selected three ligands (**A**) Amb23604659, (**B**) Amb23604132, and (**C**) Amb1153724 contact map with the desire S1-NTD protein.

2.13.4. Ligand Properties Analysis

Ligand properties were analyzed to evaluate the stabilities of the selected three compounds Amb23604659, Amb23604132, and Amb1153724 under the MD simulation. The ligands properties were analyzed based on the RMSD of the ligands, Radius of Gyration (rGyr), Intramolecular Hydrogen Bonds (intraHB), Molecular Surface Area (MolSA), Solvent Accessible Surface Area (SASA), and Polar Surface Area (PSA), which found favorable for all the three compounds shown in Figure S3. Additionally, the selected three ligands and the co-crystal ligand (folic acid) RMSD have combined analysis and been compared, which has been provided in Figure S4.

2.14. MM/GBSA Analysis

The MM/GBSA methods have been used to calculate the end-point binding free energy of the protein–ligand complex. The MM/GBSA of the complex system has been calculated from the single trajectory collected from the respective 200 ns simulation (Table S3). Analysis of MM/GBSA found higher net negative binding free energy values for the selected compounds Amb1153724, Amb23604132, and Amb23604659 in complex with MERS-CoV S1-NTD protein (Figure 12). The complex analysis of MM/GBSA found −32.47 kcal/mol, −24.75 kcal/mol, and −26.18 kcal/mol binding free energy for the compounds Amb1153724, Amb23604132, and Amb23604659, respectively, at the last stage of the MD simulation. Therefore, the screened compounds will be able to maintain a durable interaction with the desired protein. Additionally, analysis of physico-chemical components for the selected three compounds revealed a significant contribution of $G_{\text{Bind Coulomb}}$ (Coulomb energy) and $G_{\text{Bind vdW}}$ (Van der Waals interaction energy) shown in Figure 11. From the above result it can be suggested that the selected three compounds can maintain a long-term interaction with the MERS-CoV S1-NTD protein binding site and result in inhibition of the desired protein.

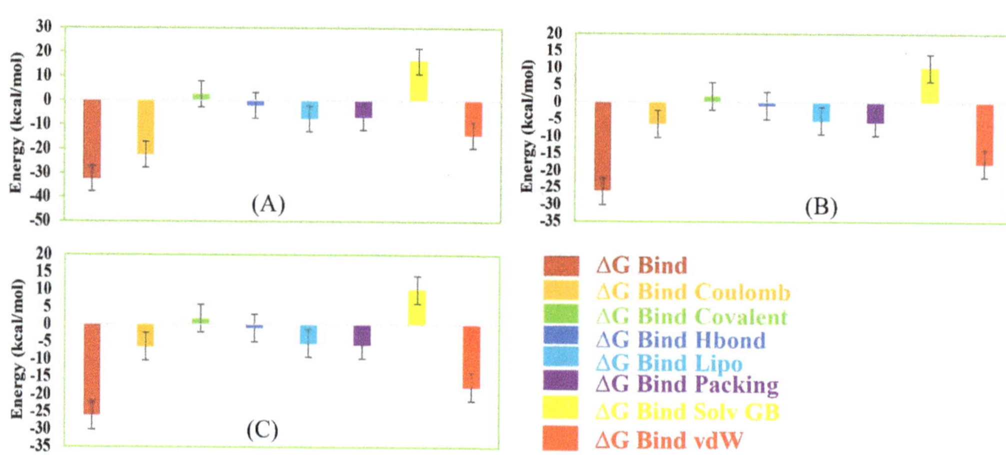

Figure 12. Representing different energy components and net MM/GBSA binding free energy (kcal/mol) with standard deviation values for extracted snapshots of MERS-CoV S1-NTD protein in complex with selected compounds, i.e., (**A**) Amb1153724, (**B**) Amb23604132, and (**C**) Amb23604659 from respective 200 ns MD simulation trajectories.

3. Discussion

Since the emergence of MERS-CoV in 2012, necessary steps to revoking the infection caused by the pathogen have become a major research focus. However, to date, no effective anti-viral drug candidates against the zoonotic pathogen have developed yet. It has been found that the distinctive NTD of the virus S1 subunit functioning as a RBD, which plays an important role to determine the host range resulting in cross-species infection [9]. Therefore, the study aimed to inhibit the function of S1-NTD of the virus to identify a novel and effective antiviral drug candidate against MERS-CoV infections.

In this study, we first identified the available experimental protein structure of MERS-CoV S1-NTD in complex with different inhibitory compounds from the protein databank. The MERS-CoV S1-NTD having a complex structure was used to generate an SBPM to the active site cavity of the protein. All the individual pharmacophore models generated from the complex structure have merged, and a combined final pharmacophore model has been generated to screen 11,295 natural compounds collected from the Ambinter natural compounds database. The pharmacophore model has validated using 12 experimentally known active compounds with their correspondence 1326 decoy compounds, where the AUC under the ROC curve indicated good discrimination ability of the model. The validated pharmacophore model has been utilized for the virtual screening process and retrieved 32 compounds as hits, which has been further screened through molecular docking simulation methods. Based on the molecular docking score, the top four compounds having a binding score of >8.0 kcal/mol have been chosen for further validation.

The selected four compounds Amb6600135, Amb23604132, Amb23604659, and Amb1153724 have been evaluated based on the ADME properties, where all compounds except the compound Amb6600135 have shown a good value of the ADME properties. The compound Amb6600135 disobeyed maximum (three) Lipinski's rule of five, on the other hand, the P-GP efflux pump was active of the compound (Figure S1), therefore the compound has skipped for further evaluation. The compound with good ADME properties has been further evaluated through the toxicity properties to measure the harmful effect on humans or animals. Analysis of toxicity found no or low toxicity of the selected three compounds.

To investigate and optimize the geometry of the compounds a computational DFT-based QM simulation has been performed. The geometry optimized through the DFT has been retrieved and re-docked with the desired protein, which exhibited substantial docking energy >−9.00 kcal/mol. The FMO based HOMO-LUMO energy gap was also calculated to evaluate the chemical reactivity of the compounds. The HOMO-LUMO gap energy found for all the compounds was high >3.50 eV which confirms the low reactivity correspondence to the bioactivity of the compounds.

The geometry optimized re-docked complex structure have been stimulated again by the MD simulation approach to identify the stability of the compounds to the binding site of the protein. The 200 ns simulation trajectories have been retrieved, and analysis based on the RMSD, RMSF, protein–ligand contact mapping, and ligand torsion properties (Figure S5) that confirm the stability of the compounds to the binding sites of the protein. Additionally, the MM/GBSA calculated from the single trajectory found a high ΔG_{bind} value, indicating the stability of the selected protein–ligand complex for long-term simulation.

4. Conclusions

To the best of our knowledge, this study offers the first compressive in-silico approaches to identify potential natural antiviral drug candidates against MERS-CoV S1-NTD. An integrative structure-based pharmacophore modeling, virtual screening, molecular docking, ADMET, QM calculation, MD simulation, and MM/GBSA approaches revealed Taiwanhomoflavone B, 2,3-Dihydrohinokiflavone, and Sophoricoside as potential drug candidates that will help to inhibit the activity of the S1-NTD of the virus. Further evaluation through different lab-based experiment techniques can help to determine the activity of the compound that will provide alternatives for MERS-CoV immunotherapy.

5. Materials and Methods

5.1. Pharmacophore Modeling

The crystallographic structure of MERS coronavirus S1-NTD submitted between 2012 to 2021 was searched in Protein Data Bank (PDB). A total of 16 S1-NTD protein structures generated through X-ray crystallographic method were identified from the PDB and filtered based on the protein resolution, having a range between 1 to 2.5 Å. After filtration six protein structures were retrieved, where two (PDB ID: 5VYH and 6PXH) protein structures were in a complex with potential inhibitors and selected as an input to generate a SBPM. This two-crystal structure corresponding to MERS coronavirus S1-NTD protein is complexed with its potential inhibitor Folic Acid (FOL409: A) and Dihydrofolic Acid (DHF428: A), were chosen to generate the pharmacophore model [10,13]. The protein structure in complex with different inhibitors was optimized by using the MMFF94 force field available at LigandScout 4.4 software [20]. The LigandScout tools were also used to generate and analyze the pharmacophore features originated from the two selected crystallographic structures based on protein–ligands interaction. Initially, two separate pharmacophore models were developed by using the protein 5VYH and 6PXH in complex with Folic Acid and Dihydrofolic Acid, respectively. The pharmacophore features generated from the complex structure were centrally coordinated for alignment perspectives. After center coordinates of all the structures, a final pharmacophore model was created by using the align and merge pharmacophore features option available at the LigandScout tool. The software generated a combined pharmacophore model and provided a fit value by aligning the molecule on the pharmacophore model. The pharmacophore features observed in the study were described based on Hydrogen Bond Donor (HBD), Hydrogen Bond Acceptor (HBA), Positive Ionizable Area (PI), Negative Ionizable Area (NI), Hydrophobic Interactions (H), and Aromatic Ring (AR) features.

5.2. *Molecule Library Preparation*

Ambinter (www.ambinter.com) is a brand and worldwide supplier of advanced chemicals that supporting the scientific community by providing active compounds for drug discovery. The Ambinter database contains over 36 million molecules including screening molecules as well as a large collection of natural compounds. Therefore, the database contains a targeted library of SARS-CoV-2 has retrieved for the further screening process. LigandScout can screen single or multi-conformational compounds from a large database for drug design and discovery. The tool can recognize and screen molecular libraries having the proprietary LDB as a file format. Hence, it is important to convert the compounds library into the LDB file format before the virtual screening. In this study, the LigandScout tools have been utilized to prepare the molecular library.

5.3. *Active Compounds Identification and Decoy Set Generation*

Experimentally validated active compounds against MERS coronavirus S1-NTD have been identified from the ChEMBL database (https://www.ebi.ac.uk/chembl, accessed on 03 April 2021) [28]. The active compounds identified from the ChEMBL database have been submitted to DUDE-E (Database of Useful Decoys: Enhanced) decoy database available at (http://dude.docking.org/, accessed on 03 April 2021) [29]. The DUDE-E decoy database identified and generated a decoys compounds list correspondence to the active compounds. The decoy compounds have been retrieved and converted into LDB file format by using the LigandScout tool for the validation of the pharmacophore model.

5.4. *Model Performance Analysis*

To determine the performance and ability of discrimination between an active set from a decoy set a receiver operating characteristics (ROC) curve has been developed in this study. From the ROC curves, the Area under (AUC) the ROC Curve has been evaluated, which helps to measure the 2D area underneath the entire ROC curve. The ROC curve also helps to evaluate the Enrichment Factor (EF) of the pharmacophore model.

5.5. *Virtual Screening*

In-silico 3D pharmacophore-based virtual screening of the molecule libraries has been performed by using the LigandScout virtual screening tools. The final pharmacophore model generated in this study has been utilized as filter criteria for the database screening process. During the multitude pharmacophore screening processes, the parameters pharmacophore-fit has been chosen as the scoring function, match all query features as the screening mode, and first matching confirmation as the retrieval mode, where a maximum of five pharmacophore features have been omitted. The excluded volume clashes generated during the pharmacophore modeling have not been checked in the screening process. Before running the pharmacophore-based virtual screening process the natural compounds database were marked as active, where decoy compounds have been marked as inactive databases. After finishing the screening process, hit compounds with the number of confirmations and geometric fit scores were investigated for further evaluation.

5.6. Protein and Ligands Preparation

The crystal structure of MERS-CoV S1-NTD has been retrieved from the RCSB (www.rcsb.org, accessed on 03 April 2021) protein data bank (PDB ID: 5VYH) consisting of 343 amino acids (AA) length with a resolution value of 2.00 Å [13]. The S1-NTD protein was prepared by removing water, metal ions, and cofactors from the complex structure. The nonpolar hydrogen atoms were merged, polar hydrogen atoms were added, and Gasteiger charges were calculated for the protein [19]. The hits generated during pharmacophore-based virtual screening have been retrieved and prepared by adding gasteiger charges and AD4 atom types to the molecules. The non-polar hydrogens were merged, and aromatic carbons were detected to setting up the 'torsion tree' of the molecules by using AutoDockTools and were saved in PDBQT format for the further screening process.

5.7. Binding Site Identification and Grid Box Generation

Binding sites can be identified through the analysis of similar pockets from known protein–ligands interaction. The known and experimental validated S1-NTD protein structure in complex with the ligand folic acid was retrieved from the PDB (PDB ID: 5VYH) and the binding site of the protein has been analyzed through BIOVIA Discovery Studio Visualizer v19.1 (BIOVIA). The binding site determined from the complex structure has been utilized for the receptor grid generation during the molecular docking simulation by using the PyRx virtual screening tool.

5.8. Molecular Docking Simulation

To identify the best hit candidates against the desired protein, a molecular docking simulation has been performed by using the PyRx tool [30]. PyRx is an open-source virtual screening tool that includes both AutoDock 4 and AutoDock Vina as a docking wizard which can screen a large compounds database against a specific biological targeted macromolecule. The AutoDock Vina wizard with default configuration parameters of PyRx has been used for molecular docking simulation. The top 10% compounds, having the highest binding affinity (kcal/mol) to the desired protein, have been chosen for further evaluation.

5.9. ADME Analysis

In the early stage of the drug design and development process assessment of ADME, it is necessary to understand the safety and efficacy of a drug candidate that will be processed by a living organism. The ADME properties describe pharmacokinetics behavior and the movement of drugs into, though, and out of the body. Traditionally, the ADME properties were evaluated at the last stage of the drug discovery process, but in-silico tools can be predicted the properties at the early stages of the drug design process and help to optimize the pharmacodynamic response. To evaluate and understand the pharmacodynamic response of selected drug candidates, the SwissADME (http://www.swissadme.ch, accessed on 03 April 2021) web tool has been used in this study [24]. The freely accessible web server helps to predict the physicochemical, pharmacokinetics, and drug-likeness properties of the selected drug candidates.

5.10. Toxicity Test

Toxicity testing in the drug design and development process is essential to evaluate the compound's toxic properties and the dose level requirements for the treatment of a specific disease. The toxicity profile of drug candidates gives an idea about the health and environmental risks and safety/toxicity of a chemical's substances. Nowadays, computer-aided in-silico toxicity testing is playing an important role in the assessment of compounds toxicity more accurately without using the experimental animal models. Therefore, to evaluate the early-stage toxicity of the selected drug candidates ProTox-II (http://tox.charite.de/protox_II, accessed on 03 April 2021) webserver has been used, which helps to determine acute toxicity, hepatotoxicity, cytotoxicity, carcinogenicity, mutagenicity, and immunotoxicity of the selected compounds [25].

5.11. Quantum Mechanics (QM)-Based Calculation

Conformation analysis of a ligand to the binding site of a protein is an essential part to identify potential active conformation, binding affinity, and strain discipline associated with the binding mechanism. This type of binding possess can be achieved through the calculation of minimum energy conformation and structural optimization, which is dependent on the solution phase and associated gas-phase energy. The classical molecular mechanics (MM) process is unable to describe the process properly due to the presentation of metal ions in a ligand–protein complex system [31]. In the last few years, QM-based calculations have helped to enhance the scoring functions that can describe the electronic structure, electronic changes, and system-specific charges during a reaction of a molecular system. Interestingly, more than 80–90% of all QM-based calculations are nowadays solve depend on density functional theory (DFT). Therefore, this study performed the DFT methods-based QM calculations of selected three compounds. Initially, the bond lengths, bond angles, and dihedral angles for potential compounds were optimized, then the DFT of the compounds has been calculated by using the Schrödinger Jaguar version 10.9 [27]. Calculation of DFT has been performed by utilizing a mix of conventional functionals Becke's three parameters with Lee-Yang-Parr functionals (B3LYP) and a dispersion correction energy term D3 combinedly known as B3LYP-D3. The conventional mix functionals B3LYP-D3 has been chosen in this study to not alter the wavefunction or any other molecular property directly, and 6-31G**, also known as 6-31G (d, p), has been chosen as a basis set to represent the electronic wave function of the molecules.

5.12. Frontier Molecular Orbital HOMO/LUMO Calculation

The highest energy occupied molecular orbital (HOMO) and lowest energy unoccupied molecular orbital (LUMO) are central to the frontier molecular orbital (FMO) theory or Fukui functions developed by Kenichi Fukui in the 1950s. The FMO of a molecule is the "frontier" of an electron that helps to determine the energy difference between two orbitals HOMO and LUMO. HOMO is mainly an electron donor (nucleophilic) and LUMO is an electron acceptor (electrophilic) in nature and the interaction between electron donor and electron acceptor pair can dominate other chemical reactivity of a molecule [32]. During the electrophilic-nucleophilic reaction, electrons from the HOMO jump to the LUMO and produce an energy difference between two molecular orbitals. The energy difference between two molecular orbitals is known as the HOMO-LUMO gap, which can explain the photochemistry and the strength and stability of transition metal complexes of organic molecules. To understand the sensitivity of atoms toward electrophilic and nucleophilic attacks, the HOMO and LUMO energy were calculated by using the Schrödinger Jaguar version 10.9 [27], and the energy difference between two molecular orbital HOMO-LUMO gaps was calculated from the following Equation (3).

$$\Delta E_{(gap)} = E_{LUMO} - E_{HOMO} \tag{3}$$

where, ΔE is the HOMO-LUMO gaps, E_{LUMO} is the lowest energy unoccupied molecular orbital energy, and E_{HOMO} is the highest energy occupied molecular orbital energy.

5.13. Re-Docking and Interaction Analysis

Geometry optimized through the DFT-based QM method of selected three compounds has been retrieved and docked again to the same binding site of the MERS-COV S1-NTD. A rigid molecule docking was conducted in this study due to rigid molecules that are not change their spatial shape during the docking process. The docking was performed through the PyRx tools AutoDock vina by using the default parameter as a setting [23]. Herein, the best binding poses with the lowest root-mean-square deviation (RMSD) have been selected for binding interaction analysis. The complex protein–ligand interaction has been analyzed through BIOVIA Discovery Studio tools.

5.14. MD Simulation

To analyze the physical movements and behavior of the selected compounds in the macromolecular environment, the protein–ligand complex structure obtained from the re-docking studies was evaluated through 200 ns MD simulations. The physical motions of atoms in the protein molecules have been observed through the Desmond module of Schrödinger (Release 2020-3) under a Linux environment [19]. Initially, the predefined simple point-charge (SPC) water model has been used to solvate the complex system for obtaining the correct density and dielectric permittivity of water. The boundary condition selected in this study was orthorhombic (box shape), and a buffer box has been chosen as a calculation method with a box distance of 15 Å. To obtain and maintain a salt concentration of 0.15 M, the system has been neutralized by adding Na+ and Cl- ions. The isothermal-isobaric (NPT) ensemble has been performed at constant pressure (1.01325 bar) and temperature (300 K) with an energy value of 1.2. The atomic movement of the molecules has been recorded for every 2 ps recording interval and OPLS-2005 is set as a force field to obtain the trajectory as an output for 200 ns simulation.

Analysis of MD Trajectory

The simulation snapshots for each atomic movement have been recorded for 2 PS intervals were rendered by using Schrödinger maestro interface v9.5. The simulation event has been analyzed through the Simulation Interaction Diagram (SID) available at the Schrödinger package. From the trajectory output, root-mean-square fluctuation (RMSF), RMSD, and protein–ligand contacts (P–L contact) have also been analyzed.

5.15. End-Point Binding Free Energy Calculation with MM/GBSA

MM/GBSA are nowadays getting more popularity for estimating ligand-binding affinities in many systems. They are typically based on the MD simulations of the receptor-ligand complex, which is more accurate than most scoring functions of molecular docking and computationally less demanding to alchemical-free energy methods [33]. Therefore, to estimate ligand-binding free energy (ΔG_{bind}) of the selected three compounds to the S1-NTD protein, the MM/GBSA methods have been performed by using the using Prime MM/GBSA module in the Schrödinger-Maestro package [26].

Supplementary Materials: The following are available online, Figure S1: Showing the blood brain barrier (BBB) and P-gp P-glycoprotein (P-GP) substrate activity of the selected four compounds, Amb6600135, Amb1153724, Amb23604132, and Amb23604659; Figure S2: Showing the contact mapping of the protein-ligands interactions for the selected three compounds found during the 200 ns simulation run. Herein, showing the selected three ligands (A) Amb23604659, (B) Amb23604132, and (C) Amb1153724 contact map with the desire S1-NTD protein; Figure S3: Depicted the RMSD ((Å), rGyr (Å), intra-HB, MolSA (Å2), SASA (Å2), and PSA (Å2 of the selected three compounds in complex with the MERS-CoV S1-NTD protein. Herein, showing the value of the compounds (A) Amb23604659, (B) Amb23604132, and (C) Amb1153724; Figure S4: The RMSD values the selected three complex structures and Folic acid in complex with the protein S1-NTD (PDB:5VYH). Herein, showing the RMSD of the compounds Amb23604659 (orange), Amb23604132 (Yellow), Amb1153724 (gray), and folic acid (blue) colors; and Figure S5: Depicted the torsion properties of the selected three compounds (A) Amb23604659, (B) Amb23604132, and (C) Amb1153724 during the 200 ns MD simulation run, Table S1: A list of 32 compounds generated as hits during pharmacophore based virtual screening process with there geometric fit score, conformation number and binding affinity (kcal/mol) with desire protein; Table S2: List of HOMO, LUMO, HOMO-LUMO gap, softness and hardness of the selected three compound; and Table S3: List of MM/GBSA component and their energy with standard error value of the selected three compound.

Author Contributions: Conceptualization, F.A., I.Q. and J.S.-G.; methodology, T.A.B., S.P., A.A. and A.M.A.; software, A.S., R.A. and F.A.; validation, T.A.B., S.P., A.A., K.A.-G., M.E.K.T. and M.S.H.; formal analysis, T.A.B., S.P., A.A., K.A.-G., M.E.K.T., A.M.A., A.S., R.A., M.S.H. and F.A.; investigation, F.A., I.Q. and J.S.-G.; resources, A.S., R.A. and F.A.; data curation, T.A.B., S.P., A.A. and A.M.A.; writing—original draft preparation, T.A.B., S.P., A.A., K.A.-G., M.E.K.T., A.M.A., A.S., R.A., M.S.H. and F.A.; writing—review and editing, A.S., R.A., I.Q., J.S.-G. and F.A.; visualization, F.A., T.A.B., S.P. and A.A.; supervision, F.A., I.Q. and J.S.-G. All authors have read and agreed to the published version of the manuscript.

Funding: This research received no external funding.

Institutional Review Board Statement: Not applicable.

Informed Consent Statement: Not applicable.

Data Availability Statement: Not applicable.

Acknowledgments: We extend our gratitude towards the Deanship of Scientific Research (DSR) at King Abdulaziz University and Biological Solution Centre (https://biosolcentre.org/) for providing technical support. We also thanks to the authority of King Abdul-Aziz University High Performance Computing Center (Aziz Supercomputer) (http://hpc.kau.edu.sa,) for providing facility in different computational work.

Conflicts of Interest: The authors declare no conflict of interest.

References

1. Fehr, A.R.; Perlman, S. Coronaviruses: An overview of their replication and pathogenesis. In *Coronaviruses: Methods and Protocols;* Springer: New York, NY, USA, 2015; Volume 1282, pp. 1–23. ISBN 9781493924387.
2. Cui, J.; Li, F.; Shi, Z.L. Origin and evolution of pathogenic coronaviruses. *Nat. Rev. Microbiol.* **2019**, *17*, 181–192. [CrossRef]
3. Cai, Z.; Lu, C.; He, J.; Liu, L.; Zou, Y.; Zhang, Z.; Zhu, Z.; Ge, X.; Wu, A.; Jiang, T.; et al. Identification and characterization of circRNAs encoded by MERS-CoV, SARS-CoV-1 and SARS-CoV-2. *Brief. Bioinform.* **2021**, *22*, 1297–1308. [CrossRef] [PubMed]
4. Samad, A.; Ahammad, F.; Nain, Z.; Alam, R.; Imon, R.R.; Hasan, M.; Rahman, M.S. Designing a multi-epitope vaccine against SARS-CoV-2: An immunoinformatics approach. *J. Biomol. Struct. Dyn.* **2020**, 1–17. [CrossRef] [PubMed]
5. Cascella, M.; Rajnik, M.; Cuomo, A.; Dulebohn, S.C.; Di Napoli, R. *Features, Evaluation and Treatment Coronavirus (COVID-19);* StatPearls Publishing, 2020.
6. Wu, J.T.; Leung, K.; Leung, G.M. Nowcasting and forecasting the potential domestic and international spread of the 2019-nCoV outbreak originating in Wuhan, China: A modelling study. *Lancet* **2020**, *395*, 689–697. [CrossRef]
7. O'Driscoll, M.; Ribeiro Dos Santos, G.; Wang, L.; Cummings, D.A.T.; Azman, A.S.; Paireau, J.; Fontanet, A.; Cauchemez, S.; Salje, H. Age-specific mortality and immunity patterns of SARS-CoV-2. *Nature* **2021**, *590*, 140–145. [CrossRef]
8. Park, Y.J.; Walls, A.C.; Wang, Z.; Sauer, M.M.; Li, W.; Tortorici, M.A.; Bosch, B.J.; DiMaio, F.; Veesler, D. Structures of MERS-CoV spike glycoprotein in complex with sialoside attachment receptors. *Nat. Struct. Mol. Biol.* **2019**, *26*, 1151–1157. [CrossRef] [PubMed]
9. Yuan, Y.; Cao, D.; Zhang, Y.; Ma, J.; Qi, J.; Wang, Q.; Lu, G.; Wu, Y.; Yan, J.; Shi, Y.; et al. Cryo-EM structures of MERS-CoV and SARS-CoV spike glycoproteins reveal the dynamic receptor binding domains. *Nat. Commun.* **2017**, *8*, 15092. [CrossRef]
10. Wang, N.; Rosen, O.; Wang, L.; Turner, H.L.; Stevens, L.J.; Corbett, K.S.; Bowman, C.A.; Pallesen, J.; Shi, W.; Zhang, Y.; et al. Structural Definition of a Neutralization-Sensitive Epitope on the MERS-CoV S1-NTD. *Cell Rep.* **2019**, *28*, 3395–3405. [CrossRef]
11. Du, L.; Yang, Y.; Zhou, Y.; Lu, L.; Li, F.; Jiang, S. MERS-CoV spike protein: A key target for antivirals. *Expert Opin. Ther. Targets* **2017**, *21*, 131–143. [CrossRef]
12. Rahman, S.M.M.; Atikullah, M.; Islam, M.N.S.; Mohaimenul, M.; Ahammad, F.; Islam, M.N.S.; Saha, B.; Rahman, M.H. Anti-inflammatory, antinociceptive and antidiarrhoeal activities of methanol and ethyl acetate extract of Hemigraphis alternata leaves in mice. *Clin. Phytoscience* **2019**, *5*, 16. [CrossRef]

13. Pallesen, J.; Wang, N.; Corbett, K.S.; Wrapp, D.; Kirchdoerfer, R.N.; Turner, H.L.; Cottrell, C.A.; Becker, M.M.; Wang, L.; Shi, W.; et al. Immunogenicity and structures of a rationally designed prefusion MERS-CoV spike antigen. *Proc. Natl. Acad. Sci. USA* **2017**, *114*, E7348–E7357. [CrossRef] [PubMed]
14. Kleine-Weber, H.; Elzayat, M.T.; Wang, L.; Graham, B.S.; Müller, M.A.; Drosten, C.; Pöhlmann, S.; Hoffmann, M. Mutations in the Spike Protein of Middle East Respiratory Syndrome Coronavirus Transmitted in Korea Increase Resistance to Antibody-Mediated Neutralization. *J. Virol.* **2018**, *93*. [CrossRef] [PubMed]
15. Trigueiro-Louro, J.; Correia, V.; Figueiredo-Nunes, I.; Gíria, M.; Rebelo-de-Andrade, H. Unlocking COVID therapeutic targets: A structure-based rationale against SARS-CoV-2, SARS-CoV and MERS-CoV Spike. *Comput. Struct. Biotechnol. J.* **2020**, *18*, 2117–2131. [CrossRef] [PubMed]
16. Leelananda, S.P.; Lindert, S. Computational methods in drug discovery. *Beilstein J. Org. Chem.* **2016**, *12*, 2694–2718. [CrossRef] [PubMed]
17. Agamah, F.E.; Mazandu, G.K.; Hassan, R.; Bope, C.D.; Thomford, N.E.; Ghansah, A.; Chimusa, E.R. Computational/in silico methods in drug target and lead prediction. *Brief. Bioinform.* **2020**, *21*, 1663–1675. [CrossRef] [PubMed]
18. Pal, R.; Kumar, A.; Misra, G. Exploring TEAD2 as a drug target for therapeutic intervention of cancer: A multi-computational case study. *Brief. Bioinform.* **2021**, *2021*, 1–10. [CrossRef]
19. Opo, F.A.D.M.; Rahman, M.M.; Ahammad, F.; Ahmed, I.; Bhuiyan, M.A.; Asiri, A.M. Structure based pharmacophore modeling, virtual screening, molecular docking and ADMET approaches for identification of natural anti-cancer agents targeting XIAP protein. *Sci. Rep.* **2021**, *11*, 4049. [CrossRef]
20. Wolber, G.; Langer, T. LigandScout: 3-D pharmacophores derived from protein-bound ligands and their use as virtual screening filters. *J. Chem. Inf. Model.* **2005**, *45*, 160–169. [CrossRef]
21. Ahammad, F.; Fuad, F.A.A. The in silico identification of potent natural bioactive anti-dengue agents by targeting the human hexokinase 2 enzyme. In Proceedings of the 5th International Electronic Conference on Medicinal Chemistry, Basel, Switzerland, 30 October 2019; p. 6342.
22. Ahammad, F.; Alam, R.; Mahmud, R.; Akhter, S.; Talukder, E.K.; Tonmoy, A.M.; Fahim, S.; Al-Ghamdi, K.; Samad, A.; Qadri, I. Pharmacoinformatics and molecular dynamics simulation-based phytochemical screening of neem plant (Azadiractha indica) against human cancer by targeting MCM7 protein. *Brief. Bioinform.* **2021**, *2021*, 1–15. [CrossRef]
23. Adyani, F.; Fuad, A.; Ahammad, F. Virtual Screening and Molecular Docking Approaches for Identification of Natural Lead Compounds for Improved Anti-Dengue Therapeutics. *Kerice* **2019**. [CrossRef]
24. Daina, A.; Michielin, O.; Zoete, V. SwissADME: A free web tool to evaluate pharmacokinetics, drug-likeness and medicinal chemistry friendliness of small molecules. *Sci. Rep.* **2017**, *7*, 42717. [CrossRef]
25. Banerjee, P.; Eckert, A.O.; Schrey, A.K.; Preissner, R. ProTox-II: A webserver for the prediction of toxicity of chemicals. *Nucleic Acids Res.* **2018**, *46*, W257–W263. [CrossRef]
26. Bharadwaj, S.; Dubey, A.; Yadava, U.; Mishra, S.K.; Kang, S.G.; Dwivedi, V.D. Exploration of natural compounds with anti-SARS-CoV-2 activity *via* inhibition of SARS-CoV-2 Mpro. *Brief. Bioinform.* **2021**, *2021*, 1–17. [CrossRef]
27. Bochevarov, A.D.; Harder, E.; Hughes, T.F.; Greenwood, J.R.; Braden, D.A.; Philipp, D.M.; Rinaldo, D.; Halls, M.D.; Zhang, J.; Friesner, R.A. Jaguar: A high-performance quantum chemistry software program with strengths in life and materials sciences. *Int. J. Quantum Chem.* **2013**, *113*, 2110–2142. [CrossRef]
28. Bento, A.P.; Gaulton, A.; Hersey, A.; Bellis, L.J.; Chambers, J.; Davies, M.; Krüger, F.A.; Light, Y.; Mak, L.; McGlinchey, S.; et al. The ChEMBL bioactivity database: An update. *Nucleic Acids Res.* **2014**, *42*, D1083–D1090. [CrossRef]
29. Mysinger, M.M.; Carchia, M.; Irwin, J.J.; Shoichet, B.K. Directory of useful decoys, enhanced (DUD-E): Better ligands and decoys for better benchmarking. *J. Med. Chem.* **2012**, *55*, 6582–6594. [CrossRef]
30. Dallakyan, S.; Olson, A.J. Small-molecule library screening by docking with PyRx. *Methods Mol. Biol.* **2015**, *1263*, 243–250. [CrossRef]
31. Friesner, R.A.; Guallar, V. Ab initio quantum chemical and mixed quantum mechanics/molecular mechanics (QM/MM) methods for studying enzymatic catalysis. *Annu. Rev. Phys. Chem.* **2005**, *56*, 389–427. [CrossRef]
32. Li, Y.; Evans, J.N.S. The Fukui Function: A Key Concept Linking Frontier Molecular Orbital Theory and the Hard-Soft-Acid-Base Principle. *J. Am. Chem. Soc.* **1995**, *117*, 7756–7759. [CrossRef]
33. Çlnaroğlu, S.S.; Timuçin, E. Comprehensive evaluation of the MM-GBSA method on bromodomain-inhibitor sets. *Brief. Bioinform.* **2020**, *21*, 2112–2125. [CrossRef]

molecules

MDPI

Article

Severe Acute Respiratory Syndrome Coronavirus-2 Inactivation Activity of the Polyphenol-Rich Tea Leaf Extract with Concentrated Theaflavins and Other Virucidal Catechins

Yohei Takeda [1,2], Kyohei Tamura [2], Dulamjav Jamsransuren [2], Sachiko Matsuda [2] and Haruko Ogawa [2,*]

1 Research Center for Global Agromedicine, Obihiro University of Agriculture and Veterinary Medicine, 2-11 Inada, Obihiro 080-8555, Japan; ytakeda@obihiro.ac.jp

2 Department of Veterinary Medicine, Obihiro University of Agriculture and Veterinary Medicine, 2-11 Inada, Obihiro 080-8555, Japan; s16110021@st.obihiro.ac.jp (K.T.); duuya.dj@gmail.com (D.J.); chaka@obihiro.ac.jp (S.M.)

* Correspondence: hogawa@obihiro.ac.jp; Tel.: +81-155-49-5893

Abstract: Since severe acute respiratory syndrome coronavirus 2 (SARS-CoV-2) is producing a large number of infections and deaths globally, the development of supportive and auxiliary treatments is attracting increasing attention. Here, we evaluated SARS-CoV-2-inactivation activity of the polyphenol-rich tea leaf extract TY-1 containing concentrated theaflavins and other virucidal catechins. The TY-1 was mixed with SARS-CoV-2 solution, and its virucidal activity was evaluated. To evaluate the inhibition activity of TY-1 in SARS-CoV-2 infection, TY-1 was co-added with SARS-CoV-2 into cell culture media. After 1 h of incubation, the cell culture medium was replaced, and the cells were further incubated in the absence of TY-1. The viral titers were then evaluated. To evaluate the impacts of TY-1 on viral proteins and genome, TY-1-treated SARS-CoV-2 structural proteins and viral RNA were analyzed using western blotting and real-time RT-PCR, respectively. TY-1 showed time- and concentration-dependent virucidal activity. TY-1 inhibited SARS-CoV-2 infection of cells. The results of western blotting and real-time RT-PCR suggested that TY-1 induced structural change in the S2 subunit of the S protein and viral genome destruction, respectively. Our findings provided basic insights in vitro into the possible value of TY-1 as a virucidal agent, which could enhance the current SARS-CoV-2 control measures.

Keywords: catechin; polyphenol; SARS-CoV-2; tea leaf extract; theaflavin; virucidal activity

Citation: Takeda, Y.; Tamura, K.; Jamsransuren, D.; Matsuda, S.; Ogawa, H. Severe Acute Respiratory Syndrome Coronavirus-2 Inactivation Activity of the Polyphenol-Rich Tea Leaf Extract with Concentrated Theaflavins and Other Virucidal Catechins. *Molecules* **2021**, *26*, 4803. https://doi.org/10.3390/molecules26164803

Academic Editors: Giovanni N. Roviello and Caterina Vicidomini

Received: 19 June 2021
Accepted: 5 August 2021
Published: 8 August 2021

1. Introduction

Severe acute respiratory syndrome coronavirus 2 (SARS-CoV-2) is an enveloped single-stranded RNA virus belonging to the genus *Betacoronavirus* and was first identified in December of 2019 in China [1,2]. As the causative agent of the coronavirus disease 2019 (COVID-19) pandemic, it is still causing a very large number of infections and deaths worldwide [3]. Restricting the spread of pathogenic viruses, reducing the number of infections and deaths, and eventually ending the pandemic are clearly important. There are three main types of virus control measures: (1) prevention of respiratory and contact transmissions by reducing the chance of virus inhalation and removing viruses from environmental surfaces and hands using disinfectants and hand washing, (2) treatment of infections through the use of medication, and (3) induction of herd immunity by vaccination.

Many disinfectants that can efficiently inactivate SARS-CoV-2 on surfaces in a short time have been reported [4–6]. However, the main transmission mode of SARS-CoV-2 is respiratory infection via droplet inhalation, and contact transmission is considered to be rare [7]; therefore, the effect of inactivating viruses on environmental surfaces and hands might be limited.

With respect to point 2, many types of approved or experimental drugs have been investigated for COVID-19 treatment. The main types of drugs are RNA-dependent

RNA polymerase blockers such as remdesivir; neutralizing monoclonal antibodies; and corticosteroids such as dexamethasone, Janus kinase inhibitor, and anti-IL-6 receptor antibody, which inhibit excessive inflammation induced by SARS-CoV-2 infection [8,9]. However, the therapeutic efficacies of these drugs are not absolute and are not effective in all COVID-19 cases.

Vaccines are the third line of defense against viruses, and several different vaccines have been approved for use in different countries. However, vaccination status differs among countries, and there remain many countries in which widespread vaccination will not be achieved for several months or more. Some variant strains, especially the B.1.351 lineage variant (Beta strain), which was first reported in South Africa, have shown a capacity to escape from the neutralizing antibodies established using the original strain [10,11]. These findings indicate the concern that herd immunization against novel variant strains which will appear in the future may not be achievable using currently available vaccines. The current SARS-CoV-2 control measures are therefore still insufficient to bring the pandemic under control, and the development of additional supportive and auxiliary approaches, especially those that are easily implemented by the general public, are required.

The application of naturally derived, especially plant-based, compounds for supportive and auxiliary treatments is attracting increasing attention. Phytochemicals have a wide variety of biological activities, including broad antimicrobial activities. Medicinal plant-derived compounds tend to be low toxic, and more than 50% of medical drugs used in Western countries are derived from plant-based compounds. Several antiviral phytochemicals have been identified, and the potential application of these compounds to the treatment of viral diseases has been suggested [12]. Phenolics, carotenoids, terpenoids, and alkaloids are well-known antiviral phytochemicals. In the phenolics family, phenolic acids, flavonoids, stilbenes, coumarins, and tannins are known to have antiviral activities [13,14]. The virucidal and antiviral activities of naturally derived polyphenols are widely known; they have been reported to prevent viral absorption and entry to host cells and inhibit viral genome and protein synthesis in infected cells [14]. Catechin is one of the flavonoids contained in the leaves of green tea (*Camellia sinensis*). Green tea catechins, including catechin, epicatechin (EC), epicatechin gallate (ECG), epigallocatechin (EGC), and epigallocatechin gallate (EGCG), have antitumor, antioxidative, antibacterial, and antiviral activities against multiple pathogenic virus species [15]. Fermented black tea contains theaflavins (TFs), which are dimers of catechins with a benzophenone structure. Although TFs are considered to show more potent biological activities than green tea catechins and the quality of black tea partially depends on TF content, TF content is low in many black teas [16,17]. The chemical structures of TFs and green tea catechins are described in an article by Liu et al. [18]. Since TFs, as well as green tea catechins, have been reported to show antiviral activity against various viral species [19], these antiviral catechin-rich components may be valuable as SARS-CoV-2 control measures. In this study, we evaluated the SARS-CoV-2 inactivation activity of a polyphenol-rich tea leaf extract, named TY-1, containing green tea catechins and concentrated TFs [20], and discuss its possible usage for the prevention or treatment of COVID-19.

2. Results

2.1. Virucidal Activity of TY-1 against SARS-CoV-2

The virucidal activities of solutions with different concentrations of TY-1 against the ancestral strain and multiple variant strains of SARS-CoV-2 were evaluated. TY-1 solution showed time- and concentration-dependent virucidal activity against the ancestral strain. At 3-h to 24-h reaction times, the viral titers in all tested concentrations of TY-1 were significantly lower than those of the dextrin group. At 6-h and 24-h reaction times, the viral titers in the 2.5 and 5.0 mg/mL TY-1 groups were almost below the detection limit (2.5 mg/mL TY-1: $\geq$ 4.38 $\log_{10}$ 50% tissue culture infective dose [$TCID_{50}$]/mL reduction; 5.0 mg/mL: $\geq$ 4.50 $\log_{10}$ $TCID_{50}$/mL reduction compared with the dextrin group at 6 h). At 10-min reaction time, the significant reductions of viral titers in the 1.3, 2.5, and 5.0 mg/mL

TY-1 groups were 0.63, 1.17, and 2.54 $\log_{10}$ $TCID_{50}$/mL, respectively. The 5.0 mg/mL TY-1 group also showed statistically significant virucidal activity at 1-min reaction time, and the reduction of the viral titer was 1.5 $\log_{10}$ $TCID_{50}$/mL (Figure 1a). The significant reductions of viral titers in the 5.0 mg/mL TY-1 group were 2.31, $\geq$ 3.34, $\geq$ 3.19, 1.50, and 1.50 $TCID_{50}$/mL against Alpha, Beta, Gamma, Delta, and Kappa strains at 10-min reaction time, respectively (Figure 1b–f).

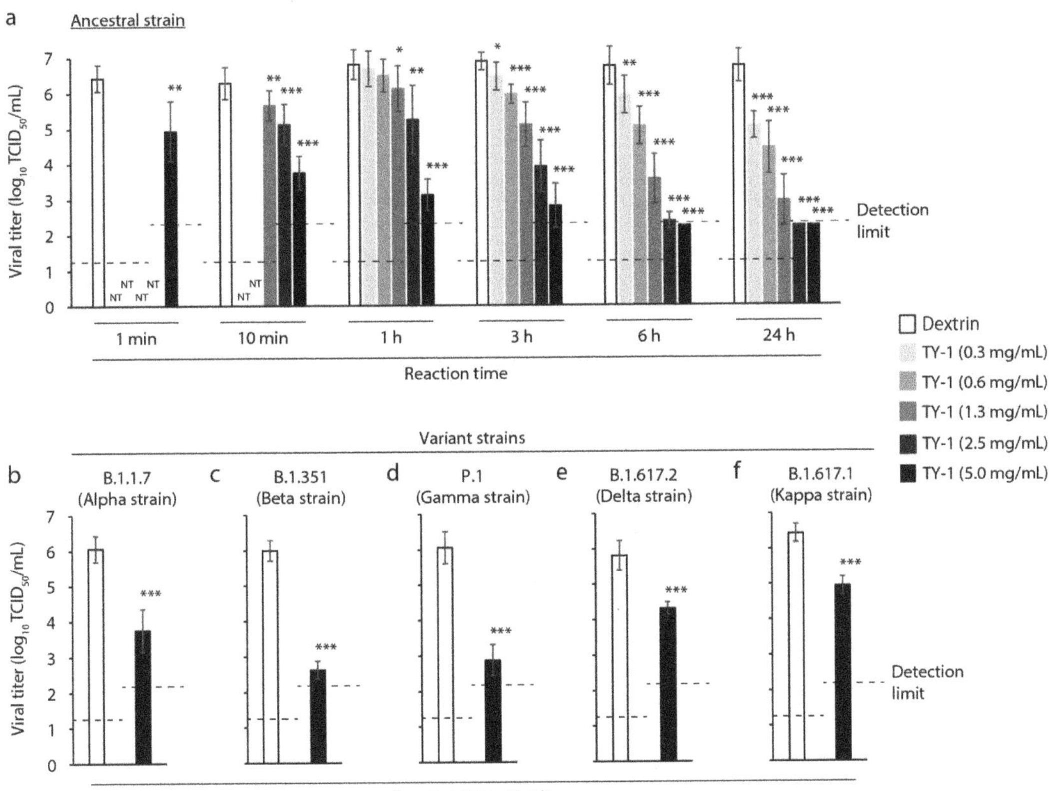

Figure 1. Virucidal activity of TY-1 against SARS-CoV-2. (**a**)–(**f**) SARS-CoV-2 solution ((**a**) ancestral strain, (**b**) Alpha strain, (**c**) Beta strain, (**d**) Gamma strain, (**e**) Delta strain, (**f**) Kappa strain) was mixed with TY-1 at several different concentrations. As a diluent control, dextrin (2.5 mg/mL) was mixed with the viral solution. The mixtures were incubated at 22–25 °C from 1 min to 24 h (**a**) or for 10 min (**b**)–(**f**); then, the viral titers were evaluated. The detection limits of the viral titer were 1.25 $\log_{10}$ $TCID_{50}$/mL in the dextrin and TY-1 (0.3–1.3 mg/mL) groups and 2.25 $\log_{10}$ $TCID_{50}$/mL in the TY-1 (2.5 and 5.0 mg/mL) groups. The results are indicated as mean ± SD (n = 4–12 per group). Student's t-tests were performed to evaluate the statistical significance of the differences between the dextrin and each TY-1 groups; * $p < 0.05$; ** $p < 0.01$; *** $p < 0.001$; NT: not tested.

2.2. Inhibitory Effects of TY-1 on SARS-CoV-2 Infection of Cells

SARS-CoV-2-added cells were incubated for 1 h in the presence of 25 µg/mL dextrin or 50 µg/mL TY-1, and inhibition by TY-1 of the viral infection in the cells was evaluated. The number of plaques was fewer in the TY-1 group than in the dextrin group (Figure 2a). The viral titer was significantly lower in the TY-1 group than in the dextrin group. The viral titer in the dextrin group was 15.68 $\times$ 10^5 plaque-forming unit (PFU)/mL and that in the TY-1 group was 8.00 $\times$ 10^5 PFU/mL (Figure 2b).

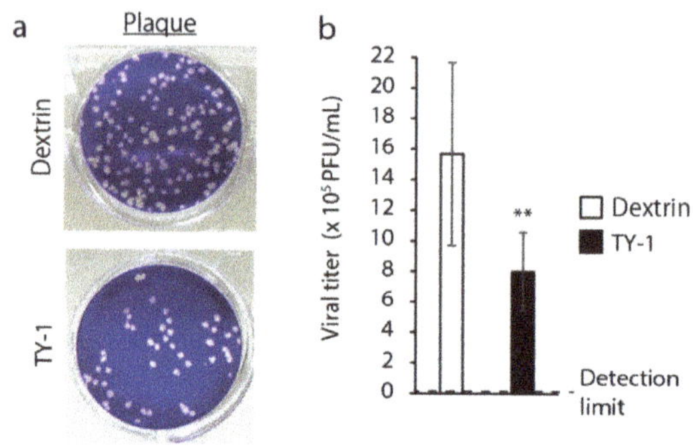

Figure 2. Inhibitory effect of TY-1 on SARS-CoV-2 infection of cells. (**a**,**b**): SARS-CoV-2 (ancestral strain)-added cells were incubated at 37 °C for 1 h in the presence of dextrin (25 μg/mL) or TY-1 (50 μg/mL); then, the cell culture media was removed. The cells were additionally incubated in 1.55% carboxymethyl cellulose sodium salt-containing virus growth medium (VGM) without dextrin or TY-1 at 37 °C for 3 d. (**a**) Plaques in representative wells, inoculated with 10^4-times diluted SARS-CoV-2 are shown. (**b**) The viral titer ($\times$ 10^5 PFU/mL) in each test group is shown. The detection limit of the viral titer was 0.00004 $\times$ 10^5 PFU/mL. The results are indicated as mean ± SD (n = 8 per group). Student's t-tests were performed to evaluate the statistical significance of the differences between the dextrin and each of the TY-1 groups; ** $p < 0.01$.

2.3. Impact of TY-1 on SARS-CoV-2 Structural Proteins

SARS-CoV-2 solution was mixed with dextrin or TY-1 solutions, and the impact on viral structural proteins expressed in viral particles was evaluated using western blotting. There were no differences between the band patterns of the S1 subunit of the S protein, the S2 subunit of the S protein, or the N protein in the dextrin and TY-1 groups at 0-h reaction time. There were also no differences in the band patterns of the S1 subunit and the N protein between these two groups at 24-h reaction time (Figure 3a, left and right). The two S2 subunit-specific bands observed in the dextrin group disappeared or were weakened in the TY-1 group at 24-h reaction time. An additional band, with a higher molecular mass, appeared in the TY-1 group (Figure 3a, middle). To more specifically evaluate the impact of TY-1 on each structural protein, each recombinant protein was mixed with dextrin or TY-1 solution. Western blotting identified no differences between the dextrin and TY-1 groups at 0-h reaction time in the band patterns of the three proteins tested. There was also no difference in the band patterns of the S1 subunit between these two groups at 24-h reaction time (Figure 3b, left). The band intensity of the S2 subunit tended to be weakened, and an additional ladder with higher molecular mass appeared in the TY-1 group at 24-h reaction time (Figure 3b, middle). The N protein-specific band disappeared in the TY-1 group at 24-h reaction time (Figure 3b, right).

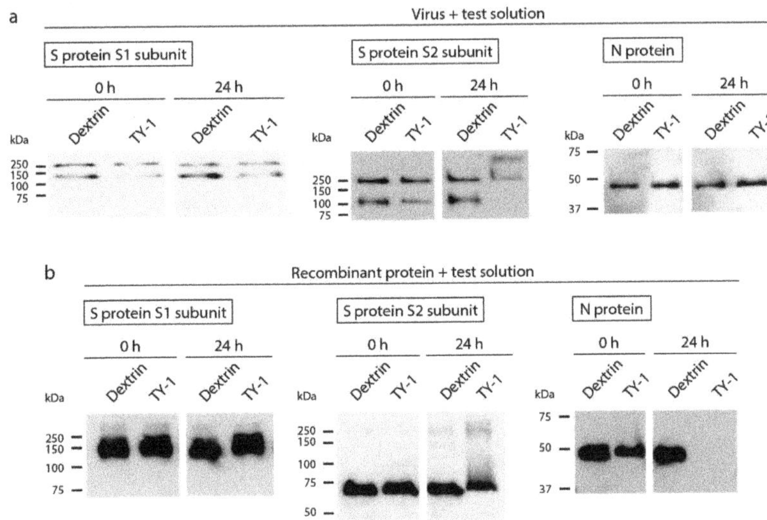

Figure 3. Impact of TY-1 on SARS-CoV-2 structural proteins. (**a**) SARS-CoV-2 (ancestral strain) was mixed with dextrin (2.5 mg/mL) or TY-1 (5.0 mg/mL). (**b**) The recombinant S1 subunit of the S protein, the S2 subunit of the S protein, or the N protein was mixed with dextrin or TY-1. (**a**,**b**) After 0-h and 24-h reaction time, respectively, western blotting was performed. The pictures on the left, middle, and right show the results of western blotting to detect the S1 subunit, S2 subunit, and N protein, respectively.

2.4. *Impact of TY-1 on the SARS-CoV-2 Genome*

SARS-CoV-2 solution was mixed with dextrin or TY-1 solution, and the impact on the viral genome was evaluated using real-time reverse transcription polymerase chain reaction (RT-PCR) analysis. There was no difference in the cycle threshold (Ct) value between the dextrin and TY-1 groups at 0-h reaction time. The Ct value was higher in the TY-1 group than in the dextrin group at 24-h reaction time. The difference in Ct value between these two groups was 12.06 (Figure 4). This result suggests that TY-1 induced the destruction of the viral genome.

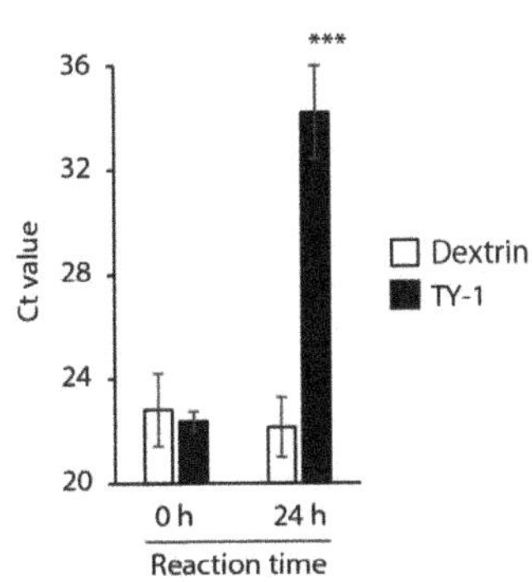

Figure 4. Impact of TY-1 on the SARS-CoV-2 genome. SARS-CoV-2 (ancestral strain) was mixed with dextrin (2.5 mg/mL) or TY-1 (5.0 mg/mL). After 0-h and 24-h reaction time, respectively, real-time RT-PCR was performed, and the Ct value was evaluated. The results are indicated as mean ± SD ($n = 4$ per group). Student's t-tests were performed to evaluate the statistical significance of differences between the dextrin and TY-1 groups; *** $p < 0.001$.

3. Discussion

In this study, we found a SARS-CoV-2 inactivation activity of TY-1, which contains abundant polyphenols, including green tea catechins and concentrated TFs. TY-1 also contains 1.8% caffeine and 1.3% theanine (Table S1, in Supplementary Materials). Although there have been several reports that caffeine has some inhibitory activity against virus proliferation in infected cells, it did not appear to have clear virucidal activity [21]. Although it has been suggested that theanine contributes to enhanced antiviral immunity [22], there are almost no reports related to its virucidal activity. A direct impact of TFs has been reported against the viral particles of both enveloped and nonenveloped virus species, including influenza virus, human immunodeficiency virus (HIV), rotavirus, enterovirus, calicivirus, and hepatitis C virus [18,23–26]. The virucidal activities of green tea catechins against SARS-CoV-2, influenza A virus, calicivirus, and many other virus species have been also shown in previous studies by ourselves and other researchers [19,27]. Hence, the SARS-CoV-2 inactivation activity of TY-1 may largely depend on virucidal tea polyphenols such as TFs and other catechins. Some reports have suggested that several plant-derived polyphenols have additive and synergistic antiviral activities [28,29]. Multiple polyphenols could have contributed to the total SARS-CoV-2 inactivation activity of TY-1.

In this study, we showed that TY-1 treatment induced structural changes in the S protein of SARS-CoV-2 (Figure 3). Polyphenols, including catechins, bind to proteins [15,30], and previous reports have shown that TFs impact HA and NA proteins, the spike proteins of the influenza A virus, resulting in the inhibition of their functions [23,31]. In this study, a high-molecular-mass band/ladder appeared in the western blots of the TY-1-treated S2 subunit of the S protein. In our previous study, in which the SARS-CoV-2-inactivating activity of olive polyphenol-rich extract was demonstrated, the appearance of similar high-molecular-mass bands/ladders was observed for both the S1 and S2 subunits of the S protein [32]. These results indicate that some types of plant-based polyphenols may induce the aggregation of S proteins. While olive polyphenols impacted both the S1 and S2 subunits, TY-1 seemed to impact only the S2 subunit. The S1 subunit is responsible for the binding of SARS-CoV-2 to the viral receptor ACE2 on host cells, and the S2 subunit is responsible for the fusion of the virus envelope and host cell membrane [33]. Liu et al. [18] showed that TFs and EGCG did not clearly block HIV Env glycoprotein gp120 subunit-CD4 interaction but blocked the six-helix bundle formation in the gp41 subunit, resulting in the inhibition of virus envelope-cell membrane fusion. In the present study, a similar mechanism of action of the catechins in TY-1 on the S2 subunit may have contributed to the inhibition of SARS-CoV-2 infection of cells (Figure 2). Susceptibility to TY-1 seemed to be slightly different in the different SARS-CoV-2 strains. The susceptibility to TY-1 was highest in the Beta and Gamma strains, followed by the ancestral and Alpha strains. The susceptibility of the Delta and Kappa strains tended to be lower (Figure 1). Since there are multiple discrepancies in the amino acid sequences of the S protein among these different strains, such amino acid substitutions, especially in the S2 subunit, may impact the sensitivity to TY-1. The S2 subunit contains the fusion peptide, heptapeptide repeat 1 and 2 (HR1, HR2), the transmembrane domain, and the cytoplasm domain [34]. In the strains tested here, there were several discrepancies in the amino acid sequence in the HR1 and HR2 regions, but not in the fusion peptide or the transmembrane/cytoplasm domains. As HR1 and HR2 are indispensable to virus envelope-cell membrane fusion and virus entry, these regions may be a possible target of the virucidal compounds contained in TY-1. Future computational docking analysis of the S2 subunit and TFs/green tea catechins may contribute to further understanding of the details of their inactivation of SARS-CoV-2.

TY-1 also impacted uncovered recombinant N protein, but not N protein in viral particles. These differing results suggest that TY-1 interacts with the uncovered N protein, but the amount of TY-1 reaching the N protein located inside viral particles was possibly not enough to have an effect.

TY-1 also impacted the SARS-CoV-2 genome (Figure 4). This result was consistent with those of our previous studies, in which olive polyphenol-rich extract and herbal plant-derived pyrogallol-enriched fraction disrupted the SARS-CoV-2 genome [27,32]. The green tea catechins generated reactive oxygen species and damaged the DNA [35,36]. Such reactive oxygen species-dependent nucleotide damage is a possible mechanism of action of viral genome destruction. Ikigai et al. [37] demonstrated that EGCG damages phosphatidylcholine liposomes, and such a finding might indicate that catechins also impacted the virus envelope. In the present study, we could not perform electron microscopic observation of SARS-CoV-2 particles because of issues of safety in handling pathogens. However, such a structural observation of viral particles will provide information about the influence of TY-1 on the virus envelope.

Multiple previous studies evaluating the virucidal activity of TFs and green tea catechins have suggested that the number of galloyl and hydroxyl groups contributed to their virucidal activity, and TF3 and EGCG tended to show more potent antiviral activities than other catechins [15,18,19,27]. Therefore, further research to increase the content of TF3 and EGCG in TY-1 may contribute to the further enhancement of its virucidal activity.

4. Materials and Methods

4.1. Viruses and Cells

SARS-CoV-2 strains (2019-nCoV/Japan/TY/WK-521/2020 [ancestral strain], hCoV-19/Japan/QHN001/2020 [B.1.1.7 lineage; Alpha strain, GISAID ID: EPI_ISL_804007], hCoV-19/Japan/TY8-612-P1/2021 [B.1.351 lineage; Beta strain, GISAID ID: EPI_ISL_1123289], hCoV-19/Japan/TY7-501/2021 [P.1 lineage; Gamma strain, GISAID ID: EPI_ISL_833366], hCoV-19/Japan/TY11-927-P1/2021 [B.1.617.2 lineage; Delta strain, GISAID ID: EPI_ISL_2158617], and hCoV-19/Japan/TY11-330-P1/2021 [B.1.617.1 lineage; Kappa strain, GISAID ID: EPI_ISL_2158613]) were provided by the National Institute of Infectious Diseases (Tokyo, Japan). VeroE6/TMPRSS2 cells [38] established by the National Institute of Infectious Diseases were purchased from the Japanese Collection of Research Bioresources (No. JCRB1819, Osaka, Japan). SARS-CoV-2-inoculated VeroE6/TMPRSS2 cells were cultured in a previously described VGM [32].

4.2. Sample Preparation

TY-1 powder was provided by Yokoyama Food Co., Ltd. (Sapporo, Japan); its components are shown in Table S1. To prepare a 10 mg/mL TY-1 solution, 1.0 g TY-1 powder was dissolved in 100 mL phosphate-buffered saline (PBS), and the water-soluble layer was collected and stored at −80 °C. When the experiments were performed, to prepare TY-1 solutions with multiple concentrations, TY-1 solution was additionally diluted with PBS. As TY-1 powder contains 50% dextrin, 5 mg/mL dextrin solution was prepared as a solvent control by dissolving 0.5 g of dextrin in 100 mL PBS.

4.3. Evaluation of the Virucidal Activity of TY-1 against SARS-CoV-2

SARS-CoV-2 solution (~7.0 $\log_{10}$ $TCID_{50}$/mL) was mixed with an equal volume of dextrin or TY-1. The final concentrations of dextrin and TY-1 in the mixture were 2.5 and 0.3–5.0 mg/mL, respectively. The mixtures were incubated at 22–25 °C from 1 min to 24 h and were then inoculated into cells; a 10-fold serial dilution of the cell culture medium, VGM, was performed. After incubation for 3 d, the cytopathic effect induced by SARS-CoV-2 infection was observed, and the viral titer ($\log_{10}$ $TCID_{50}$/mL) was calculated using the Behrens–Kärber method [39]. The virucidal activities of TY-1 solutions of different concentrations were evaluated by comparing the difference in the viral titers between the dextrin group and each concentration of TY-1. The detection limits of the viral titers in the dextrin and TY-1 groups were determined based on the cytotoxic concentrations of those test solutions in virus-free conditions (Table S2). In Table S2, no cytotoxicity was defined as $\geq$ 80% cell survival, because abnormalities of cell morphology were not observed, and susceptibility to virus infection was maintained in these cells. The detection limits of viral titers in the groups treated with 2.5 mg/mL dextrin and 0.3–1.3 mg/mL TY-1 solutions were set to 1.25 $\log_{10}$ $TCID_{50}$/mL, according to our viral titer calculation. The detection limits in the groups treated with 2.5 and 5.0 mg/mL TY-1 were set to 2.25 $\log_{10}$ $TCID_{50}$/mL.

4.4. Evaluation of the Inhibitory Effect of TY-1 on SARS-CoV-2 Infection of Cells

SARS-CoV-2 solution was added into cells, and a twofold serial dilution was performed in the cell culture medium, VGM. At the same time, dextrin or TY-1 solution was added to the cell culture media. The final concentrations of dextrin and TY-1 in the cell culture media were 25 and 50 μg/mL, respectively; these concentrations did not show any cytotoxicity under virus-free conditions (Table S3). The cells to which virus was added were incubated at 37 °C for 1 h in the presence of dextrin or TY-1; then, the cell culture medium was removed. After washing with VGM, new VGM containing 1.55% carboxymethyl cellulose sodium salt (Nacalai tesque Co. Ltd., Kyoto, Japan) without dextrin or TY-1 was added, and the cells were further incubated at 37 °C for 3 d. The plaques were then observed, and the viral titer ($\times 10^5$ PFU/mL) was calculated.

4.5. Evaluation of the Impact of TY-1 on SARS-CoV-2 Structural Proteins

Western blotting targeting the S1 subunit of the S protein, the S2 subunit of the S protein, or the N protein was performed as previously described [32]. Briefly, SARS-CoV-2 solution (~7.0 $\log_{10}$ $TCID_{50}$/mL), or each recombinant viral protein, was mixed with an equal volume of dextrin or TY-1. The final concentrations of dextrin and TY-1 in the mixture were 2.5 and 5.0 mg/mL, respectively. The final concentration of recombinant proteins in the mixture was 10 μg/mL. These mixtures were combined with sodium dodecyl sulfate (SDS) buffer with 2-mercaptoethanol (FUJIFILM Wako Pure Chemical Co., Ltd., Osaka, Japan) either immediately (0-h reaction time) or were incubated at 25 °C for 24 h prior to adding the SDS buffer (24-h reaction time). These samples were subjected to SDS–polyacrylamide gel electrophoresis using Mini-PROTEAN® 3 Cell (Bio-Rad Laboratories Inc., Hercules, CA) and Power PAC™ HC power supply (Bio-Rad Laboratories Inc.), and western blotting using Power PAC™ HC power supply and Luminescent Image Analyzer LAS-3000 (FUJIFILM Co. Ltd., Tokyo, Japan) to detect the S1 subunit of the S protein, the S2 subunit of the S protein, and the N protein.

4.6. Evaluation of the Impact of TY-1 on the SARS-CoV-2 Genome

Real-time RT-PCR analysis targeting the SARS-CoV-2 N gene was performed as previously described [32]. Briefly, SARS-CoV-2 solution (~7.0 $\log_{10}$ $TCID_{50}$/mL) was mixed with an equal volume of dextrin or TY-1. The final concentrations of dextrin and TY-1 in the mixture were 2.5 and 5.0 mg/mL, respectively. These mixtures were combined with ISOGEN-LS (Nippon Gene Co., Ltd., Tokyo, Japan) immediately (0-h reaction time) or were incubated at 25 °C for 24 h prior to adding ISOGEN-LS (24-h reaction time). These samples were subjected to real-time RT-PCR using LifeECO (Bioer Technology Co. Ltd., Hangzhou, China) and LightCycler® 96 (F. Hoffmann-La Roche, Ltd., Basel, Switzerland). The primers and probe used in this study were NIID_2019-nCoV_N_F2, NIID_2019-nCoV_N_R2, and NIID_2019-nCoV_N_P2 [40].

4.7. Statistical Analysis

Student's *t*-tests were performed to determine statistically significant differences between the dextrin and each of the TY-1 groups. p values of less than 0.05 were considered to be statistically significant.

5. Conclusions

In the present study, we showed that the tea leaf extract TY-1, which contains abundant tea polyphenols, including virucidal TFs and other green tea catechins, had time- and concentration-dependent SARS-CoV-2 inactivation activity. TY-1 induced the structural changes or destruction of the S protein S2 subunit and viral genome and inhibited the SARS-CoV-2 infection of cells. Since oral mucosal epithelial cells are one of the target cells of SARS-CoV-2 infection [41], TY-1 can be applied not only as an antiviral supplement but also as a troche, mouth rinse, or gargle. There may also be other ways of using TY-1 to prevent SARS-CoV-2 transmission. A previous report showed that green tea extract-doped masks blocked the passage of infective influenza A virus [42]. However, only the in vitro virucidal activity of TY-1 was tested in the current study. Further animal experiments and clinical tests in humans are needed to evaluate the actual use of TY-1. Nevertheless, our findings provide basic insights into the possible value of TY-1 as a highly safe antiviral agent, which may help reinforce the current SARS-CoV-2 control measures.

Supplementary Materials: Table S1: Composition of 100,000 mg of TY-1 powder, Table S2: Cytotoxicity of dextrin and TY-1 (cell culture time in the presence of dextrin or TY-1: 3 d), Table S3: Cytotoxicity of dextrin and TY-1 (cell culture time in the presence of dextrin or TY-1: 1 h).

Author Contributions: Conceptualization, Y.T.; methodology, Y.T.; software, Y.T.; validation, Y.T.; formal analysis, Y.T., K.T., D.J., S.M.; investigation, Y.T., K.T., H.O.; data curation, Y.T.; writing—original draft preparation, Y.T.; writing—review and editing, Y.T., K.T., H.O.; visualization, Y.T., K.T.; supervision, H.O.; project administration, H.O.; funding acquisition, H.O. All authors have read and agreed to the published version of the manuscript.

Funding: This research was supported in part by Grants-in-Aid for Regional R&D Proposal-Based; Program from Northern Advancement Center for Science & Technology of Hokkaido, Japan.

Institutional Review Board Statement: Not applicable.

Informed Consent Statement: Not applicable.

Data Availability Statement: Not applicable.

Acknowledgments: We are grateful to Yokoyama Food Co., Ltd. (Sapporo, Hokkaido, Japan) for providing TY-1 powder.

Conflicts of Interest: The authors declare no conflict of interest.

Sample Availability: Samples of the compounds are available from the authors.

References

1. Zhu, N.; Zhang, D.; Wang, W.; Li, X.; Yang, B.; Song, J.; Zhao, X.; Huang, B.; Shi, W.; Lu, R.; et al. A novel coronavirus from patients with pneumonia in China, 2019. *N. Engl. J. Med.* **2020**, *382*, 727–733. [CrossRef]
2. Ramanathan, K.; Antognini, D.; Combes, A.; Paden, M.; Zakhary, B.; Ogino, M.; Maclaren, G.; Brodie, D. Genomic characterisation and epidemiology of 2019 novel coronavirus: Implications for virus origins and receptor binding. *Lancet* **2020**, *395*, 565–574.
3. World Health Organization. Coronavirus Disease (COVID-2019) Situation Reports. Available online: https://www.who.int/emergencies/diseases/novel-coronavirus-2019/situation-reports (accessed on 12 May 2021).
4. Kratzel, A.; Todt, D.; V'kovski, P.; Steiner, S.; Gultom, M.; Thao, T.T.N.; Ebert, N.; Holwerda, M.; Steinmann, J.; Niemeyer, D.; et al. Inactivation of severe acute respiratory syndrome coronavirus 2 by WHO-recommended hand rub formulations and alcohols. *Emerg. Infect. Dis.* **2020**, *26*, 1592–1595. [CrossRef]
5. Takeda, Y.; Uchiumi, H.; Matsuda, S.; Ogawa, H. Acidic electrolyzed water potently inactivates SARS-CoV-2 depending on the amount of free available chlorine contacting with the virus. *Biochem. Biophys. Res. Commun.* **2020**, *530*, 1–3. [CrossRef]
6. Chan, K.-H.; Sridhar, S.; Zhang, R.R.; Chu, H.; Fung, A.Y.-F.; Chan, G.; Chan, J.F.-W.; To, K.K.-W.; Hung, I.F.-N.; Cheng, V.C.-C.; et al. Factors affecting stability and infectivity of SARS-CoV-2. *J. Hosp. Infect.* **2020**, *106*, 226–231. [CrossRef]
7. Meyerowitz, E.A.; Richterman, A.; Gandhi, R.T.; Sax, P.E. Transmission of SARS-CoV-2: A review of viral, host, and environmental factors. *Ann. Intern. Med.* **2021**, *174*, 69–79. [CrossRef]
8. Ita, K. Coronavirus disease (COVID-19): Current status and prospects for drug and vaccine development. *Arch. Med. Res.* **2021**, *52*, 15–24. [CrossRef] [PubMed]
9. Rizk, J.G.; Kalantar-Zadeh, K.; Mehra, M.R.; Lavie, C.J.; Rizk, Y.; Forthal, D.N. Pharmaco-immunomodulatory therapy in COVID-19. *Drugs* **2020**, *80*, 1267–1292. [CrossRef] [PubMed]
10. Salim, S.A.K.; Tulio, D.O. New SARS-CoV-2 variants-clinical, public health, and vaccine implications. *N. Engl. J. Med.* **2021**, *384*, 1866–1868. [CrossRef]
11. Wibmer, C.K.; Ayres, F.; Hermanus, T.; Madzivhandila, M.; Kgagudi, P.; Oosthuysen, B.; Lambson, B.E.; de Oliveira, T.; Vermeulen, M.; van der Berg, K.; et al. SARS-CoV-2 501Y.V2 escapes neutralization by South African COVID-19 donor plasma. *Nat. Med.* **2021**, *27*, 622–625. [CrossRef] [PubMed]
12. Kapoor, R.; Sharma, B.; Kanwar, S.S. Antiviral phytochemicals: An overview. *Biochem. Physiol. Open Access* **2017**, *6*. [CrossRef]

13. Ghildiyal, R.; Prakash, V.; Chaudhary, V.K.; Gupta, V.; Gabrani, R. Phytochemicals as antiviral agents: Recent updates. In *Plant-Derived Bioactives*; Springer: Singapore, 2020; pp. 279–295.
14. Musarra-Pizzo, M.; Pennisi, R.; Ben-Amor, I.; Mandalari, G.; Sciortino, M.T. Antiviral activity exerted by natural products against human viruses. *Viruses* **2021**, *13*, 828. [CrossRef]
15. Xu, J.; Xu, Z.; Zheng, W. A review of the antiviral role of green tea catechins. *Molecules* **2017**, *22*, 1337. [CrossRef] [PubMed]
16. Tong, T.; Liu, Y.J.; Kang, J.; Zhang, C.M.; Kang, S.G. Antioxidant activity and main chemical components of a novel fermented tea. *Molecules* **2019**, *24*, 2917. [CrossRef] [PubMed]
17. Jiang, Y.; Hua, J.; Wang, B.; Yuan, H.; Ma, H. Effects of variety, season, and region on theaflavins content of fermented Chinese Congou black tea. *J. Food Qual.* **2018**, *2018*, 1–9. [CrossRef]
18. Liu, S.; Lu, H.; Zhao, Q.; He, Y.; Niu, J.; Debnath, A.K.; Wu, S.; Jiang, S. Theaflavin derivatives in black tea and catechin derivatives in green tea inhibit HIV-1 entry by targeting gp41. *Biochim. Biophys. Acta Gen. Subj.* **2005**, *1723*, 270–281. [CrossRef]
19. Mhatre, S.; Srivastava, T.; Naik, S.; Patravale, V. Antiviral activity of green tea and black tea polyphenols in prophylaxis and treatment of COVID-19: A review. *Phytomedicine* **2020**, *85*, 153286. [CrossRef] [PubMed]
20. Takemoto, M. Manufacturing Method for Theaflavins Using Raw Tea Leaves. Japanese Patent No 4,817,206, 9 September 2011.
21. Utsunomiya, H.; Ichinose, M.; Uozaki, M.; Tsujimoto, K.; Yamasaki, H.; Koyama, A.H. Antiviral activities of coffee extracts in vitro. *Food Chem. Toxicol.* **2008**, *46*, 1919–1924. [CrossRef]
22. Takagi, Y.; Kurihara, S.; Higashi, N.; Morikawa, S.; Kase, T.; Maeda, A.; Arisaka, H.; Shibahara, S.; Akiyama, Y. Combined administration of L-cystine and L-theanine enhances immune functions and protects against influenza virus infection in aged mice. *J. Vet. Med. Sci.* **2010**, *72*, 157–165. [CrossRef]
23. Nakayama, M.; Suzuki, K.; Toda, M.; Okubo, S.; Hara, Y.; Shimamura, T. Inhibition of the infectivity of influenza virus by tea polyphenols. *Antiviral Res.* **1993**, *21*, 289–299. [CrossRef]
24. Mukoyama, A.; Ushijima, H.; Nishimura, S.; Koike, H.; Toda, M.; Hara, Y.; Shimamura, T. Inhibition of rotavirus and enterovirus infections by tea extracts. *Jpn. J. Med. Sci. Biol.* **1991**, *44*, 181–186. [CrossRef]
25. Ohba, M.; Oka, T.; Ando, T.; Arahata, S.; Ikegaya, A.; Takagi, H.; Ogo, N.; Zhu, C.; Owada, K.; Kawamori, F.; et al. Antiviral effect of theaflavins against caliciviruses. *J. Antibiot.* **2017**, *70*, 443–447. [CrossRef] [PubMed]
26. Chowdhury, P.; Sahuc, M.E.; Rouillé, Y.; Rivière, C.; Bonneau, N.; Vandeputte, A.; Brodin, P.; Goswami, M.; Bandyopadhyay, T.; Dubuisson, J.; et al. Theaflavins, polyphenols of black tea, inhibit entry of hepatitis C virus in cell culture. *PLoS ONE* **2018**, *13*, e0198226. [CrossRef]
27. Takeda, Y.; Murata, T.; Jamsransuren, D.; Suganuma, K.; Kazami, Y.; Batkhuu, J.; Badral, D.; Ogawa, H. *Saxifraga spinulosa*-derived components rapidly inactivate multiple viruses including SARS-CoV-2. *Viruses* **2020**, *12*, 699. [CrossRef] [PubMed]
28. Knox, Y.M.; Hayashi, K.; Suzutani, T.; Ogasawara, M.; Yoshida, I.; Shiina, R.; Tsukui, A.; Terahara, N.; Azuma, M. Activity of anthocyanins from fruit extract of Ribes nigrum L. against influenza A and B viruses. *Acta Virol.* **2001**, *45*, 209–215.
29. Amoros, M.; Simões, C.M.O.; Girre, L.; Sauvager, F.; Cormier, M. Synergistic effect of flavones and flavonols against herpes simplex virus type 1 in cell culture. Comparison with the antiviral activity of propolis. *J. Nat. Prod.* **1992**, *55*, 1732–1740. [CrossRef] [PubMed]
30. Brudzynski, K.; Maldonado-Alvarez, L. Polyphenol-protein complexes and their consequences for the redox activity, structure and function of honey. A current view and new hypothesis—A review. *Polish J. Food Nutr. Sci.* **2015**, *65*, 71–80. [CrossRef]
31. Zu, M.; Yang, F.; Zhou, W.; Liu, A.; Du, G.; Zheng, L. In vitro anti-influenza virus and anti-inflammatory activities of theaflavin derivatives. *Antiviral Res.* **2012**, *94*, 217–224. [CrossRef]
32. Takeda, Y.; Jamsransuren, D.; Matsuda, S.; Crea, R.; Ogawa, H. The SARS-CoV-2-inactivating activity of hydroxytyrosol-rich aqueous olive pulp extract (HIDROX®) and its use as a virucidal cream for topical application. *Viruses* **2021**, *13*, 232. [CrossRef]
33. Xia, S.; Liu, M.; Wang, C.; Xu, W.; Lan, Q.; Feng, S.; Qi, F.; Bao, L.; Du, L.; Liu, S.; et al. Inhibition of SARS-CoV-2 (previously 2019-nCoV) infection by a highly potent pan-coronavirus fusion inhibitor targeting its spike protein that harbors a high capacity to mediate membrane fusion. *Cell Res.* **2020**, *30*, 343–355. [CrossRef]
34. Huang, Y.; Yang, C.; Xu, X.F.; Xu, W.; Liu, S.W. Structural and functional properties of SARS-CoV-2 spike protein: Potential antivirus drug development for COVID-19. *Acta Pharmacol. Sin.* **2020**, *41*, 1141–1149. [CrossRef] [PubMed]
35. Arakawa, H.; Maeda, M.; Okuda, S.; Shimamura, T. Role of hydrogen peroxide in bactericidal action of catechin. *Biol. Pharm. Bull.* **2004**, *27*, 277–281. [CrossRef]
36. Furukawa, A.; Oikawa, S.; Murata, M.; Hiraku, Y.; Kawanishi, S. (-)-Epigallocatechin gallate causes oxidative damage to isolated and cellular DNA. *Biochem. Pharmacol.* **2003**, *66*, 1769–1778. [CrossRef]
37. Ikigai, H.; Nakae, T.; Hara, Y.; Shimamura, T. Bactericidal catechins damage the lipid bilayer. *Biochim. Biophys. Acta Biomembr.* **1993**, *1147*, 132–136. [CrossRef]
38. Nao, N.; Sato, K.; Yamagishi, J.; Tahara, M.; Nakatsu, Y.; Seki, F.; Katoh, H.; Ohnuma, A.; Shirogane, Y.; Hayashi, M.; et al. Consensus and variations in cell line specificity among human metapneumovirus strains. *PLoS ONE* **2019**, *14*, e0215822. [CrossRef]
39. Kärber, G. Beitrag zur kollektiven behandlung pharmakologischer reihenversuche. *Naunyn. Schmiedebergs. Arch. Exp. Pathol. Pharmakol.* **1931**, *162*, 480–483. [CrossRef]
40. Shirato, K.; Nao, N.; Katano, H.; Takayama, I.; Saito, S.; Kato, F.; Katoh, H.; Sakata, M.; Nakatsu, Y.; Mori, Y.; et al. Development of genetic diagnostic methods for detection for novel coronavirus 2019(nCoV-2019) in Japan. *Jpn. J. Infect. Dis.* **2020**, *73*, 304–307. [CrossRef] [PubMed]

41. Huang, N.; Pérez, P.; Kato, T.; Mikami, Y.; Okuda, K.; Gilmore, R.C.; Conde, C.D.; Gasmi, B.; Stein, S.; Beach, M.; et al. SARS-CoV-2 infection of the oral cavity and saliva. *Nat. Med.* **2021**, *27*, 892–903. [CrossRef] [PubMed]
42. Oxford, J.S.; Lambkin, R.; Guralnik, M.; Rosenbloom, R.A.; Petteruti, M.P.; DiGian, K.; LeFante, C. Preclinical in vitro activity of QR-435 against influenza a virus as a virucide and in paper masks for prevention of viral transmission. *Am. J. Ther.* **2007**, *14*, 455–461. [CrossRef]

MDPI
St. Alban-Anlage 66
4052 Basel
Switzerland
Tel. +41 61 683 77 34
Fax +41 61 302 89 18
www.mdpi.com

Molecules Editorial Office
E-mail: molecules@mdpi.com
www.mdpi.com/journal/molecules

www.ingramcontent.com/pod-product-compliance
Lightning Source LLC
Chambersburg PA
CBHW040804250926
51850CB00007BA/1076

* 9 7 8 3 0 3 6 5 6 9 6 3 5 *